Practical Ship Hydrodynamics

Practical Ship Hydrodynamics

Second edition

Volker Bertram

AMSTERDAM • BOSTON • HEIDELBERG • LONDON
NEW YORK • OXFORD • PARIS • SAN DIEGO
SAN FRANCISCO • SINGAPORE • SYDNEY • TOKYO
Butterworth-Heinemann is an imprint of Elsevier

Butterworth-Heinemann is an imprint of Elsevier
The Boulevard, Langford Lane, Kidlington, Oxford OX5 1GB, UK
225 Wyman Street, Waltham, MA 02451, USA

First edition 2000
Second edition 2012

British Library Cataloguing in Publication Data
A catalogue record for this book is available from the British Library

Library of Congress Cataloging-in-Publication Data
A catalog record for this book is availabe from the Library of Congress

ISBN–13: 978-1-4832-9971-6

For information on all Butterworth-Heinemann publications
visit our web site at books.elsevier.com

Printed and bound in the UK

11 12 13 14 15   10 9 8 7 6 5 4 3 2 1

Contents

Preface

Ten years after the 1st edition, it was time to update, extend and reorganize the material. The book still gives an introduction to modern ship hydrodynamics, which is in my opinion suitable for teaching at a senior undergraduate level or even at a postgraduate level. It is thus also suitable for engineers working in industry. The book assumes that the reader has a solid knowledge of general fluid dynamics. In teaching, general fluid dynamics and specific ship hydrodynamics are often mixed but I believe that universities should first teach a course in general fluid dynamics which should be mandatory to most engineering students. There are many good textbooks on the market for this purpose. Naval architects should then concentrate on the particular aspects of their field and cover material more suited to their needs. This book is organized to support such a strategy in teaching.

The first chapter covers basics of computational fluid dynamics and model tests, and Chapters 2 to 6 cover the four main areas of propeller flows, resistance and propulsion, ship seakeeping including ship vibrations, and maneuvering. Chapter 5 was added to cover ship vibrations from a hydrodynamic point, as a natural extension of rigid-body motions in waves in seakeeping. It is recommended that this sequence be followed in teaching. The book tries to find a suitable balance for practical engineers between facts and minimizing formula work. However, there are still formulae. These are intended to help those tasked with computations or programming. Readers with a practical interest may simply skip these passages. Readers with a more theoretical interest will find additional background, e.g. derivations of formulae, on the associated website.

The final two chapters of the 1st edition involved more extensive background on boundary element methods. They were intended for graduate and postgraduate teaching. Research is no longer active in these methods and more modern field methods are covered in standard textbooks. The original two chapters on boundary element theory are now still available but only as appendices to this book.

The book is supplemented by exercises and solutions, formula derivations and texts, intended to support teaching or self studies. The material can be obtained from the publisher upon request.

This book is based largely on lectures for German students. The nucleus of the book was formed by lectures on ship seakeeping and ship maneuvering, which I have taught for several years with Professor Heinrich Söding. I always felt that we should have a comprehensive textbook that would also cover resistance and propulsion, as ship seakeeping and maneuvering are both interwoven strongly with the steady base flow. Many colleagues helped with providing material, allowing me to pick the best from their teaching approaches. A lot of material was written and compiled in a new way, inspired by these sources, but the chapters on ship seakeeping and maneuvering use extensive existing material. For this 2nd edition, material on ship vibrations, propulsion-improving devices, and simple design approaches were added. Also, CFD has progressed significantly over the past decade and required updating of pertaining passages. Readers interested in marine CFD applications may see the latest progress in the proceedings of the Numerical Towing Tank Symposium, a conference series that I initiated for faster dissemination of research results in this field. The proceedings can be downloaded from www.uni-due.de/IST/ismt_nutts8.

Thanks are due to Seehafen-Verlag Hamburg for permission to reprint text and figures from the *Manoeuvring Technical Manual*, an excellent book unfortunately no longer in print. Thanks are due to Hansa-Verlag Hamburg for permission to reprint text and figures from German contributions in *Handbuch der Werften XXIV*. Thanks are also due to Germanischer Lloyd for permission to reprint text and figures from its GL Technology on ship vibrations.

Countless colleagues supported the endeavor of writing this book by supplying material, proof-reading, making comments, or just discussing engineering or didactic matters. Among these are (in alphabetical order) Poul Andersen, Kai Graf, Mike Hughes, Hidetsugu Iwashita, Gerhard Jensen, Meinolf Kloppenburg, Maurizio Landrini, Jochen Laudan, Eike Lehmann, Friedrich Mewis, Holger Mumm, Prasanta Sahoo, Katsuji Tanizawa, Gerhard Thiart, Michel Visonneau, and Hironori Yasukawa. Special thanks to Dodo Wagener for all the great artwork. Most of all, Professor Heinrich Söding has supported this book to an extent that he should have been named as co-author, but, typically for him, he declined the offer. He even refused to allow me to dedicate this book to him. I then dedicate this book to the best mentor I ever had, a role model as a scientist and a man, so much better than I will ever be. You know who.

Volker Bertram

CHAPTER 1

Introduction

Chapter Outline

Models now in tanks we tow.
All of that to Froude we owe.
Will computers, fast and new,
Make us alter Euler's view?
Marshall Tulin

1.1. Overview of Problems and Approaches

The prediction of ship hydrodynamic performance can be broken down into the general areas of:

- resistance and propulsion;
- seakeeping and ship vibrations;
- maneuvring.

Propeller flows and propeller design can be seen as a subtopic of resistance and propulsion, but it is so important and features special techniques that it is treated as a separate topic in its own

Practical Ship Hydrodynamics. DOI: 10.1016/B978-0-08-097150-6.10001-6

right. Morgan and Lin (1998) give a good short introduction to the historical development of these techniques to the state of the art in the late 1990s.

The basic approaches can be roughly classified into:

- *Empirical/statistical approaches.* Design engineers need simple and reasonably accurate estimates, e.g. of the power requirements of a ship. Common approaches combine a rather simple physical model and regression analysis to determine required coefficients either from one parent ship or from a set of ships. The coefficients may be given in the form of constants, formulae, or curves. Because of the success with model testing, experimental series of hull forms have been developed for varying hull parameters. Extensive series were tested in the 1940s and the subsequent two decades. These series were created around a 'good' hull form as the parent form. The effect of essential hull parameters, e.g. block coefficient, was determined by systematic variations of these parameters. Because of the expense of model construction and testing, there are no recent comparable series tested of modern hull forms and the traditional ship series must be considered as outdated by now. Rather than using model tests, today computational fluid dynamics could be used to create data for systematic series varying certain parameters for a ship type (Harries and Tillig 2011). Once such a dedicated 'numerical hull series' is set up, designers can rapidly interpolate within such a database.
- *Experimental approaches, either in model tests or in full-scale trials.* The basic idea of model testing is to experiment with a scale model to extract information that can be scaled (transformed) to the full-scale ship. Despite continuing research and standardization efforts, a certain degree of empiricism is still necessary, particularly in the model-to-ship correlation, which is a method to enhance the prediction accuracy of ship resistance by empirical means. The total resistance can be decomposed in various ways. Traditionally, model basins tend to adopt approaches that seem most appropriate to their respective organization's corporate experience and accumulated databases. Unfortunately, this makes various approaches and related aggregated empirical data incompatible. Although there has been little change in the basic methodology of ship resistance since the days of Froude (1874), various aspects of the techniques have progressed. We now understand better the flow around three-dimensional, appended ships, especially the boundary layer effects. Also non-intrusive experimental techniques like laser-Doppler velocimetry (LDV) or particle image velocimetry (PIV) allow the measurement of the velocity field in the ship wake to improve propeller design. Another more recent experimental technique is wave pattern analysis to determine the wave-making resistance. In propulsion tests, measurements include towing speed and propeller quantities such as thrust, torque, and rpm. Normally, open-water tests on the propeller alone are run to aid the analysis process as certain coefficients are necessary for the propeller design. Strictly, open-water tests are not essential for power prediction. The model propeller is usually a stock propeller (taken from a large selection (= stock) of propellers) that approximates the actual design propeller.

Propulsion tests determine important input parameters for the actual detailed propeller design, e.g. wake fraction and thrust deduction. The wake distribution, also needed for propeller design, is measured behind the ship model using pitot tubes or laser-Doppler velocimetry (LDV). For propeller design, measured nominal wakes (for the ship without a propeller) for the model must be transformed to effective wakes (for the ship with a working propeller) for the full-scale ship. While semi-empirical methods for this transformation apparently work well for most hull forms, for those with considerable flow separation at the stern, i.e. typically full hulls, there are significant scale effects on the wake between model and full scale. To some extent, computational fluid dynamics can help here in estimating the scale effects. Although the procedures for predicting full-scale resistance from model tests are well accepted, full-scale data available for validation purposes are extremely limited and difficult to obtain. The powering performance of a ship is validated by actual ship trials, ideally conducted in calm seas. The parameters usually measured are torque, rpm, and speed. Thrust is measured only as a special requirement because of the difficulty and extra expense involved in obtaining accurate thrust data. Whenever possible and appropriate, corrections are made for the effects of waves, current, wind, and shallow water. Since the 1990s, the Global Positioning System (GPS) and computer-based data acquisition systems have considerably increased the accuracy and economy of full-scale trials. The GPS has eliminated the need for 'measured miles' trials near the shore with the possible contamination of data due to shallow-water effects. Today trials are usually conducted far away from the shore. Model tests for seakeeping are often used only for validation purposes. However, for open-top container ships and ro-ro ships model tests are often performed as part of the regular design process, as IMO regulations require certain investigations for ship safety which may be documented using model tests. Most large model basins have a maneuvring model basin. The favored method to determine the coefficients for the equations of motion is through a planar motion mechanism. However, scaling the model test results to full scale using the coefficients derived in this manner is problematic, because vortex shedding and flow separation are not similar between model and full scale. Appendages generally make scaling more difficult. Also, maneuvering tests have been carried out with radio-controlled models in lakes and large reservoirs. These tests introduce additional scale effects, since the model propeller operates in a different self-propulsion point than the full-scale ship propeller. Despite these concerns, the maneuvering characteristics of ships seem generally to be predicted with sufficient accuracy by experimental approaches.

- *Numerical approaches, either rather analytical or using computational fluid dynamics (CFD).* For ship resistance and powering, CFD has become increasingly important and is now an indispensable part of the design process. Typically inviscid free-surface methods based on the boundary element approach are used to analyze the forebody, especially the interaction of bulbous bow and forward shoulder. Viscous flow codes focus on the aftbody or appendages. Flow codes modeling both viscosity and the wave-making are widely applied for flows involving breaking waves. CFD is still considered by industry as too inaccurate for resistance or power predictions. So far CFD has been used to gain insight into local flow details and derive recommendation on how to improve a given design or select a most promising candidate design for model testing. However, numerical power prediction with good accuracy has been demonstrated in research applications by 2010 for realistic ship geometries. It is expected to drift into industry applications within the next decade. For seakeeping, simple strip methods are used to analyze the seakeeping properties. These usually employ boundary element methods to solve a succession of two-dimensional problems and integrate the results into a quasi-three-dimensional result with usually good accuracy. For water impact problems (slamming and sloshing), free-surface RANSE methods are widely used in industry practice. Also, for problems involving green water on deck, free-surface RANSE methods have become the preferred tool in practice, replacing previously favored non-linear boundary element methods. A commonly used method to predict the turning and steering of a ship is to use equations of motions with experimentally determined coefficients. Once these coefficients are determined for a specific ship design – by model tests, estimated from similar ships, by empirically enhanced strip methods or CFD – the equations of motions are used to simulate the dynamic behavior of the ship. The form of the equations of motions is fairly standard for most hull designs. The predictions can be used, for example, to select rudder size and steering control systems, or to predict the turning characteristics of ships. As viscous CFD codes became more robust and efficient to use, the reliance on experimentally derived coefficients in the equations of motions has been reduced. In some industry applications, CFD has been used exclusively to compute maneuvering coefficients for ship simulators, for example.

Although a model of the final ship design is still tested in a towing tank, the testing sequence and content have changed significantly over time. Traditionally, unless the new ship design was close to an experimental series or a known parent ship, the design process incorporated many model tests. The process has been one of design, test, redesign, test, etc., sometimes involving more than ten models, each with slight variations. This is no longer feasible due to time-to-market requirements from shipowners and no longer necessary thanks to CFD developments. Combining CAD (computer-aided design) to generate new hull shapes in concert with CFD to analyze these hull shapes allows for rapid design explorations without model testing. With massive parallel computing and progress in optimization strategies (e.g. response surfaces), formal optimization of hulls, propellers, and appendages has drifted into industrial applications. CFD is increasingly used for the actual design of hull and propellers. Then often only the final design is actually tested to validate the intended performance features and to get a power prediction accepted in practice as highly accurate. As a consequence of this practice, model tests for shipyard customers have declined considerably since the 1980s. This was partially compensated for by more sophisticated and detailed tests funded from research projects to validate and calibrate CFD methods.

One of the biggest problems for predicting ship seakeeping is determining the nature of the sea: how to predict and model it, for both experimental and computational analyses. Many long-term predictions of the sea require a Fourier decomposition of the sea and ship responses with an inherent assumption that the sea and the responses are 'moderately small', while the physics of many seakeeping problems is highly non-linear. Nevertheless, seakeeping predictions are often considered to be less important or covered by empirical safety factors where losses of ships are shrugged off as 'acts of God', until they occur so often or involve such spectacular losses of life that safety factors and other regulations are adjusted to a stricter level. Seakeeping is largely not understood by shipowners and global 'sea margins' of, e.g., 15% to finely tuned (±1%) power predictions irrespective of the individual design are not uncommon.

1.2. Model Tests — Similarity Laws

Since the purely numerical treatment of ship hydrodynamics has not yet reached a completely satisfactory stage, model tests are still essential in the design process and for validation purposes. The model tests must be performed such that model and full-scale ships exhibit similar behavior, i.e. the results for the model can be transferred to full scale by a proportionality factor. We indicate in the following the full-scale ship by the index s and the model by the index m.

We distinguish between:

- geometrical similarity;
- kinematical similarity;
- dynamical similarity.

Geometrical similarity means that the ratio of a full-scale 'length' (length, width, draft, etc.) L_s to a model-scale 'length' L_m is constant, namely the model scale λ:

$$L_s = \lambda \cdot L_m \tag{1.1}$$

Correspondingly we get for areas and volumes: $A_s = \lambda^2 \cdot A_m$; $\nabla_s = \lambda^3 \cdot \nabla_m$. In essence, the model then 'appears' to be the same as the full-scale ship. While this is essential for movie makers, it is not mandatory for naval architects who want to predict the hydrodynamic performance of a full-scale ship. In fact, there have been proposals to deviate from geometrical similarity to achieve better similarity in the hydrodynamics. However, these proposals were not accepted in practice and so we always strive at least in macroscopic dimensions for geometrical similarity. In microscopic dimensions, e.g. for surface roughness, geometrical similarity is not obtained.

Kinematic similarity means that the ratio of full-scale times t_s to model-scale times t_m is constant, namely the kinematic model scale τ:

$$t_s = \tau \cdot t_m \tag{1.2}$$

Geometrical and kinematical similarity result then in the following scale factors for velocities and accelerations:

$$V_s = \frac{\lambda}{\tau} \cdot V_m; \quad a_s = \frac{\lambda}{\tau^2} \cdot a_m \tag{1.3}$$

Dynamical similarity means that the ratio of all forces acting on the full-scale ship to the corresponding forces acting on the model is constant, namely the dynamical model scale:

$$F_s = \kappa \cdot F_m \tag{1.4}$$

Forces acting on the ship encompass inertial forces, gravity forces, and frictional forces.

Inertial forces follow Newton's law $F = m \cdot a$, where F denotes force, m mass, and a acceleration. For displacement ships, $m = \rho \cdot \nabla$, where ρ is the density of water and ∇ the displacement. We then obtain for ratio of the inertial forces:

$$\kappa = \frac{F_s}{F_m} = \frac{\rho_s}{\rho_m} \cdot \frac{\nabla_s}{\nabla_m} \cdot \frac{a_s}{a_m} = \frac{\rho_s}{\rho_m} \cdot \frac{\lambda^4}{\tau^2} \tag{1.5}$$

This equation couples all three scale factors. It is called Newton's law of similarity. We can rewrite Newton's law of similarity as:

$$\kappa = \frac{F_s}{F_m} = \frac{\rho_s}{\rho_m} \cdot \lambda^2 \cdot \left(\frac{\lambda}{\tau}\right)^2 = \frac{\rho_s}{\rho_m} \cdot \frac{A_s}{A_m} \cdot \left(\frac{V_s}{V_m}\right)^2 \tag{1.6}$$

Hydrodynamic forces are often described by a coefficient c as follows:

$$F = c \cdot \frac{1}{2} \rho \cdot V^2 \cdot A \tag{1.7}$$

V is a reference speed (e.g. ship speed), A a reference area (e.g. wetted surface in calm water). The factor ½ is introduced in analogy to stagnation pressure $q = ½\rho \cdot V^2$. Combining the above equations then yields:

$$\frac{F_s}{F_m} = \frac{c_s \cdot \frac{1}{2} \rho_s \cdot V_s^2 \cdot A_s}{c_m \cdot \frac{1}{2} \rho_m \cdot V_m^2 \cdot A_m} = \frac{\rho_s}{\rho_m} \cdot \frac{A_s}{A_m} \cdot \left(\frac{V_s}{V_m}\right)^2 \tag{1.8}$$

This results in $c_s = c_m$, i.e. the non-dimensional coefficient c is constant for both ship and model. For the same non-dimensional coefficients Newton's similarity law is fulfilled and vice versa.

Gravity forces can be described in a similar fashion as inertial forces:

$$G_s = \rho_s \cdot g \cdot \nabla_s \text{ resp. } G_m = \rho_m \cdot g \cdot \nabla_m \tag{1.9}$$

This yields another force scale factor:

$$\kappa_g = \frac{G_s}{G_m} = \frac{\rho_s}{\rho_m} \cdot \frac{\nabla_s}{\nabla_m} = \frac{\rho_s}{\rho_m} \cdot \lambda^3 \tag{1.10}$$

For dynamical similarity both force scale factors must be the same, i.e. $\kappa = \kappa_g$. This yields for the time scale factor:

$$\tau = \sqrt{\lambda} \tag{1.11}$$

We can now eliminate the time scale factors in all equations above and express the proportionality exclusively in the length scale factor λ, e.g.:

$$\frac{V_s}{V_m} = \sqrt{\lambda} \quad \rightarrow \quad \frac{V_s}{\sqrt{L_s}} = \frac{V_m}{\sqrt{L_m}} \tag{1.12}$$

It is customary to make the ratio of velocity and square root of length non-dimensional with $g = 9.81$ m/s^2. This yields the Froude number:

$$F_n = \frac{V}{\sqrt{g \cdot L}} \tag{1.13}$$

The same Froude number in model and full scale ensures dynamical similarity only if inertial and gravity forces are present (Froude's law). For the same Froude number, the wave pattern in model and full scale are geometrically similar. This is only true for waves of small amplitude where gravity is the only relevant physical mechanism. Breaking waves and splashes involve another physical mechanism (e.g. surface tension) and do not scale so easily. Froude's law is kept in all regular ship model tests (resistance and propulsion tests, seakeeping tests, maneuvering tests). This results in the following scales for speeds, forces, and powers:

$$\frac{V_s}{V_m} = \sqrt{\lambda} \qquad \frac{F_s}{F_m} = \frac{\rho_s}{\rho_m} \cdot \lambda^3 \qquad \frac{P_s}{P_m} = \frac{F_s \cdot V_s}{F_m \cdot V_m} = \frac{\rho_s}{\rho_m} \cdot \lambda^{3.5} \tag{1.14}$$

Frictional forces follow yet another similarity law, and are primarily due to frictional stresses (due to friction between two layers of fluid):

$$R = \mu \cdot \frac{\partial u}{\partial n} \cdot A \tag{1.15}$$

μ is a material constant, namely the dynamic viscosity. The partial derivative is the velocity gradient normal to the flow direction. A is the area subject to the frictional stresses. Then the ratio of the frictional forces is:

$$\kappa_f = \frac{R_s}{R_m} = \frac{\mu_s(\partial u_s / \partial n_s) A_s}{\mu_m(\partial u_m / \partial n_m) A_m} = \frac{\mu_s}{\mu_m} \cdot \frac{\lambda^2}{\tau} \tag{1.16}$$

Again we demand that the ratio of frictional forces and inertial forces should be the same, $\kappa_f = \kappa$. This yields:

$$\frac{\mu_s}{\mu_m} \cdot \frac{\lambda^2}{\tau} = \frac{\rho_s}{\rho_m} \cdot \frac{\lambda^4}{\tau^2} \tag{1.17}$$

If we introduce the kinematic viscosity $\nu = \mu/\rho$ this yields:

$$\frac{\nu_s}{\nu_m} = \frac{\lambda^2}{\tau} = \frac{V_s \cdot L_s}{V_m \cdot L_m} \quad \rightarrow \quad \frac{V_s \cdot L_s}{\nu_s} = \frac{V_m \cdot L_m}{\nu_m} \tag{1.18}$$

$R_n = V \cdot L/\nu$ is the Reynolds number, a non-dimensional speed parameter important in viscous flows. The same Reynolds number in model and full scale ensures dynamic similarity if only inertial and frictional forces are present (Reynolds' law). (This is somewhat simplified as viscous flows are complicated by transitions from laminar to turbulent flows, microscopic-scale effects such as surface roughness, flow separation, etc.) The kinematic viscosity ν of seawater [m/s^2] can be estimated as a function of temperature t (°C) and salinity s (%):

$$\nu = 10^{-6} \cdot (0.014 \cdot s + (0.000645 \cdot t - 0.0503) \cdot t + 1.75) \tag{1.19}$$

Sometimes slightly different values for the kinematic viscosity of water may be found. The reason is that water is not perfectly pure, containing small organic and inorganic matter which differs regionally and in time.

Froude number and Reynolds number are related by:

$$\frac{R_n}{F_n} = \frac{V \cdot L}{\nu} \cdot \frac{\sqrt{gL}}{V} = \frac{\sqrt{gL^3}}{\nu} \tag{1.20}$$

Froude similarity is easy to fulfill in model tests, as with smaller models also the necessary test speed decreases. Reynolds' law on the other hand is difficult to fulfill as smaller models mean higher speeds for constant kinematic viscosity. Also, forces do not scale down for constant viscosity.

Ships operating at the free surface are subject to gravity forces (waves) and frictional forces. Thus in model tests both Froude's and Reynolds' laws should be fulfilled. This would require:

$$\frac{R_{ns}}{R_{nm}} = \frac{\nu_m}{\nu_s} \cdot \sqrt{\frac{L_s^3}{L_m^3}} = \frac{\nu_m}{\nu_s} \cdot \lambda^{1.5} = 1 \tag{1.21}$$

i.e. model tests should chose model scale and viscosity ratio of the test fluid such that $(\nu_m/\nu_s) \cdot \lambda^{1.5} = 1$ is fulfilled. Such fluids do not exist or at least are not cheap and easy to handle for usual model scales. However, sometimes the test water is heated to improve the viscosity ratio and reduce the scaling errors for viscous effects.

Söding (1994) proposed 'sauna tanks' where the water is heated to a temperature of 90°C. Then the same Reynolds number as in cold water can be reached using models of only half the length. Smaller tanks could be used which could be better insulated and may actually require less energy than today's large tanks. The high temperature would also allow similar cavitation numbers as for the full-scale ship. A futuristic towing tank may be envisioned that would also perform cavitation tests on propellers, eliminating the need for special cavitation tunnels. However, such 'sauna tanks' have not been established yet and there are doubts concerning the feasibility of such a concept.

For model tests investigating vibrations Froude's similarity law does not automatically also give *similarity in vibrations*. For example, for propeller blade vibrations, model propellers of the same material as the full-scale propeller are too stiff under Froude similarity. Similarity in vibrations follows Cauchy's scaling law, requiring that the Cauchy number is the same in model and full scale:

$$C_n = \frac{E \cdot I \cdot t^2}{\rho \cdot g \cdot L^6} \tag{1.22}$$

where E is the modulus of elasticity, I the moment of inertia, t the time, and L the length. The same Cauchy and Froude numbers mean that, for the same density, the modulus of elasticity is downscaled by λ from full scale to model scale.

1.3. Full-Scale Trials

Trial tests of the built ship are an important prerequisite for the acceptance of the ship by the shipowner and are always specified in the contract between shipowner and shipyard. The problem is that the trial conditions differ from both model test conditions and design conditions. The contract usually specifies a contract speed at design load at a given percentage of the maximum continuous rating of the engine, this at calm sea without wind and current on deep water. Trial conditions are usually in ballast load, natural seaways, in the presence of currents and sometimes shallow water. Only on rare occasions is it possible to perform trial tests under ideal conditions as specified in the contract. However, upper limits for the wind and sea conditions are usually defined in the contract and test trials are performed only at times or places where the actual conditions are within the specified limits.

The difference between contract and trial conditions requires various corrections to correlate trial results to contract conditions. Apart from the difficulties and margins of uncertainties in the trial measurements, the correlation procedure is plagued by many doubts. The traditional methods are partly empirical, involving curves with manual interpolation, etc. It was not uncommon that the results of various consultants, e.g. towing tank experts, differed by several tenths of a knot for the obtainable speed under contract conditions. This margin may make a difference between paying and not paying considerable penalties! Subsequently, trial evaluation was susceptible to disputes between representatives of shipowners and shipyards. The increasing demand for quality management and clearly documented procedures, preferably on an international standard, led to the formation of various panels of experts. The Japan Marine Standards Association submitted in 1998 a proposal for an ISO standard for the assessment of speed and power in speed trials. Also, the 'trial and monitoring' subcommittee of the ITTC (International Towing Tank Conference) was tasked with the development of an international standard.

Test trials were historically 'measured mile trials', as ships were tested between measured miles near the coast for different ship speeds. The ship speed can be measured 'over ground' (relative to the earth) or 'in water' (relative to the water). The speed in water includes currents and local flow changes. Historically, various logs have been developed, including devices towed behind the ship, on long rods alongside the ship, electro-acoustic devices, and pitot tubes in the ship bottom. There is still no accurate and reliable way to measure a ship's speed through water. The speed over ground was traditionally determined by electro-acoustic devices, celestial navigation, and radio navigation. The advent of satellite systems, namely GPS (global positioning system) and DGPS (differential GPS), has eliminated many of the previous uncertainties and problems. GPS allows accurate determination of the speed over ground, although the speed of interest is the speed in water. Trials are usually performed by repeatedly testing the ship on opposite courses to eliminate the effects of current. It is best to align the course with the wind and predominant wave propagation direction to make elimination of these effects in the correlation procedure easier.

Seakeeping is usually not measured in detail as a normal procedure in ship deliveries. Full-scale seakeeping tests are sometimes used in research and are discussed in more detail in Section 4.2.

1.4. Numerical Approaches (Computational Fluid Dynamics)

1.4.1. Basic Equations

For the velocities involved in ship flows, water can be regarded as incompressible, i.e. the density ρ is constant. Therefore we will limit ourselves here to incompressible flows. All equations are given in a Cartesian coordinate system with z pointing downwards.

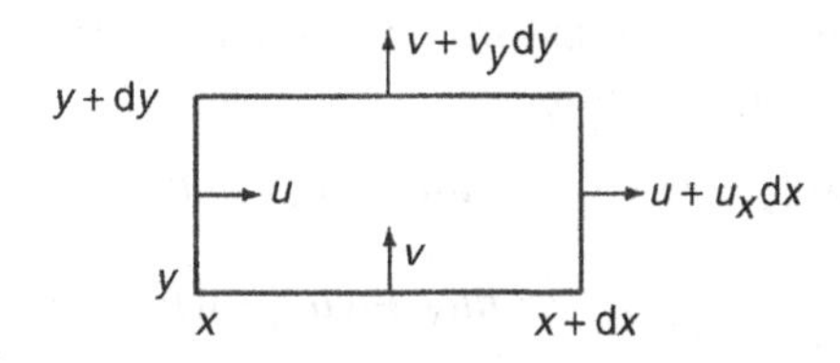

Figure 1.1:
Control volume to derive continuity equation in two dimensions

The continuity equation states that any amount flowing into a control volume also flows out of the control volume at the same time. We consider for the two-dimensional case an infinitely small control volume as depicted in Fig. 1.1. u and v are the velocity components in x resp. y direction. The indices denote partial derivatives, e.g. $u_x = \partial u/\partial x$. Positive mass flux leaves the control volume; negative mass flux enters the control volume. The total mass flux has to fulfill:

$$-\rho \, \mathrm{d}y \, u + \rho \, \mathrm{d}y \, (u + u_x \, \mathrm{d}x) - \rho \, \mathrm{d}x \, v + \rho \, \mathrm{d}x \, v + \rho \, \mathrm{d}x \, (v + v_y \, \mathrm{d}y) = 0 \tag{1.23}$$

$$u_x + v_y = 0 \tag{1.24}$$

The continuity equation in three dimensions can be derived correspondingly to:

$$u_x + v_y + w_z = 0 \tag{1.25}$$

where w is the velocity component in z direction.

The Navier–Stokes equations together with the continuity equation suffice to describe all real flow physics for ship flows. The Navier–Stokes equations describe conservation of momentum in the flow:

$$\begin{aligned}
\rho(u_t + uu_x + vu_y + wu_z) &= \rho f_1 - p_x + \mu(u_{xx} + u_{yy} + u_{zz}) \\
\rho(v_t + uv_x + vv_y + wv_z) &= \rho f_2 - p_y + \mu(v_{xx} + v_{yy} + v_{zz}) \\
\rho(w_t + uw_x + vw_y + ww_z) &= \rho f_3 - p_z + \mu(w_{xx} + w_{yy} + w_{zz})
\end{aligned} \tag{1.26}$$

where f_i is an acceleration due to a volumetric force, p the pressure, μ the viscosity, and t the time. Often the volumetric forces are neglected, but gravity can be included by setting $f_3 = g$ ($= 9.81 \text{ m/s}^2$) or the propeller action can be modeled by a distribution of volumetric forces f_1. The l.h.s. of the Navier–Stokes equations without the time derivative describes convection, the time derivative describes the rate of change ('source term'), the last term on the r.h.s. describes diffusion.

The Navier–Stokes equations in the above form contain on the l.h.s. products of the velocities and their derivatives. This is a non-conservative formulation of the Navier–Stokes equations. A conservative formulation contains unknown functions (here velocities) only as first derivatives. Using the product rule for differentiation and the continuity equation, the non-conservative formulation can be transformed into a conservative formulation, e.g. for the first of the Navier–Stokes equations above:

$$\begin{aligned}(u^2)_x + (uv)_y + (uw)_z &= 2uu_x + u_y v + uv_y + u_z w + uw_z \\ &= uu_x + vu_y + wu_z + u\underbrace{(u_x + v_y + w_z)}_{=0} \\ &= uu_x + vu_y + wu_z\end{aligned} \tag{1.27}$$

Navier–Stokes equations and the continuity equation form a system of coupled, non-linear partial differential equations. An analytical solution of this system is impossible for ship flows. Even if the influence of the free surface (waves) is neglected, today's computers are not powerful enough to allow a numerical solution either. Even if such a solution may become feasible in the future, it is questionable if it is really necessary for engineering purposes in naval architecture.

Velocities and pressure may be divided into a time average and a fluctuation part to bring the Navier–Stokes equations closer to a form where a numerical solution is possible. Time averaging yields the Reynolds-averaged Navier–Stokes equations (RANSE). u, v, w, and p are from now on time averages. u', v', w' denote the fluctuation parts. For unsteady flows (e.g. maneuvering), high-frequency fluctuations are averaged over a chosen time interval (assembly average). This time interval is small compared to the global motions, but large compared to the turbulent fluctuations. Most computations for ship flows are limited to steady flows where the terms u_t, v_t, and w_t vanish. The RANSE have a similar form to the Navier–Stokes equations:

$$\begin{aligned}\rho(u_t + uu_x + vu_y + wu_z) &= \rho f_1 - p_x + \mu(u_{xx} + u_{yy} + u_{zz}) \\ &\quad - \rho((\overline{u'u'})_x + (\overline{u'v'})_y + (\overline{u'w'})_z) \\ \rho(v_t + uv_x + vv_y + wv_z) &= \rho f_2 - p_y + \mu(v_{xx} + v_{yy} + v_{zz}) \\ &\quad - \rho((\overline{u'v'})_x + (\overline{v'v'})_y + (\overline{v'w'})_z) \\ \rho(w_t + uw_x + vw_y + ww_z) &= \rho f_3 - p_z + \mu(w_{xx} + w_{yy} + w_{zz}) \\ &\quad - \rho((\overline{u'w'})_x + (\overline{v'w'})_y + (\overline{w'w'})_z)\end{aligned} \tag{1.28}$$

They contain as additional terms the derivatives of the Reynolds stresses:

$$\begin{matrix} -\rho\overline{u'u'} & -\rho\overline{u'v'} & -\rho\overline{u'w'} \\ -\rho\overline{u'v'} & -\rho\overline{v'v'} & -\rho\overline{v'w'} \\ -\rho\overline{u'w'} & -\rho\overline{v'w'} & -\rho\overline{w'w'} \end{matrix} \tag{1.29}$$

The time averaging eliminated the turbulent fluctuations in all terms except the Reynolds stresses. The RANSE require a turbulence model that couples the Reynolds stresses to the average velocities. There are whole books and conferences dedicated to turbulence modeling. Recommended for further studies is, e.g., Ferziger and Peric (1996). Turbulence modeling will not be treated in detail here, except for a brief discourse in Section 1.5.1. It suffices to say that, despite considerable progress in turbulence modeling, none of the present models is universally convincing and research continues to look for better solutions for ship flows. Because we are so far from being able to solve the actual Navier–Stokes equations, we often say 'Navier–Stokes' (as in Navier–Stokes solver) when we really mean RANSE.

'Large-eddy simulations' (LES) are located between Navier–Stokes equations and RANSE. LES let the grid resolve the large vortices in the turbulence directly and only model the smaller turbulence structures. Depending on what is considered 'small', this method lies closer to RANSE or actual Navier–Stokes equations. So far few researchers have attempted LES computations for ship flows and the grid resolution was often too coarse to allow any real progress compared to RANSE solutions. However, many experts see LES as a key technology for maritime CFD applications that may drift into industry application within the next two decades.

Neglecting viscosity – and thus of course all turbulence effects – turns the Navier–Stokes equations (also RANSE) into the Euler equations, which still have to be solved together with the continuity equations:

$$\begin{aligned}
\rho(u_t + uu_x + vu_y + wu_z) &= \rho f_1 - p_x \\
\rho(v_t + uv_x + vv_y + wv_z) &= \rho f_2 - p_y \\
\rho(w_t + uw_x + vw_y + ww_z) &= \rho f_3 - p_z
\end{aligned} \tag{1.30}$$

Euler solvers allow coarser grids and are numerically more robust than RANSE solvers. They are suited for computation of flows about lifting surfaces (foils) and are thus popular in aerospace applications. They are not so well suited for ship flows and generally not recommended because they combine the disadvantages of RANSE and Laplace solvers without being able to realize their major advantages: programming is almost as complicated as for RANSE solvers, but the physical model offers hardly any improvements over simple potential flow codes (Laplace solvers).

A further simplification is the assumption of irrotational flow:

$$\nabla \times \vec{v} = \begin{Bmatrix} \partial/\partial x \\ \partial/\partial y \\ \partial/\partial z \end{Bmatrix} \times \vec{v} = 0 \tag{1.31}$$

A flow that is irrotational, inviscid and incompressible is called potential flow. In potential flows the components of the velocity vector are no longer independent from each other. They are coupled by the potential ϕ. The derivative of the potential in arbitrary direction gives the velocity component in this direction:

$$\vec{v} = \begin{Bmatrix} u \\ v \\ w \end{Bmatrix} = \nabla\phi \tag{1.32}$$

Three unknowns (the velocity components) are thus reduced to one unknown (the potential). This leads to a considerable simplification of the computation.

The continuity equation simplifies to Laplace's equation for potential flow:

$$\Delta\phi = \phi_{xx} + \phi_{yy} + \phi_{zz} = 0 \tag{1.33}$$

If the volumetric forces are limited to gravity forces, the Euler equations can be written as:

$$\nabla\left(\phi_t + \frac{1}{2}(\nabla\phi)^2 - gz + \frac{1}{\rho}p\right) = 0 \tag{1.34}$$

Integration gives Bernoulli's equation:

$$\phi_t + \frac{1}{2}(\nabla\phi)^2 - gz + \frac{1}{\rho}p = \text{const.} \tag{1.35}$$

The Laplace equation is sufficient to solve for the unknown velocities. The Laplace equation is linear. This offers the big advantage of combining elementary solutions (so-called sources, sinks, dipoles, vortices) to arbitrarily complex solutions. Potential flow codes are still the most commonly used CFD tools in ship and propeller design.

Boundary layer equations represent a special branch in the development of hydrodynamics (see Schlichting 1979), which are historically important. The boundary layer equations introduce many simplifications in the physical model: diffusion in the predominant flow direction is neglected, the thickness of the boundary layer is taken as small, and the pressure is constant over the thickness. These assumptions are violated near separating boundary layers. Therefore separation cannot be predicted properly. Of course, neither is any evaluation of the separated flow possible. But this is the area of interest for improving aftbodies and designing the propeller. One of the last doctoral theses on boundary layer methods for ship flows concluded in 1993: 'With the present method the practically interesting velocities at the propeller plane cannot be determined because there is no wall. In order to compute all the velocity components in a thick boundary layer and at the propeller plane, the Navier–Stokes equations have to be solved.'

Boundary layer methods had been substituted almost completely by RANSE solvers by the end of the 1980s. A series of validation workshops demonstrated that the solution of the equations for thin boundary layers failed in the stern region because of the rapid thickening of the boundary layer in this zone. The limited success of generalizations of thin boundary layer equations involving high-order corrections was subsequently demonstrated so that the tendency towards computing the full solution of the Navier–Stokes equations became stronger and stronger because increased computer resources became more and more available at continuously decreasing costs.

Basic equations (and flows) are sometimes classified as elliptic, hyperbolic or parabolic. Consider a two-dimensional differential equation of second order:

$$A\frac{\partial^2 f}{\partial x^2}+2B\frac{\partial^2 f}{\partial x\partial y}+C\frac{\partial^2 f}{\partial y^2}+a\frac{\partial f}{\partial x}+b\frac{\partial f}{\partial y}+cf+d=0 \tag{1.36}$$

For $\delta = AC - B^2 > 0$ the equation is 'elliptic', for $\delta = 0$ 'parabolic' and for $\delta < 0$ 'hyperbolic'. The names are derived from an analogy to the algebraic equation:

$$Ax^2+2Bxy+Cy^2+ax+by+d=0 \tag{1.37}$$

This equation describes for $\delta > 0$ an ellipse, for $\delta = 0$ a parabola, and for $\delta < 0$ a hyperbola. Behind these rather abstract mathematical definitions lies a physical meaning (Fig. 1.2):

- *Elliptic.* disturbances propagate in all directions. RANSE and the Laplace equation are in general elliptic.
- *Hyperbolic.* Disturbances are limited in their propagation to a conical (or in two dimensions a wedge-shaped) region. Supersonic flow with a Mach cone follows a hyperbolic field equation. The Kelvin angle in the wave pattern results in a radiation condition of 'hyperbolic' character.
- *Parabolic.* The extreme case of a hyperbolic flow is a parabolic flow. Here the angle of the cone/wedge opens up to 90°. Disturbances propagate only downstream. 'Parabolic' RANSE solvers allow faster solution with reduced storage requirements. They start the computation at the upstream end and solve strip after strip marching downstream. Instead of considering the whole domain at one time, only two adjacent strips have to be considered at any time. However, local flow reversals could never be captured by such a method because they violate the assumed parabolic character of the flow. Parabolic RANSE solvers thus appeared only shortly in the 1980s and were replaced by fully elliptic solvers when more computer power became widely available. All unsteady equations are parabolic in time.

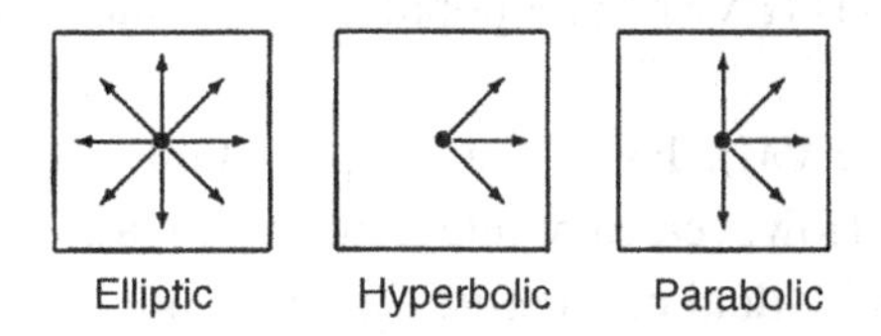

Figure 1.2:
A disturbance propagates differently depending on the type of field equation

1.4.2. Basic CFD Techniques

CFD comprises methods that solve the basic field equations subject to boundary conditions by approaches involving a large number of (mathematically simple) elements. These approaches lead automatically to a large number of unknowns.

Basic CFD techniques are:

- *Boundary element methods (BEM).* BEM are used for potential flows. For potential flows, the integrals over the whole fluid domain can be transformed to integrals over the boundaries of the fluid domain. The step from space (3-d) to surface (2-d) simplifies grid generation and often accelerates computations. Therefore practical applications for potential flows about ships (e.g. wave resistance problems) use exclusively BEM, which are called panel methods. Panel methods divide the surface of a ship (and often part of the surrounding water surface) into discrete elements (panels). Each of these elements automatically fulfills the Laplace equation. Indirect methods determine the element strengths so that at the collocation points (usually centers of the panels) a linear boundary condition (e.g. zero normal velocity) is fulfilled. This involves the solution of a full system of linear equations with the source strengths as unknowns. The required velocities are computed in a second step, hence 'indirect' method. Bernoulli's equation then yields the pressure field. Direct methods determine the potential directly. They are less suited for boundary conditions involving higher derivatives of the potential, but yield higher accuracy for lifting flows. Most commercially used codes for ship flows are based on indirect methods. BEM cannot be used to solve RANSE or Euler equations. Fundamentals of BEM can be found in, e.g., Hess (1986, 1990).
- *Finite element methods (FEM).* FEM dominate structural analysis. For ship hydrodynamics they play only a minor role. Unlike in structural analysis, the elementary functions cannot also be used as weight functions to determine the weighted error integrals (residuals) in a Galerkin method. This reduces the elegance of the method considerably. Fundamentals of FEM can be found in, e.g., Chung (1978).
- *Finite difference methods (FDM).* FDM discretize (like FEM) the whole fluid domain. The derivatives in the field equations are approximated by finite differences. Discretization errors can lead to a violation of conservation of mass or momentum, i.e. in the course of a simulation the amount of water might diminish continuously. While FDM lose popularity and finite volume methods (FVM) gain popularity, in many cases FDM give results of comparable quality.
- *Finite volume methods (FVM).* FVM also employ finite differences for the spatial and temporal discretization. However, they integrate the equations for mass and momentum conservation over the individual cell before variables are approximated by values at the cell centers. This ensures conservativeness, i.e. mass and momentum are conserved because errors at the exit face of a cell cancel with errors at the entry face of the neighbor

cell. Most commercial RANSE solvers today are based on FVM. Fundamentals of FVM can be found in Versteeg and Malalasekera (1995), and Ferziger and Peric (1996).

FEM, FDM, and FVM are called 'field methods', because they all discretize the whole fluid domain (field) as opposed to BEM, which just discretize the boundaries.

Some textbooks on CFD also include spectral methods which use harmonic functions as elementary solutions. Spectral methods have no practical relevance for ship flows. The interested reader will find an introduction in Peyret and Taylor (1985).

1.4.3. Applications

Practical CFD applications for ship flows concentrate mainly on the ship moving steadily ahead. A 1994 survey at ship model basins showed inviscid BEM computations for wave-resistance and offshore seakeeping as still the most important CFD application for commercial projects (ca. 40–50% of the turnover), followed by RANSE applications (30–40%) and computations for propellers (10–20%). All other applications combined contribute less than 5% of the turnover in the commercial sector. This global decomposition has remained remarkably stable despite an overall increase in CFD volume. Besides global aspects like resistance, sometimes local flow details are the focus of attention, e.g. the design of shaft brackets, stabilizing fins, or sonar domes (noise reduction), e.g. Larsson et al. (1998) and Larsson (1997).

The most important applications are briefly discussed in the following.

- *'Resistance and propulsion'*. CFD applications are mainly concerned with steadily advancing ships. For a double-body potential flow, where the wave-making at the free surface and the effects of viscosity are neglected, the flow computation is relatively simple, quick, and accurate. The name 'double-body flow' comes from an interpretation that the ship's hull is reflected at the waterline at rest. Then the flow in an infinite fluid domain is computed and the lower half of the flow gives automatically the flow about a ship with an undeformed (rigid) water surface. The double-body potential flow is only used as an approximate solution for other methods (boundary layer, wave resistance, seakeeping). The simultaneous consideration of viscosity and wave-making drifted into industry applications after the year 2000. But even a decade later, many viscous flow computations in practice still neglected wave-making. For steady free-surface flows ('wave-resistance problem'), inviscid BEM codes are still the workhorse. The propeller is almost always neglected in BEM computations for the steady flow ('resistance problem'). RANSE computations include the propeller action ('propulsion problem'), usually by applying an equivalent body force in the r.h.s. of the RANSE. The body forces were traditionally prescribed based on experience or experimental results. More sophisticated applications used integrated propeller models. The body forces in both thrust and rotative directions are then estimated, e.g. by a panel method. The distributions obtained by this approach depend on the propeller inflow and are determined iteratively using the RANSE data as input for the propeller computation and vice versa. The approach usually converges quickly.

- *Maneuvering.* Aspects of maneuvering properties of ships gain in importance, as public opinion and legislation are more sensitive concerning safety issues after spectacular accidents of tankers and ferries. IMO regulations concerning the (documented) maneuverability of ships increased the demand for CFD methods in this field. Model tests as an alternative method are expensive and time-consuming. Traditional simple simulation methods with hydrodynamic coefficients gained from analytical approaches or regression analysis (of usually small databases) are often considered as too inaccurate. CFD applications to simulate maneuvering model tests have progressed considerably over the last decade. However, it is difficult to assess the state of the art. By 2010, several research groups presented full maneuvers simulated in RANSE computations. Some companies used by 2010 CFD rather than model tests to furnish maneuvering models to nautical simulators. Yet at the same time, validation workshops showed disappointingly large scatter of results between different simulations. CFD for ship maneuvering appears to be a threshold technology, where we may need another decade before wider confidence in CFD as preferred technique will be established in the industry. Predicting the flow around the hull and appendages (including propellers and rudders) is much more complicated than predicting the steady flow in resistance and propulsion problems. Often, both viscosity and free-surface effects (e.g. dynamic trim and sinkage) play an important role. The rudder is most likely in the hull boundary layer, often operating in the propeller wake. The hull forces themselves are also difficult to predict computationally, because sway and yaw motions induce considerable cross-flows with shedding of strong vortices. Both BEM and field methods have been employed for selected maneuvering problems. Selected problems like side forces and moments in steady yaw are well predicted, but longitudinal forces and some flow details still showed considerable errors for real ship geometries.
- *Ship seakeeping.* The 1990s saw the advent of Rankine panel methods for seakeeping. The approaches are similar to those used for the steady wave-resistance problem, but failed to reach a similar level of acceptance. Most properties of practical relevance are calculated accurately enough for most ship types by strip methods, although the underlying physical models are generally considered as crude. The two-dimensional flow calculations for the individual strips today are based almost always on BEM, namely close-fit methods. RANSE methods for global ship motions have matured over the past decade to industry application. These simulations are applied when strong non-linearities are involved, e.g. green water on deck.
- *Slamming/water-entry problems.* Using suitable space–time transformations, the water entry of a two-dimensional wedge can also be used to model the hydrodynamics of

planing hulls. We will focus here on the seakeeping aspect of modeling water-entry problems. Slamming involves local loads changing rapidly in time and space. Hydroelastic effects, interaction between trapped air pockets and water, velocities that require consideration of water compressibility with shockwaves forming and the complex shapes of the water surface forming jets, make slamming problems already in two dimensions very challenging. Traditional approaches work well for wedges of suitable deadrise angle and two-dimensional flows. But usually ship cross-sections do not have suitable deadrise angles and the phenomena are three-dimensional. CFD has brought substantial progress in this field. Free-surface RANSE simulations are today standard industry practice to predict slamming loads. The focus lies here on forces and deformations, not local pressures. Similarly, sloshing analyses (= internal impact problem) are based on free-surface RANSE simulations.

- *Zero-speed seakeeping.* For offshore applications, global loads and motions in seakeeping can be computed quite well by BEM. For zero speed, the steady wave system vanishes and various diffraction and radiation wave systems coincide. If the geometries of offshore structure and waves are of the same order of magnitude BEM can successfully capture three-dimensional effects and complex interactions. The employed three-dimensional BEM determine forces and motions either in the time or the frequency domain. First-order forces and motions are calculated reliably and accurately. For practically required accuracy of first-order quantities, 1000–2000 elements are typically deemed sufficient. Commercial program packages (WAMIT, TIMIT, AQWA, or DIODORE) are widely accepted and used for offshore applications.
- *Propeller flows.* Inviscid flow methods have long been used in propeller design as a standard tool yielding information comparable to experiments. Lifting-surface methods and BEMs are equally popular. Lifting-surface methods (quasi-continuous method, vortex-lattice method) allow the three-dimensional modeling of the propeller. They discretize the mean camber surface of the propeller blade by individual vortex panels. In addition, the free vortices are modeled by elements of given strength. Other than the BEM described below, lifting-surface methods do not fulfill exactly the boundary conditions at the blade's lower and upper surfaces. However, the resulting errors are small for thin blades. BEM represent an improvement concerning the treatment and modeling of the geometry. BEM model both lift and displacement of the propeller blades by surface panels and/or dipoles. They can also model the propeller hub. Despite the theoretical superiority, BEM results were not clearly better than lifting-surface method results in benchmark tests. BEM codes for propeller applications often use only dipole panels which are distributed over hub, blade surfaces, and the wakes of each blade. Viscous flow CFD methods are applied by industry for complex configurations. Considerable progress in propulsive efficiency is expected when propellers are designed modeling ship, propeller and rudder together, for full scale, with CFD. This is expected to become industry standard practice by 2030.

Further, less frequently found applications of CFD in naval architecture include:

- *Air flow.* CFD has been applied to air flows around the upper hull and superstructure of ships and offshore platforms. Topics of interest are:
  - Wind resistance (especially of fast ships)
    For fast ships the wind resistance becomes important. For example, for one project of a 50 knot SES (surface effect ship = air-cushion catamaran), the wind resistance constituted ca. 25% of the total resistance. Hull changes limited to the bow decreased the wind resistance by 40%.
  - Wind-over-the-deck conditions for helicopter landing
    This application concerns both combatants and offshore platforms.
  - Wind loads
    Wind loads are important for ships with large superstructures and relatively small lateral underwater area, e.g. car transporters, car ferries, container ships, SES, and air-cushion vehicles.
  - Tracing of funnel smoke
    This is important for passenger vessels (passengers on deck, paintwork) and for offshore platforms (safety of helicopter operation). Formal optimization has been combined with CFD to minimize smoke dispersion on the deck of a yacht (Harries and Vesting 2010).

  The comparison of CFD, wind-tunnel tests, and full-scale measurements shows an overall good agreement, even if large discrepancies appear at some wind directions. The differences between CFD and model-test results are not generally larger than between full-scale and model-scale results. In fact, the differences are not much larger than often found when the same vessel is tested in different wind tunnels. The determination of wind loads on ships and offshore structures by CFD is a realistic alternative to the experimental methods.
- *Interior flows.* Sloshing in partially filled tanks is a standard CFD application, and required by classification societies for some cases. Sloshing computations may be coupled to the outer (global) motions of a ship, but industry practice uses only weak coupling: the global ship motions are prescribed for the tank, but the effect of the fluid motion in the tank on the global ship motions is neglected. Related problems are flows in a roll damping tank and water flowing into a damaged ship.

Table 1.1 summarizes an assessment of the maturity of the various CFD applications.

1.4.4. Cost and Value Aspects of CFD

The value of any product (or service) can be classified according to time, cost, and quality aspects. For CFD this means:

- *Time benefits* (How does CFD accelerate processes in ship design?). In the shipbuilding industry, we see the same trends towards ever-decreasing times for product development

Table 1.1: Maturity of CFD application on a scale from – (not applicable, no applications known) to •••• (very mature)

	Viscous	Inviscid
'Resistance test'	•••	•••
'Propulsion test'	•••	–
Maneuvering	•	•
Ship seakeeping	••	•••
Offshore	•	•••
Propeller	••	••••
Others	••	–

as in other manufacturing industries. In some cases, delivery time is the key factor for getting the contract. CFD plays a special role in this context. A numerical pre-optimization can save time-consuming iterations in model tests and may thus reduce total development time. The speed of CFD allows applications already in preliminary design. Early use thus reduces development risks for new ships. This is especially important when exploring niche markets for unconventional ships where design cannot be based on experience. In addition, another aspect related to turnover has to be realized: CFD improves chances of successful negotiations by supplying hydrodynamic analyses. It has become standard for all high-tech shipbuilders to apply at least inviscid CFD analyses to proposed hull designs when entering negotiations to obtain a contract for building a ship.

- *Quality benefits* (How does CFD enable superior ships or reduce risks in new designs?). Model tests are still more accurate for power prognosis than CFD. We see occasionally good agreement of CFD power prediction with measured data, but these cases may just benefit from fortunate error cancellation or tuning of parameters to fit a posteriori the experimental data. No 'blind' benchmark test has yet demonstrated the ability of CFD codes to predict, at least with 5% accuracy, consistently the power of ship hulls at design speed. I expect this to remain so for some more years. In the long run, CFD should outperform model tests, as with growing computational power, accurate simulations at full scale will become available overcoming current uncertainties in correlating model tests to full-scale predictions. For some projects, it is only important to ensure that a given installed power will enable the ship to achieve contract speed. In these cases, CFD is of little interest. However, CFD should be considered in cases where model test results show problems or the shipowner is willing to pay a higher price for lower operating costs (due to improved hull). CFD allows insight in flow details not offered by the standard model tests. Insight in flow details is especially important in cases where initial model tests show that power requirements for a given hull are far more than expected. Here CFD also allows the investigation of the flow far below the waterline and modifications can be quickly analyzed to see if major improvements are to be expected. The model tests and experience of a towing tank mainly indicate the potential for improvement; CFD indicates where and how to improve the design. CFD also allows formal optimization, using hundreds of 'numerical towing tanks' in parallel.

- *Cost benefits* (How does CFD reduce costs in ship designs?). While the influence of certain decisions and actions on the turnover can be estimated only qualitatively, costs can usually be quantified directly. This explains why management prefers investments with a short payback due to cost reductions even though there is general consent that cost reductions alone do not ensure the economic future of a company. However, CFD's potential for direct cost reductions is small. CFD is still considered widely as not accurate enough to substitute the model test for power prognosis. Therefore, *one* model test is always performed. In three out of four projects of the Hamburg Ship Model Basin this was sufficient already in 1990. By 2010, a single model test trial had become standard practice. Thus, there is little direct cost reduction potential. Indirect cost savings in other departments are difficult to quantify. Time benefits of CFD will also affect costs. It is possible to determine 40–60% of the total production costs of a ship in the first weeks of design. Costs for modifications in later stages are higher by order of magnitudes than those necessary at the conceptual phase. Various decisions concerning production costs can be made earlier and involve lower risks if CFD is employed consistently to determine the final hull form at an earlier time.

The benefits discussed so far only cover one-half of a cost–benefit analysis for a CFD strategy. Understanding the cost structure of CFD is at least as important and some general management guidelines can be deduced. This requires a closer look at the work process in CFD. The work process is split into:

- preprocessing (generation and quality control of grids);
- computation;
- postprocessing (graphical displays, documentation).

The individual steps sometimes have to be performed several times in iterations. Cost structures will be discussed separately for each step:

1. *Preprocessing.* Preprocessing requires staff familiar with the special programs for grid generation, especially on the hull. This requires at least a basic understanding of the subsequent CFD computation. Grid generation is best performed on workstations or fast PCs with high-resolution screens. User experience and a degree of automation mainly determine time (and labor) costs in this step. Progress in grid generation and more robust CFD has reduced the man time involved in CFD analyses. Largely automatic procedures are now available for many applications. Staff training and grid generation software are the main fixed costs in this step.
2. *Computation.* The computation involves almost no man time. Computations for inviscid CFD can usually run on PCs; viscous CFD and formal optimization require more powerful computer environments with parallel computing. Computing costs usually

account for less than 1% of total costs. Also, software licenses for the flow code are often negligible compared to other costs, especially those for training (Bertram and Couser 2010).

3. *Postprocessing*. Postprocessing is generally based on commercial software. Postprocessing requires some time (typically 10–30% of the total time). Increasingly, videos are used for unsteady CFD applications. Interpretation of results still requires expertise. You pay thus for the skilled interpretation, not the number of color plots.

The high fixed costs for training and user-defined macros to accelerate the CFD process lead to considerable economies of scale. This is often not realized by management. Experience shows that many shipyards buy CFD software, because the hardware is available or not expensive, and the software license costs may be as much as a few CFD analyses at a consulting company. The vendors are naturally only too happy to sell the software. Then the problems and the disillusion start. Usually no initial training is given by the vendor (or bought by the shipyard). Typical beginners' mistakes are the consequence:

- Time is lost in program handling
- Unsuitable models and grids are used requiring repeated analyses or leading to useless results.

By the time the management realizes the problems, it is usually too late. The software licenses are all bought, the design engineer has now already invested (lost) so much time struggling with the code. Nobody wants to admit defeat. So the CFD analyses are continued in-house with the occasional outsourcing when problems and time pressures become too large. As a general rule, outsourcing is recommended for shipyards and design offices with fewer than five projects per year. In-house CFD makes sense starting from ten projects per year. For finite-element analyses of structures we have seen a development that after an initial period where shipyards performed the analyses in-house the pendulum swung the other way with shipyards now using almost exclusively outsourcing as the sensible option. A similar development is expected for most specialized CFD applications.

Model generation plays a vital role for costs, response time, and quality of results. In this respect, CFD analyses have benefited considerably from the progress in guided or automated grid generation since the year 2000.

- *By making grid generation more user-friendly.* Grid generation was largely a matter of experience. The logical deduction was to incorporate this experience in the grid generation codes to improve user-friendliness. A fundamental dilemma found in model generation is that the procedures should be flexible to cope with a variety of problems, yet easy to handle with a minimum of input. Many flow codes offer a lot of flexibility, often at the cost of having many options which in turn leave inexperienced (i.e. occasional) users frustrated and at risk to choose exactly the wrong options for their problems. Incorporation of expert knowledge in the model generation program offering reasonable default options is a good solution to this dilemma. In the extreme case, a user may choose the 'automatic mode' where the program proceeds completely on its own knowledge. On the other hand, default values may be overruled at any stage by an experienced user.

- *By making the computation more robust.* A simple grid is cheap and fast to generate, but unsuitable for most marine problems. Therefore modern, marine CFD applications use:
  - block-structured grids, sometimes with sliding block interfaces;
  - non-matching boundaries between blocks;
  - unstructured grids;
  - chimera grids (overlapping, non-matching blocks).
- *By generating grids only once.* Time for grid generation means *total* time for all grids generated. The philosophy is to 'Get it right the first time', i.e. the codes are robust enough or the grid generators good enough that grids need to be created only once. This should also favor the eventual development of commercial codes with adaptive grid techniques. First industrial applications to ships appeared around 2010.

Standard postprocessing could save time and would also help customers in comparing results for various ships. However, at present we have at best internal company standards on how to display CFD results.

1.5. Viscous Flow Computations

Fundamentals of viscous flows are covered in detail in Ferziger and Peric (1996). Fundamentals of potential flow methods are found on the website. I will therefore limit myself here to a naval architect's view of the most important issues for applications of these methods. This is intended to raise the understanding of the matter to a level sufficient to communicate and collaborate with a CFD expert.

1.5.1. Turbulence Models

The RANSE equations require external turbulence models to couple the Reynolds stresses (terms from the turbulent fluctuations) to the time-averaged velocities. Turbulence is in general not fully understood. All turbulence models used for ship flows are semi-empirical. They use some theories about the physics of turbulence and supply the missing information by empirical constants. None of the turbulence models used so far for ship flows has been investigated for its suitability at the free surface. On the other hand, it is not clear whether an exact turbulence modeling in the whole fluid domain is necessary for engineering purposes. There are whole books on turbulence models and we will discuss here only the most popular turbulence models, which are standard options in commercial RANSE solvers. ITTC (1990) gives a literature review of turbulence models as applied to ship flows.

Turbulence models may be either algebraic (0-equation models) or based on one or more differential equations (one-equation models, two-equation models, etc.). Algebraic models compute the Reynolds stresses directly by an algebraic expression. The other models require the parallel solution of additional differential equations which is more time consuming, but also more accurate.

The six Reynolds stresses (or more precisely their derivatives) introduce six further unknowns. Traditionally, the Boussinesq approach has been used in practice which assumes isotropic turbulence, i.e. the turbulence properties are independent of the spatial direction. (Detailed measurements of ship models have shown that this is not true in some critical areas in the aftbody of full ships. It is unclear how the assumption of isotropic turbulence affects global properties like the wake in the propeller plane.) The Boussinesq approach then couples the Reynolds stresses to the gradient of the average velocities by an eddy viscosity μ_t:

$$-\rho\begin{bmatrix}\overline{u'u'} & \overline{v'u'} & \overline{w'u'}\\ \overline{u'v'} & \overline{v'v'} & \overline{w'v'}\\ \overline{u'w'} & \overline{v'w'} & \overline{w'w'}\end{bmatrix} = \mu_t\begin{bmatrix}2u_x & u_y+v_x & u_z+w_x\\ u_y+v_x & 2v_y & w_y+v_z\\ u_z+w_x & w_y+v_z & 2w_z\end{bmatrix} - \begin{bmatrix}\frac{2}{3}\rho k & 0 & 0\\ 0 & \frac{2}{3}\rho k & 0\\ 0 & 0 & \frac{2}{3}\rho k\end{bmatrix} \tag{1.38}$$

where k is the (average) kinetic energy of the turbulence:

$$k = \frac{1}{2}(\overline{u}^2 + \overline{v}^2 + \overline{w}^2) \tag{1.39}$$

The eddy viscosity μ_t has the same dimension as the real viscosity μ, but unlike μ it is not a constant, but a scalar depending on the velocity field. The eddy viscosity approach transforms the RANSE to:

$$\begin{aligned}\rho(u_t + uu_x + vu_y + wu_z) &= \rho f_1 - p_x - \frac{2}{3}\rho k_x + (\mu + \mu_t)(u_{xx} + u_{yy} + u_{zz})\\ &\quad + \mu_{tx}2u_x + \mu_{ty}(u_y + v_x) + \mu_{tz}(u_z + w_x)\\ \rho(v_t + uv_x + vv_y + wv_z) &= \rho f_2 - p_y - \frac{2}{3}\rho k_y + (\mu + \mu_t)(v_{xx} + v_{yy} + v_{zz})\\ &\quad + \mu_{tx}(u_y + v_x) + \mu_{ty}2v_y + \mu_{tz}(w_y + v_z)\\ \rho(w_t + uw_x + vw_y + ww_z) &= \rho f_3 - p_z - \frac{2}{3}\rho k_z + (\mu + \mu_t)(w_{xx} + w_{yy} + w_{zz})\\ &\quad + \mu_{tx}(u_z + w_x) + \mu_{ty}(w_y + v_z) + \mu_{tz}2w_z\end{aligned} \tag{1.40}$$

Turbulence models generally use a reference length scale and reference velocity scale. Alternatively, the velocity scale can be expressed as the fraction of length scale and a time scale. To obtain the proper dimension, the eddy viscosity is expressed proportional to the product of length scale and velocity scale. The length scale is characteristic for the larger turbulence structures which are mainly contributing to the momentum transfer in the fluid. The velocity scale is characteristic for the intensity of the turbulent fluctuations.

All commonly used turbulence models are plagued by considerable uncertainties. Internationally renowned fluid dynamicists have described turbulence models as follows:

- 'Turbulence models are voodoo. We still don't know how to model turbulence.'
- 'The word "model" is a euphemism for an uncertain, but useful postulated regularity. In the last few decades, scientists have learned to simulate some aspects of turbulence effects by the invention of "turbulence models" which purport to represent the phenomena by postulated laws of conservation, transport and sources for supposed "properties of turbulence" such as its "energy", its "frequency" or its "length scale". These "laws" are highly speculative.'

Researchers have succeeded in direct numerical simulation of turbulence for Reynolds numbers several orders of magnitude smaller than ship model Reynolds numbers and for very simple geometries. These simulations allow one at best to understand phenomena of turbulence better and to test engineering turbulence models.

The usefulness of a turbulence model for ship flows can only be evaluated in benchmark tests for similar ships. Sometimes simple models work surprisingly well; sometimes the same model fails for the next ship. The most popular turbulence model for ship flow applications in practice remains the standard k–ε model, although its results were not convincing in benchmark tests for several ship geometries.

By the late 1990s, k–ω models were proposed for ship flows. These models are like the k–ε two-equation models and can be seen as a further evolution of them. ω is proportional to ε/k and can be interpreted as a 'turbulence frequency'. k–ω models yield better agreement with experiments in many cases; however, they react more sensitively to grid quality.

Reynolds stress models calculate the individual Reynolds stresses from their modeled transport equations without recourse to an eddy viscosity hypothesis. These models require more computational effort than, e.g., the two-equation k–ε model, but showed superior results in several ship flow applications. However, it is not yet decided if similarly good results can be obtained by simple turbulence models with properly adjusted coefficients.

Large-eddy simulations may eventually solve the current debate over turbulence modeling for engineering applications, but for ship flows we will have to wait at least two more decades before realistic LES solutions become available in practice.

Probably the most widely used turbulence model in engineering applications is the (standard) $k-\varepsilon$ model (Launder and Spalding 1974). k is the kinetic energy of the turbulence, ε the dissipation rate of k. The $k-\varepsilon$ model expresses the eddy viscosity μ_t as a simple function of k and ε:

$$\mu_x = 0.09\rho\,\frac{k^2}{\varepsilon} \tag{1.41}$$

where 0.09 is an empirical constant. k and ε are expressed by two partial differential equations involving further empirical constants:

$$\frac{\mathrm{D}k}{\mathrm{D}t} = \frac{1}{\rho}\left(\left(\frac{\mu_t}{1.0}k_x\right)_x + \left(\frac{\mu_t}{1.0}k_y\right)_y + \left(\frac{\mu_t}{1.0}k_z\right)_z\right) + P_k - \varepsilon \tag{1.42}$$

$$\frac{\mathrm{D}\varepsilon}{\mathrm{D}t} = \frac{1}{\rho}\left(\left(\frac{\mu_t}{1.2}\varepsilon_x\right)_x + \left(\frac{\mu_t}{1.2}\varepsilon_y\right)_y + \left(\frac{\mu_t}{1.2}\varepsilon_z\right)_z\right) + 1.44\,\frac{\varepsilon}{k}P_k - 1.92\,\frac{\varepsilon^2}{k} \tag{1.43}$$

P_k is the production rate of k:

$$P_k = \frac{\mu_t}{\rho}\,(2u_xu_x + (u_y + v_x)u_y + (u_z + w_x)u_z + (v_x + u_y)v_x + 2v_yv_y \tag{1.44}$$

$$+(v_z + w_y)v_z + (w_x + u_z)w_x + (w_y + v_z)w_y + 2w_zw_z)$$

The substantial derivative is defined as usual:

$$\frac{\mathrm{D}}{\mathrm{D}t} = \frac{\partial}{\partial t} + u\,\frac{\partial}{\partial x} + v\,\frac{\partial}{\partial y} + w\,\frac{\partial}{\partial z} \tag{1.45}$$

These equations contain four empirical constants (1.0, 1.2, 1.44, and 1.92) which were determined (in a best fit approach) for very simple flows in physical and numerical experiments. The applicability to other turbulent flows (e.g. around ship geometries) was never explicitly validated.

The $k-\varepsilon$ model cannot be applied directly at a wall (ship hull) as it assumes inherently high (local) Reynolds numbers. If a no-slip condition (zero relative speed at the hull) is to be enforced directly at the wall, the ε differential equation must be substituted by an algebraic equation near the wall. This is the so-called low-Re $k-\varepsilon$ model. However, more popular is the introduction of a wall function coupled to the standard $k-\varepsilon$ model. The wall function is empirically determined for two-dimensional flows. One assumes that the velocity increases logarithmically with distance from the wall:

$$\frac{u}{u_\tau} = \begin{cases} y^+ & y^+ \le y_m^+ \\ \dfrac{1}{0.42}\ln(9.0y^+) & y^+ > y_m^+ \end{cases} \tag{1.46}$$

where y_m + is implicitly given by:

$$y_m^+ = \frac{1}{0.42} \ln(9.0 y_m^+) \tag{1.47}$$

0.42 and 9.0 are empirical constants. $y^+ = y\rho u_\tau/\mu$ is a non-dimensional distance from the wall, u the velocity in longitudinal (parallel to the wall) direction, $u_\tau = \sqrt{\tau_w/\rho}$ with τ_w the wall shear stress.

The centers of the innermost cells should lie within a distance from the wall where the logarithmic law of the wall function applies, i.e. $100 < y^+ < 1000$. However, y^+ contains the wall shear stress, which is part of the solution and not a priori known. It is thus only possible to judge a posteriori if a chosen wall distance was appropriate. Higher Reynolds numbers require generally smaller y^+.

The fundamental assumptions for the wall function are:

- velocity gradient in normal direction is much larger than in other directions;
- pressure gradient and gravity influence are so small that shear stresses in the boundary layer are constant;
- shear stresses and velocity vectors have the same direction in the whole boundary layer;
- equilibrium of turbulence generation and dissipation;
- linear variation of the reference length for turbulence.

These assumptions are questionable for complex flows as found in the aftbodies of ships. The standard k–ε model usually over-predicts the turbulent kinetic energy in the stern region. Also, the model cannot properly account for the reduction of the turbulent kinetic energy near the wall when the viscous layer becomes thick over the stern. The wall function approach usually yields worse results for the wall shear stresses than turbulence models that apply a no-slip condition directly at the wall. However, the wall function saves many cells (and thus computational time).

The k–ε model appears suitable for flows with a predominant boundary-layer character. Problems with defining a reference length, as in many algebraic models, are avoided and at least the important physical aspect of turbulence transport is explicitly reflected in the model. The wall function makes the approach numerically efficient, but the model is in principle not capable of predicting flow separation for curved surfaces (e.g. ships!).

1.5.2. Boundary Conditions

The computational grid can only cover part of the real fluid domain. This introduces artificial boundaries of the computational domain in addition to the physical boundaries of the hull and the free surface.

For ships moving straight ahead (as in simulations of resistance or propulsion tests), the midship plane is generally treated as a symmetry plane. The usual symmetry of the ship would intuitively suggest that this indeed reflects physical reality. However, viscous flows with symmetric inflow to symmetric bodies do not automatically result in symmetric flow patterns at all times. Vortex shedding results in asymmetric flow patterns which only in the time average are symmetric again. This may result in considerable differences in the resistance. The following example may illustrate the problem (Fig. 1.3). Behind a circular cylinder in uniform inflow one would assume intuitively a symmetrical flow which would hardly be disturbed by a flat plate behind the cylinder. However, experiments yield a considerably smaller resistance coefficient for the cylinder with a flat plate. The reason is vortex shedding behind the cylinder with large vortices oscillating from one side to the other. These large-vortex oscillations are blocked by the flat plate.

The boundary condition on the hull is a no-slip condition (zero relative speed) which is either enforced directly or via a wall function depending on the turbulence model employed.

The side and bottom boundaries may correspond to an actual physical boundary as in a model tank. In this case, the boundaries may be treated similar to the ship hull with a no-slip condition. However, one should remember that the outer boundaries then have a relative velocity to the ship. Usually, the physical boundaries would be too far away to be considered. Then the side and bottom boundaries should be sufficiently removed from the ship. Often the side and bottom boundaries form part of a cylinder (quarter for double-body flow with symmetry in y), as a cylinder usually leads to better grids (for single-block grids) than a block-type grid. A typical cylinder radius is one ship length.

At the inlet all unknowns are specified. If the inlet is chosen sufficiently upstream of the ship, uniform flow with corresponding hydrostatic pressure can be assumed. If the $k-\varepsilon$ model is employed, the distributions of k and ε at the inlet also have to be specified. The influence of the specified values for k and ε decays rapidly downstream, such that errors have decayed by some

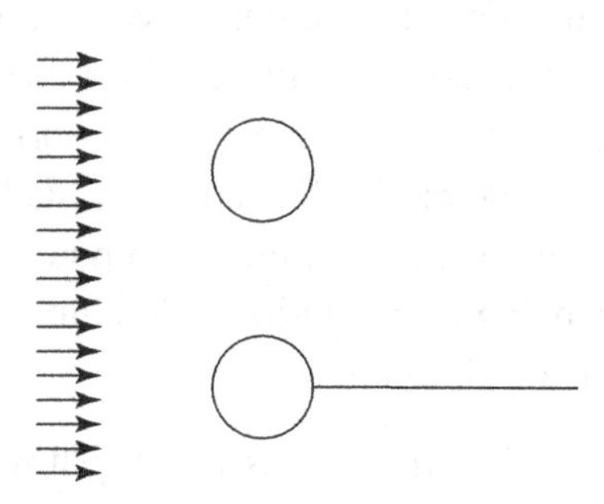

Figure 1.3:
A cylinder with a flat plate in the wake has a considerably lower resistance coefficient than a cylinder without a plate. Care is required when assuming symmetry planes for viscous flow computations

orders of magnitude after several cells. One may then simply specify zero k and ε. A slightly more sophisticated approach estimates the turbulence intensity I (turbulent fluctuation component of a velocity component made non-dimensional with the ship speed V) at the inlet. For isotropic turbulence we then get:

$$k = \frac{3}{2}(VI)^2 \tag{1.48}$$

In the absence of experimental data for I in a specific case, $I = 5\%$ has often been assumed. The dissipation rate is typically assumed as:

$$\varepsilon = 0.164\frac{k^{1.5}}{\ell} \tag{1.49}$$

where 0.164 is an empirical constant and ℓ a reference length. For ship flows, there are few indications as to how to choose this reference length. Our own computations have used 1/100th of the radius of the cylinder forming the computational domain. However, the initial choice of the quantities does not influence the final result, but 'only' the required number of iterations to obtain this result. If only the aftbody is considered, then the inlet may be placed, for example, amidships. In this case all unknowns must be specified from experiments (for validation studies) or simpler computations (e.g. coarse grid computations for the whole domain, inviscid flow computations coupled with boundary layer computations, etc.).

At the outlet usually the derivatives in the longitudinal direction are set for all unknowns to zero and the flow leaving the domain is determined such that continuity is preserved for the whole fluid domain. The longitudinal derivatives are in reality not zero, but this boundary condition prevents upstream propagation of any reflections of disturbances created by the numerical method. Numerical experiments show that these boundary conditions affect results only in a small local region near the outlet.

At symmetry planes normal velocity and all derivatives in the normal direction are set to zero. Since the normal derivatives of the tangential velocities vanish, the shear stresses are zero. The outer boundary of the computational domain (side and bottom) may be treated as 'symmetry plane', i.e. on each outer cell face normal velocity and all normal derivatives are set zero. In this case, the outer boundary must be far away from the ship such that virtually uniform flow is encountered. Another possibility is to specify values computed by inviscid codes at the outer boundary, which allows much smaller computational domains, but not many fewer cells as most cells are concentrated near the ship hull.

If a propeller is modeled in RANSE computations for ship flows (propulsion test condition), it is generally simplified by specifying the propeller effect as body forces. This simulates the acceleration of the flow by the propeller. The sum of all axial body forces yields the thrust. The distribution is often assumed to be parabolic in the radial direction and constant in the circumferential direction. Alternatively, the distribution of the body forces for the propeller may be specified from:

- experience with similar ships;
- experiments for the actual ship;
- alternating computations for the propeller loading with non-uniform inflow from the RANSE computation. The propeller loading is then computed every 10 or 20 iterations in the RANSE computation as the propeller loading converges much faster than the other properties of the RANSE computation. Convergence for the propeller loading is usually obtained with five or seven iterations.

1.5.3. Free-Surface Treatment

Most viscous flow computations for ships in design projects in the 1990s still assumed the free surface to be a symmetry plane. In reality this is not true. The free surface forms waves which break locally at the bow, and the ship changes trim and sinkage (squat) due to the free surface. The problem of turbulence models (and their specific boundary conditions) near the free surface has not been addressed in ship flows and generally the same conditions as for symmetry planes are used.

A variety of methods exists to capture wave-making with various degrees of success. The difficulty with the unknown free-surface position is usually resolved by considering the problem transient, starting from rest. The hull is thus accelerated to the requested Froude number and the time integration is continued until steady state conditions have been achieved. (This procedure corresponds to usual practice in towing tank experiments.) The free-surface position is updated as part of the iterative process.

The methods for computing flows with a free surface can be classified into two major groups:

- Interface-tracking methods define the free surface as a sharp interface whose motion is followed. They use moving grids fitted to the free surface and compute the flow of liquid only. Problems are encountered when the free surface starts folding or when the grid has to be moved along walls of a complicated shape (like a real ship hull geometry).
- Interface-capturing methods do not define a sharp boundary between liquid and gas and use grids which cover both liquid- and gas-filled regions. Either marker particles or a transport equation for the void fraction of the liquid phase are used to capture the free surface.

Interface tracking has become the standard approach as the ability to model complex geometries of hull and water surface is essential for real ship flows. Interface-tracking methods may also solve the flow in the air above the water, but for most ship flows this is not necessary.

A typical approach uses an extended volume-of-fluid (VOF) formulation introducing an additional scalar function, which describes the volume concentration of water, to identify the position of the free surface.

Initial problems with numerical damping of the ship wave propagation have been overcome by better spatial resolution and using higher-order schemes.

1.5.4. Further Details

The vector equations for conservation of momentum yield three scalar equations for three-dimensional computations. These determine the three velocity components for a given pressure. Usually these velocities do not fulfill the continuity equation. The introduction of a pressure correction equation derived from the continuity equation allows a correction of pressure and velocities. Popular methods for such pressure–velocity coupling are:

- SIMPLE (*s*emi-*im*plicit *p*ressure *l*inked *e*quations) and related methods;
- PISO (*p*ressure *i*mplicit with *s*plitting of *o*perators).

In the 1990s most RANSE codes used for ship flows employed SIMPLE or related pressure–velocity coupling. The SIMPLE method is fast, but tends to slow convergence for suboptimal grids. Figure 1.4 gives a simple flow chart for the SIMPLE algorithm. PISO, like SIMPLE, is based on a predictor–corrector method, but employs several corrector steps while SIMPLE uses only one. This makes the PISO method more stable, but less efficient. In personal experience, the computation time for one tanker was increased by a factor of 5 switching from SIMPLE to PISO. For unsteady problems, however, the PISO method is generally preferred to the SIMPLE method due to its better stability. The discretization of the fundamental differential equations leads to very large systems of linear equations which are usually sparse, i.e. most of the elements of the matrix are zero. (This is fundamentally different from boundary element methods where full matrices with an often not dominant main diagonal need to solved.) Direct solvers like Gauss elimination or matrix inversion have prohibitively excessive computational time and storage requirements. In addition, the solution of the system of equations is embedded in an outer iteration which requires only an approximate solution, because the coefficients due to the non-linearity of the differential equations and the pressure–velocity coupling require further corrections. Therefore field methods generally employ iterative solvers:

- Gauss–Seidel method (point iterative);
- LSOR (line successive overrelaxation), ADI (alternating direction implicit) (line iterative);
- ILU (incomplete lower upper) decomposition, e.g. SIP (strong implicit procedure);
- CG (conjugate gradient) method.

The various iterative methods differ in their prerequisites (dominant main diagonal, symmetry, etc.), convergence properties, and numerical effort per iteration. Strongly implicit schemes

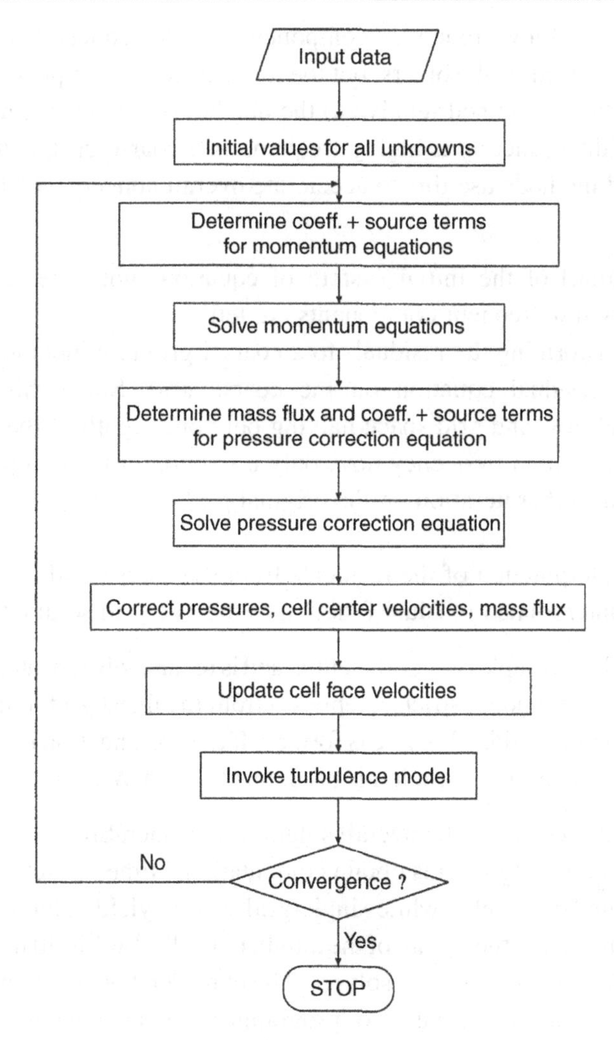

Figure 1.4:
Flow chart for SIMPLE algorithm

such as SIP feature high convergence rates. The convergence is especially high for multigrid acceleration which today is almost a standard choice.

1.5.5. Multigrid Methods

Multigrid methods use several grids of different grid size covering the same computational fluid domain. Iterative solvers determine in each iteration (relaxation) a better approximation to the exact solution. The difference between the exact solution and the approximation is called residual (error). If the residuals are plotted versus the line number of the system of equations, a more or less wavy curve appears for each iterative step. A Fourier analysis of this curve then yields high-frequency and low-frequency components. High-frequency components of the residual are quickly reduced in all solvers, but the low-frequency components are reduced only slowly. As the frequency is defined relative to the number of unknowns, respectively the grid fineness, a given residual function is highly frequent on a coarse grid, and low frequency on a fine grid. Multigrid methods use this to accelerate overall convergence by the following general procedure:

1. Iteration (relaxation) of the initial system of equations until the residual is a smooth function, i.e. only low-frequent components are left.
2. 'Restriction': transforming the residuals to a coarser grid (e.g. double the grid space).
3. Solution of the residual equation on the coarse grid. Since this grid contains for three-dimensional flow and grid space halving only one-eighth of the unknowns and the residual is relatively high frequency now, only a fraction of the computational time is needed because a further iteration on the original grid would have been necessary for the same accuracy.
4. 'Prolongation': interpolation of the residuals from the coarse grid to the fine grid.
5. Addition of the interpolated residual function to the fine-grid solution.

This procedure describes a simple two-grid method and is recursively repeated to form a multigrid method. If the multigrid method restricts (stepwise) from the finest grid to the coarsest grid and afterwards back to the finest grid, a V-cycle is formed. If the prolongation is only performed to an intermediate level, again before restriction is used, this forms a W-cycle (Fig. 1.5).

The multigrid method accelerates the overall solutions considerably, especially for grids with many unknowns. Multigrid algorithms obtain computational times which are almost proportional to the number of cells, while single-grid solvers yield computational times proportional approximately to the square of the number of cells. Multigrid methods are relatively easy to combine with all major iterative solvers. The considerable speed-up of computations more than justifies the additional expense of programming and storage requirements.

1.5.6. Numerical Approximations

Finite-volume methods require values (and derivatives) of various variables at the cell faces, when they are originally only known at the cell centers. The flow direction is often considered

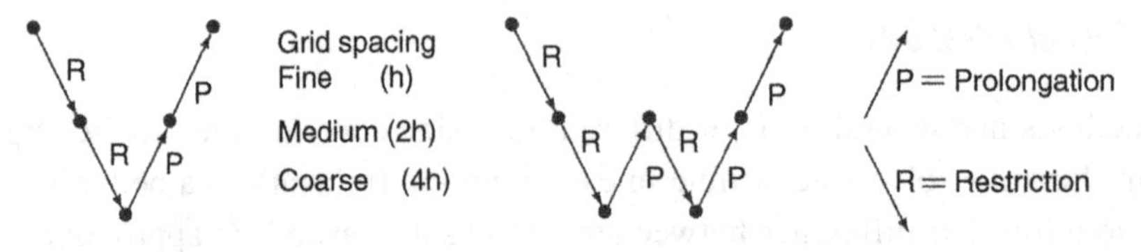

Figure 1.5:
Multigrid cycles: V-cycle (left), W-cycle (right); h: grid spacing computations

when these quantities are determined for convective terms. Time derivatives are also numerically approximated. Consider, for example, the convective fluxes in the x direction. One determines in general the value of a variable (e.g. pressure or velocity) at the location x by employing an interpolation polynomial through the neighboring cell centers x_i:

$$f(x) = a_1 + a_2(x - x_1) + a_3(x - x_1)(x - x_2) + \ldots \tag{1.50}$$

The coefficients a_i are determined by inserting the known function values $f_i = f(x_i)$. The simplest case uses just the value of the next cell centerupstream (*upwind differencing scheme, UDS*):

$$f(x) = \begin{cases} f_{i-1} & u > 0 \\ f_i & u < 0 \end{cases} \tag{1.51}$$

where u is the flow velocity in the x direction. This is a first-order approximation, i.e. (for fine grids) halving the grid size should halve the error in approximating the derivative. The order of an approximation is derived from a Taylor expansion for equidistant grids. For non-equidistant grids, an additional error appears that depends on the ratio of adjacent cell lengths. This error may dominate for coarse to moderately coarse grids, but vanishes in the theoretical limit of infinitely fine grids. The approximation thus depends on the direction of the velocity in the cell center. UDS is unconditionally stable, but plagued by large numerical diffusion. Numerical diffusion smoothes the derivatives (gradients) and may thus lead to wrong results. The numerical diffusion becomes maximal for an angle of 45° between grid lines and flow direction. Grid refinement reduces the numerical diffusion, but increases, of course, the computational effort.

The *central differencing scheme (CDS)* uses the adjacent upstream and downstream points:

$$f(x) = \frac{f_{i-1} + f_i}{2} \tag{1.52}$$

This is a second-order approximation, i.e. halving the grid size will reduce the error by one-quarter for fine grids. The approximation is independent of the sign of the flow direction. CDS tends to numerical instabilities and is therefore (for usual discretizations and speeds) unsuited for the approximation of the convective fluxes; the diffusive terms are usually approximated by CDS.

The *linear upwind differencing scheme (LUDS)* uses the cell centers of the next two upstream points:

$$f(x) = \begin{cases} \dfrac{f_{i-2} - f_{i-1}}{x_{i-2} - x_{i-1}}(x - x_{i-1}) + f_{i-1} & u > 0 \\[2ex] \dfrac{f_i - f_{i+1}}{x_i - x_{i+1}}(x - x_i) + f_i & u < 0 \end{cases} \tag{1.53}$$

This second-order approximation considers again the flow direction. LUDS is more stable than CDS, but can yield unphysical results. This is sometimes referred to as 'numerical dispersion'. Three points allow quadratic interpolation. The *QUICK* (*quadratic upstream interpolation for convective kinematics*) uses two adjacent points upstream and one downstream:

$$f(x) = \begin{cases} \dfrac{f_i + f_{i-1}}{2} - \dfrac{1}{8}(f_{i-2} + f_i - 2f_{i-1}) & u > 0 \\ \dfrac{f_i + f_{i-1}}{2} - \dfrac{1}{8}(f_{i+1} + f_{i-1} - 2f_i) & u < 0 \end{cases} \tag{1.54}$$

This third-order approximation may also produce unphysical results due to overshoots and requires a higher computational effort than the other schemes presented so far.

Blended schemes combine the basic schemes in weighted averages. Optimum weight factors depend on the problem. Blending combines the advantages (stability, accuracy) of the individual schemes, but requires more effort in each iteration. For an optimum weight the reduced number of iterations should more than compensate for this. For ship flows our experience is still insufficient to give general recommendations for blending schemes. Ideally, the weighting factors are chosen depending on the local flow. This usually involves the Peclet number, i.e. a local Reynolds number based on the local velocity and the grid size. Even more sophisticated techniques use a basic scheme (e.g. CDS) unless local instabilities (wiggles) are diagnosed automatically. These instabilities are then smoothed or filtered. These schemes do not require (error-prone) user input as do the simpler blending schemes. Blending may also be time-dependent. Then a more robust blend is used in the beginning to ensure numerical stability, and a more accurate blend is used later to obtain accurate converged results.

For the approximation of time derivatives implicit or explicit schemes may be used. In explicit schemes, the variables (e.g. derivatives of velocities) depend at each point in space only on known values of previous time steps. They can thus be computed directly (explicitly). Implicit schemes couple the unknowns to neighboring values (in time) and require the solution of a system of equations. Explicit schemes cannot usually be employed for ship flows, because they require very small time steps for the necessary very fine spatial discretization. A popular implicit scheme is the Crank–Nicholson scheme.

1.5.7. Grid Generation

Model set-up, especially grid generation, is decisive for time, cost, and quality of results in any CFD project. Grids must capture the changes in the geometries of hull and free surface (if included), but also all changes in the flow, with sufficient accuracy. For reasons of computational accuracy and efficiency (convergence rate), one should try to avoid extreme cell side ratios and skewed angles in individual cells. However, for ship flows, the flow changes drastically in the normal direction to the hull and little in the tangential longitudinal direction.

One would like to have a similar resolution for all changes in the flow direction. This automatically forces us to use cells with extreme side ratios, e.g. 1:1000. Grids should be rather fine in regions of high velocity or pressure gradients. The curvature of the ship hull and the experience with similar ship hull forms give some indications where such regions are to be expected, but often one identifies only after computations regions where the grid should have been finer. Ideally the computation should refine the grid in these regions automatically during a computation. Such adaptive grid techniques are subject to research for ship flows. They should bring considerable progress in accuracy without increasing computational effort excessively, but usually require unstructured grid capabilities of the code.

Numerical (non-physical) diffusion can be reduced by aligning grid lines along streamlines. However, flow separation and flow recirculation in ship flows allow this only to a limited extent.

Cartesian grids consist of elements with cell edges parallel to the axes of a Cartesian coordinate system. They are thus easy to generate, but unsuited for capturing complex geometries like ship hulls.

Therefore, in practice, generally *curvilinear grids* (body-fitted grids) are employed. These grids may be orthogonal or non-orthogonal. *Orthogonal grids* employ grid lines which intersect orthogonally. Since real ship geometries do not intersect the water surface orthogonally, at least some non-orthogonal grid lines have to be accepted. Otherwise, orthogonal grids are preferred since they facilitate the description of the discretized equations.

Curvilinear orthogonal grids require considerable effort in grid generation, but keep the complexity of the discretized equations relatively low. The velocity components may be grid oriented (local) or Cartesian (global). A formulation in Cartesian coordinates seems to react less sensitively to small deviations from smoothness in the grid.

Grid generation starts with specifying the cell faces on the boundaries (hull, water surface, inlet, outlet, outer boundary). Then the internal cell nodes are interpolated. Various techniques exist for this interpolation:

- *Algebraic grid generation* uses algebraic transformation and interpolation functions to create the grid geometry. For complex geometries (like real ship hulls), the resulting grid is often not smooth enough.
- *Conformal mapping* has been used for ships where the original mapping was enhanced by additional transformations to ensure that for real ship geometries the grids (within each two-dimensional section) were (nearly) orthogonal. However, this technique is fundamentally limited to two dimensions, i.e. for cross-sections. Smoothness and orthogonality in the longitudinal direction cannot be ensured automatically. Therefore, these grid generation techniques have been replaced largely by methods that solve a (simple) three-dimensional differential equation.

- *Grid generation based on differential equations* solves first a (relatively simple) differential equation subject to certain user-specified control functions or boundary conditions. The most popular choice is the Poisson equation, i.e. the Laplace equation with a specified non-zero function on the r.h.s. Thompson et al. (1985) describe such a method in detail which allows the user to control distance and orientation of the grid lines by specifying control functions. Poisson solvers create automatically smooth and orthogonal grids. Solving the Poisson equation can be interpreted physically as determining lines of constant temperature in the fluid where the ship is a heat source with heat distribution specified by the control functions.

Staggered grids specify, for example, the pressure at the cell center and the velocities at the cell faces. This improves automatically the numerical stability of the scheme, but is particularly unsuited for multigrid acceleration. Therefore staggered grids have become unpopular. Instead, other numerical techniques are employed to avoid pressure oscillations from cell to cell.

Grid generation is vital for the economic success of a CFD method. Grid techniques have been successively developed to allow more flexibility and faster grid generation:

- *Single-block structured grids.* Structured grids arrange cells in a simple $n_x \cdot n_y \cdot n_z$ array where each cross-section has the same number of cells, even though the cell shape and size may differ arbitrarily. Structured grids allow easy automation of grid generation and can easily be coupled with multigrid methods. They were traditionally employed because they allow simple program structures. Neighboring cells can be determined by a simple mathematical formula, avoiding the necessity for storing this information. However, this approach to grid generation does not allow the arrangement of additional cells in areas where the flow is changing rapidly. The choice is then either to accept insufficient accuracy in some areas or unnecessarily many cells (and thus computational effort) for areas where the flow is of little interest. In addition, complex ship geometries involving appendages are virtually impossible to model with such a grid. At least, the resulting grid is usually not smooth or involves highly skewed cells. As a result convergence problems are frequent.
- *Block-structured grids.* Block-structured grids combine various single-block grids. Each block is then structured and easily generated. But the block-structured approach allows some areas to discretize finer and others coarser. Blocks are also more easily adapted to local geometries, allowing smoother grids with largely block-like cells which improve convergence. The interpolation of results at each block requires some care, but techniques have been developed that allow accurate interpolation even for non-matching block interfaces, i.e. block interfaces where grid lines do not coincide. Block-structured grids are still the most popular choice in industry projects, typically involving 20–50 blocks for ship geometries.
- *Chimera grids.* Chimera was a fire-breathing (female) mythological monster that had a lion's head, a goat's body, and a serpent's tail. Chimera grids are arbitrarily assembled

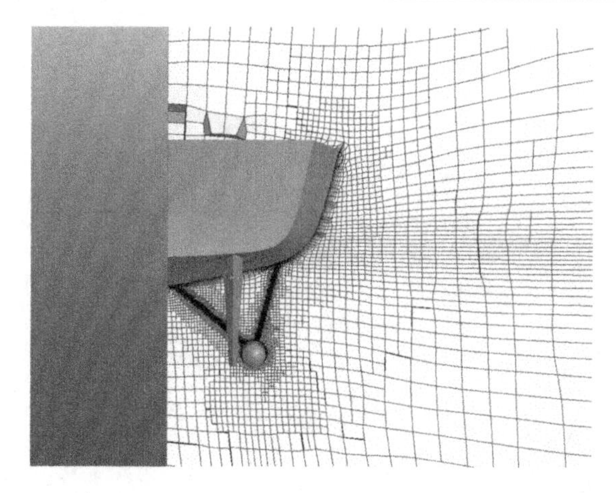

Figure 1.6:
Modern unstructured RANSE grid for ship with all appendages (one cross-section). *Source: NUMECA International.*

blocks of grids that overlap. They thus pose even fewer restrictions on grid generation and appear to be a very good choice for grid generation in ships, even though the interpolation between blocks is more complicated than for block-structured grids.

- *Unstructured grids.* Unstructured grids allow the largest flexibility in grid generation (Fig. 1.6), but require more effort. Unstructured grid programs can also handle structured or block-structured grids. One may then generate a simple grid and use adaptive grid techniques which automatically generate unstructured grids in the computation. Unstructured grids are popular, e.g. for aerodynamic flow analyses around superstructures of ships. Here the complexity of the boundary geometry makes other grid generation approaches at least very tedious. Adaptive grids and formal optimization procedures (requiring fully automated grid generation) are trends that will make unstructured grids the long-term preferred choice in many practical applications.

CHAPTER 2

Propellers

Chapter Outline

2.1. Introduction

Ships are predominantly equipped with 'simple' screw propellers. Special means of propulsion are covered towards the end of this chapter.

We will limit ourselves in the following to ships equipped with propellers. Propellers turning clockwise seen from aft are 'right-handed'. In twin-screw ships, the starboard propeller is usually right-handed and the port propeller left-handed. The propellers are then turning outwards. The propeller geometry is given in technical drawings following a special convention, or in thousands of offset points or spline surface descriptions, similar to the ship geometry. The complex propeller geometry is usually characterized by a few parameters. These include (Fig. 2.1):

- Propeller diameter D.
- Boss (or hub) diameter d.

Practical Ship Hydrodynamics. DOI: 10.1016/B978-0-08-097150-6.10002-8

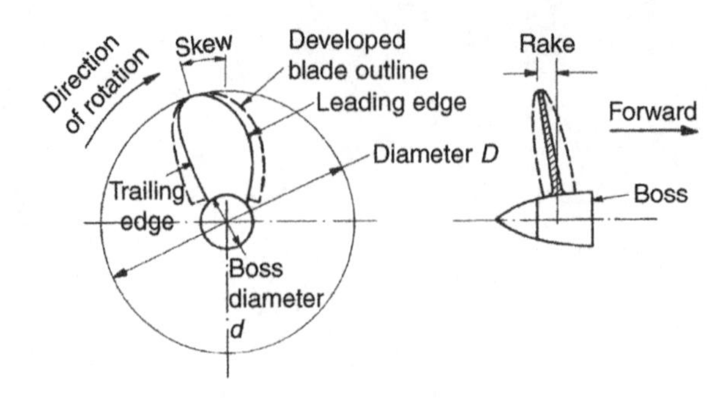

Figure 2.1:
Propeller geometry

- Propeller blade number Z.
- Propeller pitch P.

A propeller may be approximated by a part of a helicoidal surface which in rotation screws its way through the water. A helicoidal surface is generated as follows. Consider a line AB perpendicular to a line AA′ as shown in Fig. 2.2. AB rotates around the axis of AA′ with uniform angular velocity while moving along AA′ with uniform speed. AB then forms a helicoidal surface. Its pitch is the distance AA′. A propeller with a flat face and radially constant pitch would trace out a helicoidal surface. In reality, ship propellers often have neither a radially constant pitch nor a flat face. Then averaging in the circumferential direction creates a flat reference line to define the pitch as a function of the radius. Again averaging in a radial direction may define an average pitch P used to describe the propeller globally. Alternatively, the pitch at one radial position, typically $0.7R = 0.35D$, is taken as a single value to represent the radial pitch distribution.

- Disc area $A_0 = \pi D^2/4$.
- Projected area A_P.
- The blade area can be projected on to a plane normal to the shaft yielding the projected outline. Usually the area of the boss is not included.

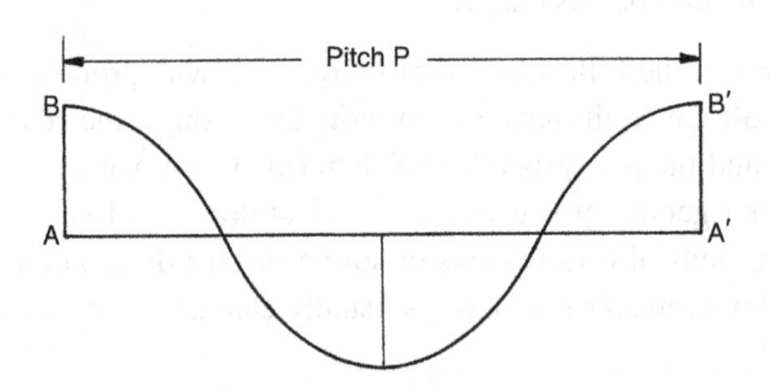

Figure 2.2:
Helicoidal surface defining pitch

- Expanded blade area A_E. The expanded outline is obtained if the circumferential chord of the blade is set out against the radius. The area of the formed outline is A_E.
- Skew (back). The line of the half chord length of each radial cross-section of the propeller is usually not a straight line, but curved back relative to the rotation of the blade. Skew is usually expressed as the circumferential displacement of the propeller tip made non-dimensional by the propeller diameter. Skew back evens out (to some extent) the influence of a highly non-uniform wake field and reduces peak values of propeller-induced vibrations. Modern ship propellers always have some skew back.
- Rake i_G. The face of the propeller may be tilted versus the normal plane on the propeller shaft. The tilt is usually backwards to increase the clearance between the blade tip and the hull.
- Profile shape. A propeller section at a given radius is the intersection between the blade and a circular cylinder of that radius. The section is then laid out flat (developed) and displayed as a two-dimensional profile. Historically, the early propeller designs had a flat face and circular cross-sections, which were then completely described by the blade width and maximum thickness. Today's profiles are far more complicated, but again usually characterized by a few parameters. The camber line is the line through the mid-thickness of the profile. If this line is curved, the profile is 'cambered'. The chord is the line joining the leading edge and the trailing edge. The camber is the maximum distance between the camber line and the chord. Profile sections are often defined by specifying the ordinates of the face and back as measured from the camber line.

Some of these data are often given as non-dimensional ratios:

- d/D;
- A_E/A_0;
- P/D;
- i_G.

The blade number Z is an important parameter for propeller-induced vibration. In general, odd numbers Z feature better vibration characteristics than even numbers. High blade numbers reduce vibration problems (due to less pronounced pressure peaks), but increase manufacturing costs. For large ships, blade numbers of four to seven are typical. For small boats, blade numbers of two to four are typical. The propellers for large ships are always tailored towards the specific ship and involve extensive hydrodynamic analyses. The propellers for boats are often mass-produced.

Typical extended blade area ratios are $0.3 < A_E/A_0 < 1.5$. Area ratios above 1 mean overlapping blades which are expensive to manufacture. A_E/A_0 is chosen such that the blade load is kept low enough to avoid unacceptable cavitation. Therefore A_E/A_0 increases with propeller load (thrust per propeller area A_0). The propeller efficiency decreases with A_E/A_0

since the increased area also increases frictional losses. Larger A_E/A_0 also often demands higher blade numbers to avoid too small side ratios for the blades.

2.2. Propeller Curves

Thrust T and torque Q are usually expressed as functions of rpm n in non-dimensional form as:

$$K_T = \frac{T}{\rho \cdot n^2 \cdot D^4} \tag{2.1}$$

$$K_Q = \frac{Q}{\rho \cdot n^2 \cdot D^5} \tag{2.2}$$

The force T is made non-dimensional by the propeller disk area times the stagnation pressure based on the circumferential speed, omitting a factor $\pi^2/8$. The moment Q is made non-dimensional by the additional length D, i.e. the propeller diameter.

The advance number J is defined as $J = V_A/(nD)$. V_A is the average inflow speed to the propeller. The propeller open-water efficiency is derived from thrust and torque coefficients and the advance number:

$$\eta_0 = \frac{T \cdot V_A}{2\pi \cdot n \cdot Q} = \frac{K_T \cdot \rho \cdot n^2 \cdot D^4}{K_Q \cdot \rho \cdot n^2 \cdot D^4} \cdot \frac{V_A}{2\pi \cdot n} = \frac{K_T}{K_Q} \cdot \frac{J}{2\pi} \tag{2.3}$$

K_T, K_Q, and η_0 are displayed over J. The curves are mainly used for propeller optimization and to determine the operation point (rpm, thrust, torque, power) of the ship. While the use of diagrams in education is still popular, in practice computer programs are almost exclusively used in propeller design. These represent traditionally the curves as polynomials in the form:

$$K_T = \sum C_T \cdot J^s \cdot \left(\frac{P}{D}\right)^t \cdot \left(\frac{A_E}{A_0}\right)^u \cdot Z^v \tag{2.4}$$

with tables of coefficients:

C_T	s	t	u	v
0.00880496	0	0	0	0
−0.20455403	1	0	0	0
...	...	...	...	...
−0.00146564	0	3	2	2

For standard Wageningen propellers the table consists of 49 coefficients for K_T and 56 coefficients for K_Q. While this may appear tedious, it is easy to program and fast to evaluate either by higher programming languages or spreadsheets. Diagrams are still popular in practice for documentation and visualization of tendencies.

Another important open-water parameter is the thrust loading coefficient:

$$C_{Th} = \frac{T}{\rho \cdot V_A^2 \cdot D^2 \cdot \frac{\pi}{8}} \tag{2.5}$$

This coefficient makes the thrust non-dimensional with the propeller disk area times stagnation pressure based on the propeller inflow velocity. Sometimes C_{Th} is also plotted explicitly in propeller characteristics diagrams, but sometimes it is omitted as it can be derived from the other quantities.

Figure 2.3 shows a typical propeller diagram. K_T and K_Q decrease monotonously with J. The efficiency η_0 has one maximum.

The open-water diagrams are based on stationary flow. They are only suitable for the case when the ship moves steadily ahead. For cases where the speed is changed, so-called four-quadrant diagrams are used. The name derives from a classification into four possible combinations:

- ship has forward speed, propeller delivers forward thrust;
- ship has forward speed, propeller delivers reverse thrust;
- ship has reverse speed, propeller delivers forward thrust;
- ship has reverse speed, propeller delivers reverse thrust.

The results of corresponding open-water tests are displayed in diagrams as shown in Fig. 2.4. The abscissa is the effective advance angle β defined by:

$$\tan\beta = \frac{V_A}{0.7 \cdot \pi \cdot n \cdot D} \tag{2.6}$$

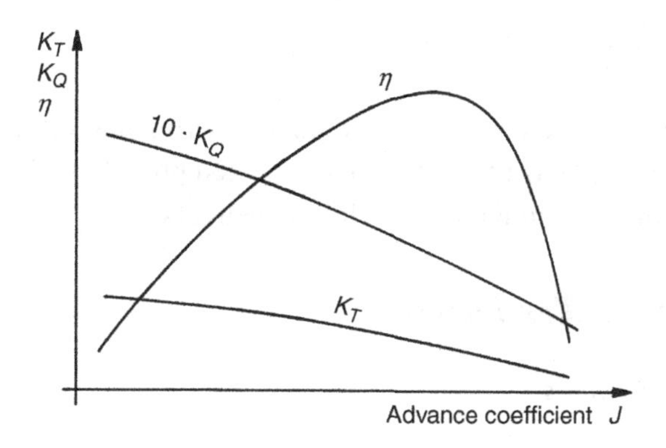

Figure 2.3:
Propeller diagram

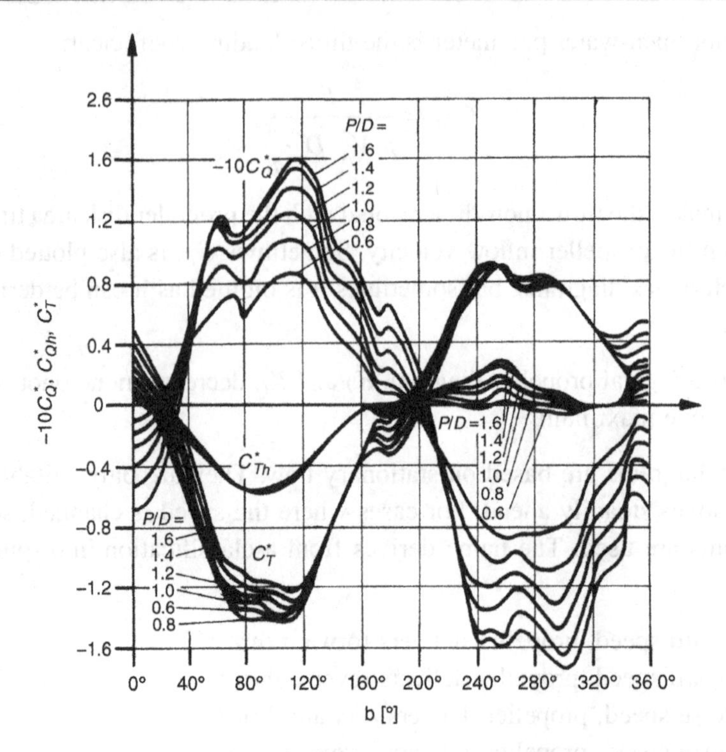

Figure 2.4:
Four-quadrant diagram for propellers

Displayed are non-dimensional modified thrust and torque coefficients:

$$C^*_{Th} = \frac{T}{\rho \cdot V_R^2 \pi D^2/8} \tag{2.7}$$

$$C^*_Q = \frac{Q}{\rho \cdot V_R^2 \pi D^3/8} \tag{2.8}$$

with $V_R = \sqrt{V_A^2 + (0.7\pi nD)^2}$.

Four-quadrant diagrams require considerably higher experimental effort than regular open-water diagrams. They are only available for some selected propellers. Four-quadrant diagrams are mainly used in computer simulations of ship maneuvers.

2.3. Analysis of Propeller Flows

2.3.1. Overview of Methods

Propellers create thrust as each of the blades is subject to local lift forces. Ideally, this lift is created with minimum drag losses. This basic goal is the same for other foil flows, e.g. airfoils, ship rudders, etc. Each propeller section resembles a cross-section of a foil. However, ship propellers feature short and stubby blades with a much smaller span-to-chord ratio than in aeronautical foils. The reason is that the limited diameter and the danger of cavitation impose more severe restrictions on ship propellers.

The small span-to-chord ratio of a ship propeller blade is one of the reasons why ship propeller flows are so complex. All two-dimensional approaches to model the flow around a propeller blade (like lifting-line theories) introduce considerable errors that must be corrected afterwards. Lifting-line approaches are still popular in propeller design as a preliminary step, before more powerful, but also more expensive, three-dimensional methods are employed. Many lifting-line codes in use today can be traced back to a fundamental formulation given by Lerbs (1952, 1955).

The advent of high-skew propellers necessitated truly three-dimensional theories to model the flow around the propeller. Empirical corrections for the lifting-line method could no longer be applied satisfactorily to the new and more complex propeller geometries. The approach was then to use lifting-line methods for an initial design serving as a starting point for more sophisticated methods which could then serve to answer the following questions:

- Will the propeller deliver the design thrust at the design rpm?
- What will be the propeller (open-water) efficiency?
- How will the propeller perform at off-design conditions?
- Will the pressure distribution be such that the propeller features favorable cavitation characteristics?
- What are the time-dependent forces and moments from the propeller on the propeller shaft and ultimately the shaft bearings?
- What are the propeller-induced pressures at the ship hull (exciting vibrations and noise)?

These more sophisticated three-dimensional propeller theories used in practical propeller design today are lifting-surface methods, namely vortex-lattice methods, which do not consider the blade thickness, and boundary element methods or panel methods, which do consider the blade thickness. Field methods are increasingly employed for advanced propulsors, particularly for off-design conditions involving cavitation.

The main methods in increasing complexity are listed below with their respective advantages and disadvantages:

- *Momentum theory.* The propeller is reduced to an actuator disk which somehow creates a pressure jump in the flow. Thrust and corresponding delivered power are expressed by increased velocities in the propeller plane and the contracted wake downstream of this plane. This simple model is unsuitable for propeller design, but popular as a simple

propeller model in RANSE ship computations and useful in understanding some basic concepts of propeller flows.
+ Simple and fast; yields ideal efficiency η_i
– Rotative and viscous losses not modeled; momentum theory is no method to design propellers or analyze given propeller designs

- *Lifting-line method.* Propeller blade is reduced to radially aligned straight vortices (lifting lines). The vortex strength varies over the radius. Free vortices are shed in the flow.
  + Proven in practice; suitable for initial design of propellers, rotative losses reflected in model; viscous losses incorporated by empirical corrections
  – Does not yield complete propeller geometry; cross-sections found, but angle of incidence and camber require corrections; no simple way to consider skew
- *Lifting-surface method, especially vortex-lattice method.* Propeller blade is reduced to a grid of horseshoe vortices; pressure distribution on the blade follows from Bernoulli's law from the induced velocities; pressure distribution yields forces and moments for the whole propeller.
  + Blade modeled three-dimensionally; corrections only necessary for viscous effects; good convergence to grid-independent solutions with grid refinement
  – More complex programming; pressure distribution must be corrected at the propeller hub
- *Boundary element method/panel method.* Exact formulation of the potential theory problem with source or dipole panels.
  + No simplifications besides the potential flow assumption; finite velocities in the hub region
  – Programming complex, especially for the Kutta condition; relatively large number of dipole and/or source panels necessary; flow near propeller tip still not well captured
- *RANSE method.* Field method formulation of the three-dimensional viscous flow.
  + Effective wake easily incorporated; viscous effects decreasing propeller efficiency directly captured; flow well captured also near hub and tip of propeller
  – Grid generation expensive; computation expensive; turbulence model questionable, possibly requiring LES.

Dedicated treatment of propeller flow analysis methods, predominantly based on lifting-surface and panel methods, can be found in, e.g., Breslin and Andersen (1994), Kinnas (1996), Streckwall (1993, 1999), and Carlton (2007).

2.3.2. Momentum Theory

Momentum theory models the propeller as a simple actuator disk accelerating the flow in the axial direction by somehow creating a pressure jump in the propeller plane. The propeller is then seen as a continuous circular disk with infinite blades and $A_E/A_0 = 1$. The

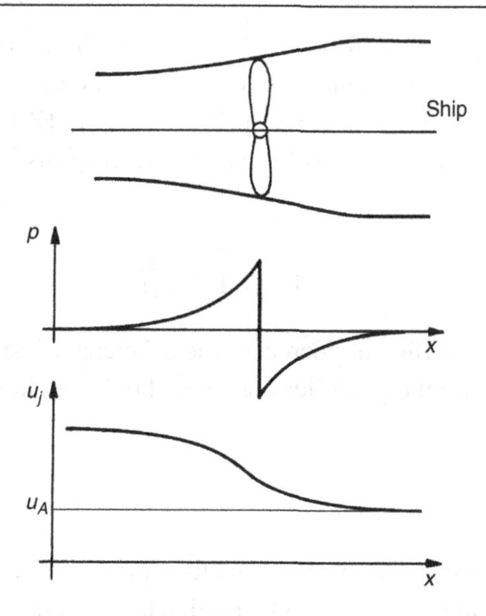

Figure 2.5:
Momentum theory considers propeller flow as one-dimensional flow with sudden pressure jump accelerating velocity from u_A to u_j

model is too crude to be of any value in propeller design, but allows some valuable insight into the global mechanisms of a propeller. The momentum theory regards inflow and outflow of the propeller plane as the flow through a tube of varying cross-section, but always of circular shape. Only the longitudinal velocity component is considered, i.e. the velocity is a scalar quantity.

The inflow to the propeller is given by $\rho \cdot u_A \cdot A_A$, where A_A is the cross-sectional area of the considered propeller plane. The propeller induces a velocity jump to the outflow velocity u_j and the cross-sectional area of the 'flow tube' is A_j. The thrust T is the change in the momentum:

$$T = \rho \cdot u_A \cdot A_A \cdot (u_j - u_A) \tag{2.9}$$

Continuity requires $A_j \cdot u_j = A_A \cdot u_A$, i.e. the flow contracts after the propeller due to the higher velocity (Fig. 2.5).

The velocity in the propeller plane is the average between the velocities far upstream and far downstream of the propeller in this model. Bernoulli's law couples the pressure to the velocity yielding qualitatively the distribution shown in Figure 2.5.

The actuator disk yields an ideal efficiency for the propeller of:

$$\eta_i = \frac{2u_A}{u_j + u_A} \tag{2.10}$$

This formula can be interpreted as follows. The smaller the increase in velocity due to the propeller, the better is the efficiency. If the velocity downstream is the same as the velocity upstream, the efficiency would be an ideal $\eta_i = 1$. (But no thrust would be produced.) The ideal efficiency can also be expressed in terms of the thrust loading coefficient C_{Th} as:

$$\eta_i = \frac{2}{1 + \sqrt{1 + C_{Th}}} \tag{2.11}$$

Thus a large thrust loading coefficient decreases the efficiency. The conclusion for practical propeller design is that usually the propeller diameter should be chosen as large as possible to increase the efficiency.

2.3.3. Lifting-Line Methods

Lifting-line methods still form a vital part of practical propeller design. They find the radial distribution of loading optimum with respect to efficiency as a first step to determine the corresponding blade geometry. Alternatively, the radial distribution of loading may be specified to determine the corresponding blade geometry (Lerbs 1952, 1955). Of course, this approach works only within limits. If unrealistic or too-demanding pressure distributions are specified, either no solution is found or the error in the framework of the theory is so large that the solution does not reflect reality.

Lifting-line methods for propellers were adapted from lifting-line theory for straight foils. We shall therefore briefly review the lifting-line theory for straight foils.

A straight line of vorticity creates lift orthogonal to the direction of the vortex line and the direction of the inflow (Fig. 2.6). Helmholtz's first and second laws state:

1. The strength of a vortex line is constant along its length.
2. A vortex line must be closed; it cannot end in the fluid.

As a consequence, the vortex lines on a foil are bent downstream at the end of the foil. Far downstream these vortex lines are closed again, but often 'far downstream' is interpreted as 'at infinity', i.e. the vortex line forms a semi-infinite horseshoe vortex. The vortex segment representing the foil is called the 'bound' vortex, as it always stays with the foil. The two vortex segments swept downstream are the 'trailing vortices', also denoted as axial vortices or tip vortices. The closing vortex segment far downstream is the 'starting' vortex.

In reality, the lift and thus the vorticity (vortex strength) are not constant over the foil span. This can be considered by approximating the continuous lift by a number of discrete, piecewise constant vortex segments. Each of these will then produce trailing vortices (Fig. 2.7). In sum, the vortex segments form a 'lifting line' of (stepwise) variable vorticity.

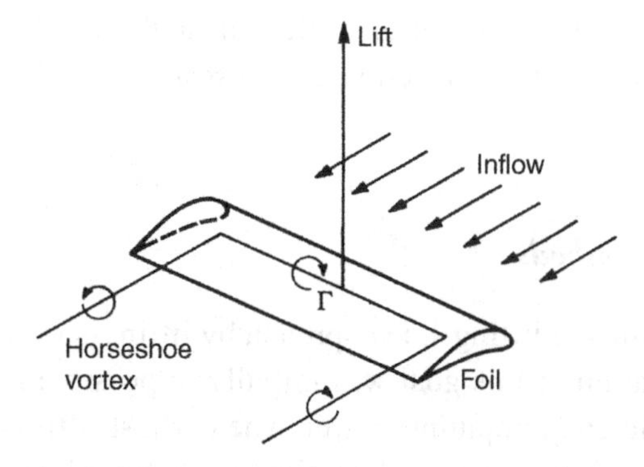

Figure 2.6:
Lifting-line theory is based on representing the foil by bound vortex and trailing vortices

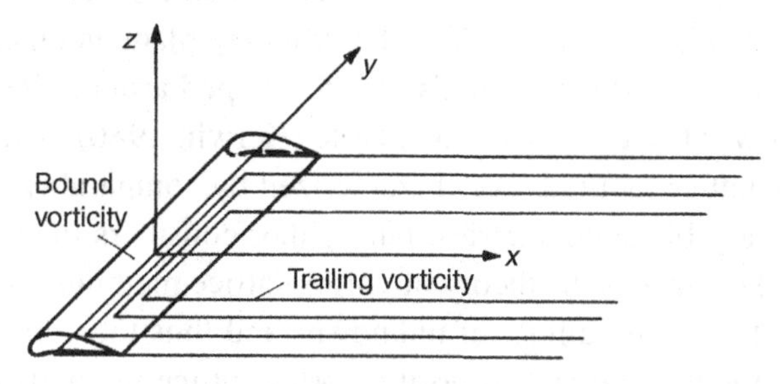

Figure 2.7:
A better model represents the foil by a distribution of horseshoe vortices

The trailing vortices induce a flow at the foil which is downward for positive lift. This velocity is therefore called downwash and changes the effective inflow angle experienced by each section of the foil.

The strengths of the individual vortex elements (each forming a closed or semi-infinite loop) are determined by requiring that there is no flow through the foil at a corresponding number of collocation points. This results in a system of linear equations which is solved numerically. Once all vortex strengths are known, the velocities and pressures can be evaluated everywhere. Lift and drag can then be computed.

For propellers, each blade is represented by one lifting line extending from hub to blade tip. Typically the lifting lines are straight with skew and rake being neglected at this point in the analysis. The proper end condition for the lifting line at the hub is unclear. Usually, the hub is neglected and the vorticity is required to go to zero as at the blade tip. This is called the 'hubless propeller assumption'. Lerbs argued that, near the hub, the blades are close enough together such that the positive pressure on the face of one blade is canceled by the negative pressure on the back of the adjacent blade. However, in practice the lifting-line results near the hub, but also often near the blade, are unrealistic and are then manually corrected (smooth connection to the rest of the lift distribution based on human insight).

2.3.4. Lifting-Surface Methods

The discussion to substitute the lifting-line approach by lifting-surface theories dates back to the 1950s, but the realization of this goal was initially impossible for real ship propeller geometries due to insufficient computing power. The earliest lifting-surface attempts were based on mode functions which prescribed continuous distributions of surface singularities. At that time, the mode function approach was the ordinary procedure for solving lifting-line or two-dimensional wing section problems and naturally it was tried first. For lifting surfaces that had to fit propeller blades these mode functions needed a careful and complicated mathematical treatment. Their ability to describe arbitrary blade geometries was poor. The second generation of lifting-surface methods was developed around the late 1970s when sufficient computer power became widely available (Kerwin 1986). These methods used vortex lattices. Vortex-lattice methods are characterized by comparably simple mathematics. They can handle arbitrary blade geometries, but neither considers the true blade thickness, nor the propeller hub. This makes the theory of vortex-lattice methods more complicated than panel methods, but reduces the number of unknowns and thus the computational effort considerably. Despite the theoretical inferiority, vortex-lattice methods gave in benchmark tests of the ITTC for propellers with moderate skew-back results of comparable quality to panel methods. Figure 2.8 shows a typical discretization of the propeller blades and the wake. The hub is not modeled, which leads to completely unrealistic results in the immediate vicinity of the hub.

Vortex-lattice methods were extended in the 1990s to rather complicated propeller geometries, e.g. contra-rotating propellers, and unsteady propeller inflow (nominal wake computations). Cavitation may be simulated by additional singularities of both source and vortex type, but this remains a rather coarse approximation of the real phenomenon.

A complete vortex-lattice method (VLM) can be established on the basis of the lifting-line method just described. The lifting-line model was used to find a circulation Γ that corresponds to a given resultant flow direction at the lifting line and is able to provide the predetermined (design) thrust. With a vortex lattice instead of a lifting line, a model for the material blade is inserted. One can now really investigate whether a given geometry corresponds to a desired thrust, a task that is beyond the scope of a lifting-line theory.

Figure 2.9 shows a vortex-lattice system. The flow is generated by spanwise and (dependent) streamwise line vortices. Control points are positioned inside the loops of the vortex system. For steady flow, the vortex elements in the wake have the same strength in

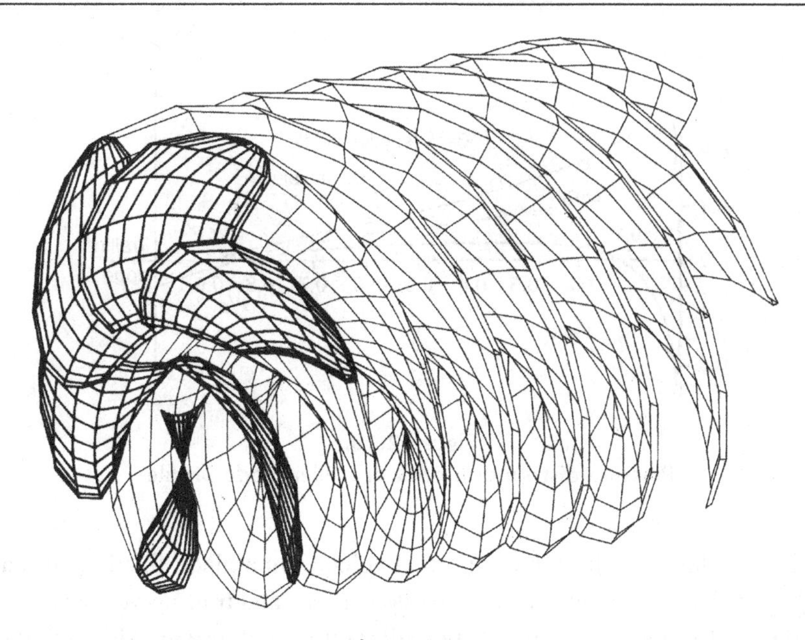

Figure 2.8:
Vortex-lattice model of a propeller and trailing wakes

each spanwise segment. The vertical vortex lines then cancel each other and a semi-infinite horseshoe vortex results. The most downstream control point is located at the trailing edge behind the last streamwise vortex, which is a very robust measure to enforce the Kutta condition.

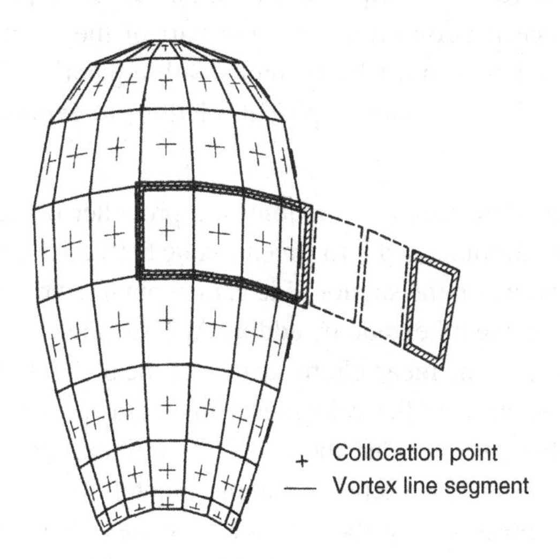

Figure 2.9:
Allocation of vortex-lattice elements on propeller blade

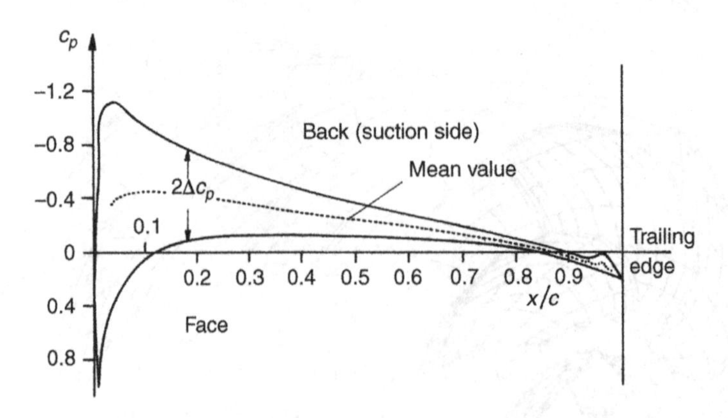

Figure 2.10:
Pressure distribution on a propeller blade profile

The kinematic boundary condition (zero normal velocity in a blade-fixed coordinate system), together with some basic relations between blade vortices and trailing vortices, is sufficient to calculate blade surface pressures and thus propeller thrust and torque. Although the kinematic condition is fulfilled on a zero-thickness blade, the influence of the blade thickness is not excluded. The thin-wing theory provides a simple formula to derive a source system from the slope of the section contours. This source system already enters the kinematic conditions and serves to correct the angle of attack of the blade sections for the displacement effect of the neighboring blades.

In most applications a 'frozen' vortex wake is used, i.e. the trailing vortex geometry is fixed from the start. A more or less empirical relation serves to prescribe the pitch of the helical lines. Since surface friction effects are not part of the solution, the forces and moments from the vortex lattice must be corrected subsequently. This is usually achieved by local section drag coefficients using empirical relations to express the Reynolds number dependence.

Figure 2.10 shows a typical pressure distribution for a propeller blade cross-section. The pressure coefficient can be decomposed into a mean value between both sides of the profile and a difference Δc_p. The pressure on the suction side is then obtained by subtracting Δc_p from the mean value, the pressure on the other side by adding Δc_p. Lifting-surface methods arrange the vortex and source elements on the mean chord surface of the blade. Following Bernoulli's law, the pressure can be computed from the velocities. This yields pressure distributions which usually reproduce the actual pressure distributions quite well except for a narrow region at the leading edge, which may extend to a length of approximately twice the nose radius. The sources yield the average pressure distribution and the vortex elements induce the pressure difference Δc_p. As the source strengths are explicitly derived from the change of the profile thickness in the longitudinal direction, the main problem is to determine the vortex strengths.

2.3.5. Boundary Element Methods

Panel methods were developed to overcome the disadvantage of an incomplete geometry model. Panel methods also model the blade thickness and include the hub in the numerical model. The development of panel methods for propellers was apparently not an easy task. After the ship hull flow could be treated by panel methods it took another decade until the late 1980s before the first successful panel approaches were established for propellers. The implementation of a robust Kutta condition is a decisive element of each propeller panel code, since it controls torque and thrust. In principle, there exist many possibilities to create panel codes, depending on panel type and the formulation of the problem (e.g. Kerwin et al. 1987). The following panel types are found:

- dipole panels;
- source panels;
- mix of dipole panels and source panels.

The problem may be formulated as:

- direct formulation (potential formulation); potential itself is the unknown;
- indirect formulation (velocity formulation); source or dipole strength is unknown.

For indirect formulations, Kerwin et al. (1987) show how a dipole-based formulation can be transformed to an equivalent vortex-based formulation.

The majority of the panel codes used for propellers follows Morino's approach (Morino and Kuo 1974, Morino 1975). Morino's approach is a direct formulation, i.e. it solves directly for the potential and determines velocities by numerical differentiation. The approach uses exclusively dipole panels, which discretize the surfaces of the propeller blades, the hub, and part of the wakes of each blade. The Kutta condition demands that at the trailing edge the pressure difference between face and back should vanish. This couples the dipoles on the wake to the dipoles on the propeller. The panels in the wake all have the same strength for steady flow conditions. The pitch of the wake is either specified by largely empirical relations or determined iteratively as part of the solution. The Kutta condition enforcing a vanishing pressure jump at the trailing edge is a non-linear condition requiring an iterative solution. The numerical implementation of the Kutta condition requires great care, since simplifications or conceptual errors in the physical model may strongly affect the computed lift forces.

The main problems of these methods lie in:

- numerical realization of the Kutta condition (stagnation point at the trailing edge);
- numerical (accurate) determination of velocity and pressure fields.

In the 1990s, panel methods were presented that were also capable of solving the problem for time-dependent inflow and ducted propellers (e.g. Kinnas 1996).

2.3.6. Field Methods

The common procedure to run unsteady propeller vortex-lattice or panel methods contains an inherent weakness. The ship is usually represented by the velocity field measured without the propeller at the propeller plane, i.e. the nominal wake. But in a real ship, the propeller rearranges the streamlines that reach the propeller plane, i.e. the propeller operates in the effective wake. There are measures to correct the nominal wake, but it is doubtful if these treat the details of the wake correctly.

No such complications arise in theory if viscous flow computations are employed. It is possible to interactively couple viscous flow computations for the ship based on RANSE solvers with potential flow computations for the propeller, e.g. vortex-lattice or panel methods. But increasingly the preferred option is to solve the flow around hull, propeller, and rudder using one RANSE model. The viscous flow representation for the propeller embedded in a viscous model for the ship makes all problems from decoupling ship flow and propeller flow vanish. By 2010, only selected analyses have appeared, but eventually this approach is expected to become standard for ship and propeller design.

Viscous flow computations are also able to deliver accurate flow details in the tip region of the propeller blade. Typical propeller geometries require careful grid generation to assure converged solutions. The warped propeller geometry makes grid generation particularly difficult, especially for high-skew propellers.

2.4. Cavitation

High velocities result in low pressures. If the pressure falls sufficiently low, cavities form and fill up with air coming out of solution and by vapor. This phenomenon is called cavitation. The cavities disappear when the pressure increases again. They grow and collapse extremely rapidly, especially if vapor is filling them. Cavitation involves highly complex physical processes with highly non-linear multi-phase flows which are subject to dedicated research by specialized physicists. We will cover the topic only to the extent that any naval architect should know. For a detailed treatment of cavitation for ship propellers, the reader is referred to the book by Isay (1989).

For ship propellers, the velocities around the profiles of the blade may be sufficiently high to decrease the local pressures to trigger cavitation. Due to the hydrostatic pressure, the total pressure will be higher on a blade at the 6 o'clock position than at the 12 o'clock position. Consequently, cavitating propellers will then have regions on a blade where alternately cavitation bubbles are formed (near the 12 o'clock position) and collapse again. The resulting rapid succession of explosions and implosions on each blade will have various negative effects:

- vibration;
- noise (especially important for navy ships like submarines);

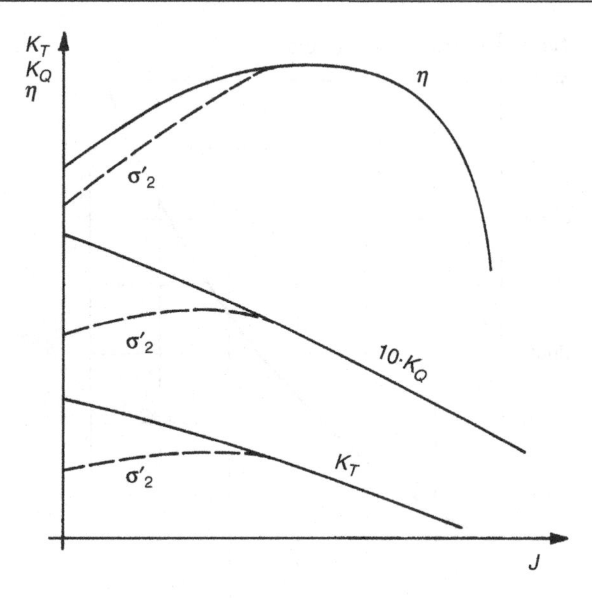

Figure 2.11:
Influence of cavitation on propeller characteristics

- material erosion at the blade surface (if the bubble collapse occurs there);
- thrust reduction (Fig. 2.11).

Cavitation occurs not only at propellers, but everywhere where locally high velocities appear, e.g. at rudders, shaft brackets, sonar domes, hydrofoils, etc.

Cavitation may be classified by:

- Location: tip cavitation, root cavitation, leading edge or trailing edge cavitation, suction side (back) cavitation, face cavitation, etc.
- Cavitation form: sheet cavitation, cloud cavitation, bubble cavitation, vortex cavitation.
- Dynamic properties of cavitation: stationary, instationary, migrating.

Since cavitation occurs in regions of low pressures, it is most likely to occur towards the blade tips where the local inflow velocity to the cross-sections is highest. But cavitation may also occur at the propeller roots near the hub, as the angle of incidence for the cross-sections is usually higher there than at the tip. The greatest pressure reduction at each cross-section profile usually occurs between the leading edge and mid-chord, so bubbles are likely to form there first.

In ideal water with no impurities and no dissolved air, cavitation will occur when the local pressure falls below vapor pressure. Vapor pressure depends on the temperature (Fig. 2.12). At 15°C it is 1700 Pa. In real water, cavitation occurs earlier, as cavitation nuclei like microscopic

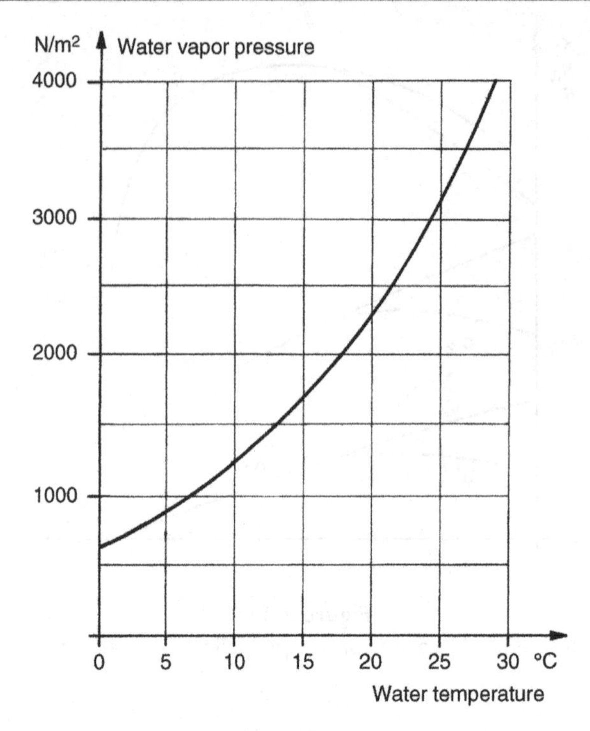

Figure 2.12:
Vapor pressure as a function of temperature

particles and dissolved gas facilitates cavitation inception. The cavitation number σ is a non-dimensional parameter to estimate the likelihood of cavitation in a flow:

$$\sigma = \frac{p_0 - p}{\frac{1}{2}\rho V_0^2} \tag{2.12}$$

p_0 is an ambient reference pressure and V_0 a corresponding reference speed. p is the local pressure. For $\sigma < \sigma_v$ (the cavitation number corresponding to vapor pressure p_v) the flow will be free of cavitation in an ideal fluid. In reality, one introduces a safety factor and sets a higher pressure than vapor pressure as the lower limit.

Cavitation is predominantly driven by the pressure field in the water. Cavitation avoidance consequently strives to control the absolute pressure minimum in a flow. This is achieved by distributing the thrust on a larger area, either by increasing the diameter or the blade area ratio A_E/A_0.

The most popular approach to estimate the danger of cavitation at a propeller uses Burill diagrams. These diagrams can only give a rough indication of cavitation danger. For well-designed, smooth propeller blades they indicate a lower limit for the projected area. Burill uses the coefficient τ_c:

$$\tau_c = \frac{T}{q_{0.7R}^2 A_p} \tag{2.13}$$

$$q_{0.7R} = \frac{1}{2}\rho V_R^2 \tag{2.14}$$

$$V_R = \sqrt{V_A^2 + (0.7\pi n D)^2} \tag{2.15}$$

V_R is the absolute value of the local velocity at 0.7 of the propeller radius. V_A is the inflow velocity to the propeller plane. A_p is the projected propeller area. Burill uses as reference pressure the atmospheric pressure plus the hydrostatic pressure at the propeller shaft:

$$p_0 = p_{atm} + \rho g h \tag{2.16}$$

The Burill diagram then yields limiting curves (almost straight) to avoid cavitation (Fig. 2.13). The curves have been transformed into algebraic expressions and are also included in propeller design programs. The upper limit for σ_c yields indirectly a minimum A_p which yields (for Wageningen B-series propellers) approximately the expanded blade area:

$$A_E \approx \frac{A_p}{1.067 - 0.229(P/D)} \tag{2.17}$$

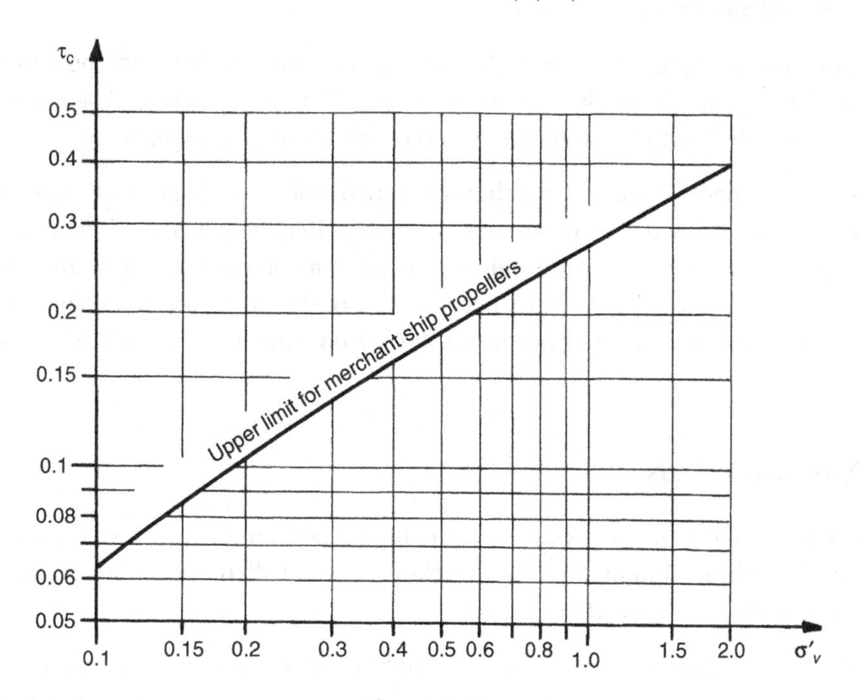

Figure 2.13:
Burill diagram

2.5. Experimental Approach

2.5.1. Cavitation Tunnels

Propeller tests (open-water tests, cavitation tests) are usually performed in cavitation tunnels. A cavitation tunnel is a closed channel in the vertical plane recirculating water by means of an impeller in the lower horizontal part. This way the high hydrostatic pressure ensures that even for reduced pressure in the tunnel, the impeller itself will not cavitate. The actual test section is in the upper horizontal part. The test section is provided with observation glass ports. The tunnels are designed to give (almost) uniform flow as inflow to the test section. If just the propeller is tested (with the driving shaft downstream), it is effectively tested in open water. Larger circulation tunnels also include ship models, thus testing the propeller in the ship wake. The ship models are sometimes shortened to obtain a thinner boundary layer in the aftbody (which thus resembles more the boundary layer in a large-scale model). Alternatively, sometimes grids are installed upstream to generate a flow similar to that of a full-scale ship wake. This requires considerable experience and is still at best a good guess at the actual wake field.

Vacuum pumps reduce the pressure in the tunnel and usually some devices are installed to reduce the amount of dissolved air and gas in the water. Wire screens may be installed to generate a desired amount of turbulence.

Cavitation tunnels are equipped with stroboscopic lights that illuminate the propeller intermittently such that propeller blades are always seen at the same position. The eye then perceives the propeller and cavitation patterns on each blade as stationary.

Usual cavitation tunnels have too much background noise to observe or measure the noise-making or hydro-acoustic properties of a propeller, which are of great interest for certain propellers, especially for submarines or antisubmarine combatants. Several dedicated hydro-acoustic tunnels have been built worldwide to allow acoustical measurements. The HYKAT (hydroacoustic cavitation tunnel) of HSVA is one of these.

2.5.2. Open-Water Tests

Although in reality the propeller operates in the highly non-uniform ship wake, a standard propeller test is performed in uniform flow yielding the so-called open-water characteristics, namely thrust, torque, and propeller efficiency.

The model scale λ for the model propeller should be the same as for the ship model in the propulsion tests. For many propulsion tests, the ship model scale is determined by the stock propeller, i.e. the closest propeller to the optimum propeller on stock at a model basin. The similarity laws (see Section 1.2, Chapter 1) determine for geometrical and Froude similarity:

$$\left(\frac{V_A}{n \cdot D}\right)_s = \left(\frac{V_A}{n \cdot D}\right)_m \tag{2.18}$$

In other words, the advance number $J = V_A/(nD)$ is the same for model and full scale. J has thus a similar role for the propeller as the Froude number F_n has for the ship. V_A is the average inflow speed to the propeller, n the propeller rpm, and D the propeller diameter. $\pi \cdot n \cdot D$ is the speed of a point at the tip of a propeller blade in the circumferential direction.

The Reynolds number for a propeller is usually based on the chord length of one blade at 0.7 of the propeller radius and the absolute value of the local velocity V_R at this point. V_R is the absolute value of the vector sum of inflow velocity V_A and circumferential velocity:

$$V_R = \sqrt{V_A^2 + (0.7\pi nD)^2} \tag{2.19}$$

Propeller model tests are performed for geometrical and Froude similarity. It is not possible to keep Reynolds similarity at the same time. Therefore, as in ship model tests, corrections for viscous effects are necessary in scaling to full scale. ITTC 1978 recommends the following empirical corrections:

$$K_{Ts} = K_{Tm} - 0.3 \cdot Z \cdot \left(\frac{c}{D}\right)_{r=0.7} \cdot \frac{P}{D} \cdot \Delta C_D \tag{2.20}$$

$$K_{Qs} = K_{Qm} + 0.25 \cdot Z \cdot \left(\frac{c}{D}\right)_{r=0.7} \cdot \Delta C_D \tag{2.21}$$

c is the propeller blade chord length at $0.7R$, R the propeller radius, $\Delta C_D = C_{Dm} - C_{Ds}$ is a correction for the propeller resistance coefficient with:

$$C_{Dm} = 2 \cdot \left(1 + 2\frac{t}{c}\right) \cdot \left(\frac{0.044}{R_n^{1/6}} - \frac{5}{R_n^{2/3}}\right) \tag{2.22}$$

$$C_{Ds} = 2 \cdot \left(1 + 2\frac{t}{c}\right) \cdot \left(1.89 + 1.62 \log\left(\frac{c}{k_p}\right)\right)^{-2.5} \tag{2.23}$$

Here t is the (maximum) propeller blade thickness and R_n is the Reynolds number based on V_R, both taken at $0.7R$. k_p is the propeller surface roughness, taken as $3 \cdot 10^{-5}$ if not known otherwise.

2.5.3. Cavitation Tests

Cavitation tests investigate the cavitation properties of propellers. Experiments usually observe the following similarity laws:

- Geometrical similarity, making the propeller as large as possible while still avoiding tunnel wall effects.

- Kinematical similarity, i.e. the same advance number in model and ship, $J_m = J_s$.
- Dynamical similarity would require that model and full-scale ship have the same Froude and Reynolds numbers. Reynolds similarity is difficult to achieve, but the water speed is chosen as high as possible to keep the Reynolds number high and reduce scaling effects for the friction on the blades. Gravity effects are negligible in propeller flows, i.e. waves usually play no role. Thus the Froude number may be varied.
- Cavitation similarity requires the same cavitation numbers in model and full-scale ships. The tunnel pressure is adjusted to give the same cavitation number at the propeller shaft axis to approximate this condition.
- For similarity in bubble formation in cavitation, the Weber number should also be the same in model and full scale:

$$\left(\frac{\rho \cdot V_A^2 \cdot D}{T_e}\right)_m = \left(\frac{\rho \cdot V_A^2 \cdot D}{T_e}\right)_s \tag{2.24}$$

where T_e is the surface tension and D the propeller diameter. This similarity law is usually violated.

The cavitation tests are performed for given inflow velocity and cavitation number, varying the rpm until cavitation on the face and back of the propeller is observed. This gives limiting curves $\sigma = \sigma(J)$ for cavitation-free operation. The tests are often performed well beyond the first inception of cavitation and then the extent and type of cavitation is observed, as often designers are resigned to accept some cavitation, but individual limits of accepted cavitation differ and are often subject to debate between shipowners, ship designers, and hydrodynamic consultants. The tests are usually based entirely on visual observation, but techniques have been developed to automatically detect and visualize cavitation patterns from video recordings. These techniques substitute the older practice of visual observation and manual drawings, making measurements by various persons at various times more objectively comparable.

2.6. Propeller Design Procedure

Traditionally, propeller design was based on design charts. These charts were created by fitting theoretical models to data derived from actual model or full-size tests and therefore their number was limited. By and large, propeller design was performed manually. In contrast, contemporary propeller design relies heavily on computer tools. Some of the traditional propeller diagrams, such as for the Wageningen B-series of propellers, have been transformed into polynomial expressions allowing easy interpolation and optimization within the traditional propeller geometries. This is still a popular starting point for modern propeller design. Then, a succession of ever more sophisticated analysis programs is employed to modify and fine-tune the propeller geometry.

Propeller design is an iterative process to optimize the efficiency of a propeller subject to more or less restrictive constraints concerning cavitation, geometry, strength, etc. The severity of the constraints depends on the ship type. For example, submarine propellers have strict constraints concerning cavitation-induced noise. Subsequently the efficiencies of these propellers are lower than for cargo ships, but the primary optimization goal is still efficiency. A formal optimization is virtually impossible for modern propellers if the description of the geometry is based on (some hundreds of) offsets, as the evaluation of the efficiency based on CFD requires considerable time. Thus, while the word 'optimization' is often used, the final design is rather 'satisficing', i.e. a good solution satisfying the given constraints. However, parametric description coupled with efficient CFD and optimization schemes had allowed formal propeller optimization in industry by 2010.

Additional constraints are inherently involved in the design process, but often not explicitly formulated. These additional constraints reflect the personal 'design philosophy' of a designer or company and may lead to considerably different 'optimal' propellers for the same customer requirements. An example of such a 'design philosophy' could be the constraint that no cavitation should occur on the pressure side of the propeller. The following procedure will reflect the design philosophy of HSVA as detailed in Reich et al. (1997). The overall procedure will, however, be similar to any other state-of-the-art propeller design process. The main engine influences the propeller design primarily through the propeller rpm and delivered power. Modern turbo-charged diesels, almost exclusively used for cargo ships today, are imposing a rather narrow bandwidth for the operating point (rpm/power combination) of the propeller. We therefore limit ourselves to such cases where the rpm, the ship's speed, and an estimated delivered power P_D are specified by requirement. This covers more than 90% of the cases in practice.

The procedure follows a few main steps which involve model tests, analytical tools of successive sophistication and power, and some experience in deciding trade-offs in conflict situations:

1. Preparation of model
   Known at this stage: experiments
   - rpm of the full-scale propeller n_s
   - ship speed V_s
   - estimate of delivered power for the ship P_D
   - ship hull form (lines plan)
   - classification society
   - often: number of blades Z
   - often: diameter of propeller D

Generally, the customer specifies within small margins what power P_D has to be delivered at what speed V_s and what is the rpm of the (selected) main engine. While in theory such a combination may be impossible to realize, in practice the shipyard engineers (i.e. the customers) have sufficient experience to estimate a realistic power for a shipowner-specified speed and rpm. The shipyard or another department in the model basin will specify a first proposal for the ship lines. Often, the customer will also already determine the number of blades for the propeller. A few simple rules gained from experience will guide this selection, e.g. if the engine has an even number of cylinders, the propeller should have an odd number of blades. The propeller of optimal efficiency can then be determined automatically based on the Wageningen B-series by computer codes. The performance of these older propellers is insufficient for today's expectations and the propeller thus determined will only be used as a starting point for the actual design. This procedure yields the average (or representative) pitch-to-diameter ratio P_m/D and the diameter D. An upper limit for the diameter is specified from the ship geometry. Sometimes the customer already specifies the diameter, otherwise it is a result of the optimization. The expanded area ratio A_E/A_0 is usually part of the optimization result, but may be restricted with respect to cavitation if problems are foreseen. In this case, a limiting value for A_E/A_0 is derived from Burill diagrams.

Then, from a database of stock propellers, the most suitable propeller is selected. This is the propeller with the same number of blades, closest in P_m/D to the optimized propeller. If several stock propellers coincide with the desired P_m/D, the propeller closest in A_E/A_0 among these is selected. A selection constraint comes from upper and lower limits for the diameter of the stock propeller which are based on experience for the experimental facilities. For example, for HSVA, the ship models may not exceed 11 meters in length to avoid the influence of canal restrictions, but should be larger than 4 meters to avoid problems with laminar flow effects. As the ship length is specified and the model scale for propeller and ship must be the same, this yields one of the constraints for upper and lower values of the diameter of the stock propeller. Usually, the search of the database is limited to the last 300 stock propellers, i.e. the most recent designs.

The selected stock propeller then determines the model scale and the ship model may be produced and tested. The output of the model tests relevant for the propeller designer is:

- nominal wake distribution (axial, tangential, and radial velocities in the propeller plane)
- thrust deduction fraction t
- effective wake fraction w
- relative rotative efficiency η_R
- delivered power P_D.

The delivered power P_D is of secondary importance (assuming that it is close to the customer's estimate). It indicates how much the later propeller design has to strive for a high efficiency. If the predicted P_D is considerably too high, then the ship form has to be changed and the tests repeated.

2. Estimate effective wake distribution full scale
   Known at this stage: all of the above and …
   number of blades Z
   diameter of propeller D
   blade area ratio A_E/A_0
   thrust deduction fraction t
   effective wake fraction w
   relative rotative efficiency η_R
   nominal wake field (axial, tangential, radial velocity components)

   Ship–propeller interaction is difficult to capture. The inflow is taken from experiments and based on experience modified to account for scale effects (model/full-scale ship). The radial distribution of the axial velocity component is transformed from the nominal (without propeller action) value for the model to an effective (with propeller) value for the full-scale ship. The other velocity components are assumed not to be affected. Several methods have been proposed to perform this transformation. To some extent, the selection of the 'appropriate' method follows rational criteria, e.g. one method is based on empirical data for full ships such as tankers, another method for slender ships such as container ships. But still the designer expert usually runs several codes, looks at the results and selects the 'most plausible' based on 'intuition'. The remaining interaction effects such as thrust deduction fraction t and relative rotative efficiency η_R are usually taken as constant with respect to the results of ship model tests with propellers.
3. Determine profile thickness according to classification society
   Known at this stage: all of the above
   Classification societies have simple rules to determine the minimum thickness of the foils. The rules of all major classification societies are implemented in programs that adjust automatically the (maximum) thickness of all profiles to the limit value prescribed by the classification society.
4. Lifting-line and lifting-surface calculations
   Known at this stage: all of the above and . . .
   max. thickness at few radii

   As additional input, default values are taken for profile form (NACA series), distribution of chord length and skew. If this step is repeated at a later stage, the designer may deviate from the defaults. At this stage, the first analytical methods are employed. A lifting-line method computes the flow for a two-dimensional profile, i.e. the three-dimensional flow is approximated by a succession of two-dimensional flows. This is numerically stable and effective. The method needs an initial starting value for the circulation distribution. This is taken as a semi-elliptic distribution. The computation then yields the optimal radial distribution of the circulation. These results are directly used for a three-dimensional lifting-surface program. The lifting-surface code yields as output the radial distribution of profile camber and pitch.

5. Smoothing results of Step 4
   Known at this stage: all of the above and . . .
   radial distribution of profile camber (estimate)
   radial distribution of pitch (estimate)
   The results of the three-dimensional panel code are generally not smooth and feature singularities at the hub and tip of the propeller. The human designer deletes 'stray' points (point-to-point oscillations) and specifies values at hub and tip based on experience.
6. Final hydrodynamic analysis
   Known at this stage: all of the above (updated)
   The propeller is analyzed in all operating conditions using a lifting-surface analysis program and taking into account the complete wake distribution. The output can be broadly described as the cavitational and vibrational characteristics of the propeller. The work sometimes involves the inspection of plots by the designer. Other checks are already automated. Based on his 'experience' (sometimes resembling a trial-and-error process), the designer modifies the geometry (foil length, skew, camber, pitch, profile form and even, as a last resort, diameter). However, the previous steps are not repeated and this step can be treated as a self-contained module.
7. Check against classification society rules
   Known at this stage: all of the above (updated)
   A finite-element analysis is used to calculate the strength of the propeller under the pressure loading. The von Mises stress criterion is plotted and inspected. As the analysis is still limited to a radially averaged inflow, a safety margin is added to account for the real inflow. In most cases, there is no problem. But if the stress is too high in some region (usually the trailing edge), the geometry is adjusted and Step 6 is repeated. The possible geometry modifications at this stage are minor and local; they have no strong influence on the hydrodynamics and therefore one or two iterations usually suffice to satisfy the strength requirements.

2.7. Propeller-Induced Pressures

Due to the finite number of blades the pressure field of the propeller is unsteady if taken at a fixed point on the hull. The associated forces induce vibrations and noise. An upper limit for the maximum pressure amplitude that arises on the stern (usually directly above the propeller) is often part of the contract between shipyard and owner.

For many classes of ships the dominant source for unsteady hull pressures is the cavitation on the propeller blades. The effect of cavitation in computations of propeller-induced pressures is usually modeled by a stationary point source positioned in the propeller plane. To assure similarity with the propeller cavitation, the source must be given an appropriate volume amplitude, a frequency of oscillation, and a suitable position in the propeller plane specified by a radius and an angle. As the propeller frequency is rather high, the dominant term in the Bernoulli equation is the time-derivative term. If mainly fluctuating forces from propeller-induced hull pressures are of interest, the pressure is therefore usually sufficiently well approximated by the term $-\rho\phi_t$, where ϕ denotes the potential on the hull due to the perturbations from the propeller.

If pressures and forces induced by a fluctuating source on solid boundaries are to be considered, the point source may be positioned underneath a flat plate to arrive at the simplest problem of that kind. The kinematic boundary condition on the plate is ensured via an image source of the same sign at the opposite side of the plate. For the pressure field on a real ship, this model is too coarse, as a real ship aftbody does not look like a flat plate and the influence of the free surface is neglected. Potential theory is still sufficient to solve the problem of a source near a hull of arbitrary shape with the free surface present. A panel method (BEM) easily represents the hull, but the free surface requires special treatment. The high frequency of propeller rpm again allows a simplification of the treatment of the free surface. It is sufficient to specify then at the free surface $z = \zeta$:

$$\phi(x, y, z = \zeta, t) = 0 \tag{2.25}$$

If the free surface is considered plane ($\zeta = 0$), $\phi = 0$ can be achieved by creating a hull image above the free surface and changing the sign for the singularities on the image panels. An image for the source that represents the cavity (again of opposite sign in strength) has to be introduced as well. The free surface can be considered in good approximation as a plane for low Froude numbers, such as typically encountered for tankers and bulkers, but it is questionable for moderate and high Froude numbers. A pronounced stern wave will have a significant effect on the wetted areas at the stern.

The main problem of the above procedure is the reliability of calculated cavity volumes.

2.8. Unconventional Propellers

Special means of propulsion are covered in greater detail in Schneekluth and Bertram (1998) and Carlton (2007). Unconventional propulsors have developed in various forms, for various special applications:

- Nozzled propellers. The (Kort) nozzle is a fixed annular forward-extending duct around the propeller. The propeller operates with a small gap between blade tips and nozzle internal wall, roughly at the narrowest point. The nozzle ring has a cross-section shaped as a hydrofoil or similar section. Nozzled propellers have the following advantages and disadvantages:

+ At high thrust-loading coefficients, better efficiency is obtainable. For tugs and pusher boats, efficiency improvements of around 20% are frequently achievable. Bollard pull can be raised by more than 30%.
+ The reduction of propeller efficiency in a seaway is lower for nozzle propellers than for non-ducted propellers.
+ Course stability is substantially improved by the nozzle.
– Course-changing ability during astern operation is somewhat impaired.
– Owing to circulation in shallow water, the nozzle propeller tends to draw into itself shingle and stones. Also possible is damage due to operation in ice. This explains the rare application on seagoing ships.
– Due to the pressure drop in the nozzle, cavitation occurs earlier.

Nozzled propellers have been fitted frequently on tugs, fishing vessels, and inland water vessels.

- Contra-rotating propellers (CRP). Rotational exit losses amount to about 8–10% in typical cargo ships. Coaxial contra-rotating propellers (Fig. 3.15) can partially compensate these losses, increasing efficiency by up to 6% (Isay 1964). To avoid problems with cavitation, the after-propeller should have smaller diameter than the forward propeller. Contra-rotating propellers have the following advantages and disadvantages:

+ The propeller-induced heeling moment is compensated (this is negligible for larger ships).
+ More power can be transmitted for a given propeller radius.
+ The propeller efficiency is usually increased.
– The mechanical installation of coaxial contra-rotating shafts is complicated, expensive and requires more maintenance.
– The hydrodynamic gains are partially compensated by mechanical losses in shafting.

Contra-rotating propellers are used on torpedoes due to the natural torque compensation. They are also found in some motorboats. For normal ships, the task of boring out the outer shafts and the problems of mounting the inner shaft bearings were not considered to be justified by the increase in efficiency (in times of relatively cheap fuel), although in the early 1990s some large tankers were equipped with contra-rotating propellers. The CRPs should not be confused with the Grim wheel, where the 'aft' propeller is not driven by a shaft. Unlike a CRP, the Grim wheel turns in the same direction as the propeller.

- Controllable-pitch propeller (CPP). CPPs with three to five blades are often used in practice and feature the following advantages and disadvantages:

+ Fast stop maneuvers are possible.
+ The main engine does not need to be reversible.
+ The CPP allows one to drive the main generator with the main engine which is efficient and cheap. Thus the electricity can be generated with the efficiency of the main engine

and using heavy fuel. Different ship speeds can be driven with constant propeller rpm as required by the generator.

– Fuel consumption is higher. The higher propeller rpm at lower speed is hydrodynamically suboptimal. The CPP requires a bigger hub (0.3–0.32D). The pitch distribution is suboptimal. The usual radial direction, almost constant, pitch would cause negative angles of attack at the outer radii for reduced pitch, thus slowing the ship down. Therefore CPPs usually have higher pitch at the outer radii and lower pitch in the inner radii. The higher pitch in the outer radii necessitates also a larger propeller clearance.

– Higher costs for propeller.

- Azimuthing propellers. Azimuthing propellers (a.k.a. rudder propellers, slewable propellers), usually equipped with nozzles, are not just a derivative of the well-known outboarders for small boats. Outboarders can only slew the propeller by a limited angle to both sides, while azimuthing propellers can cover the full 360°. Turning the propeller by 180° allows reversing the thrust. This astern operation is much more efficient than for conventional propellers turning in the reverse direction. By 2003, rudder propellers were available for up to 4000 kW permanent power. Podded drives housing an electric motor in the pod which drives the propeller(s) can be seen as a special case of azimuthing propellers.
- Podded drives. Podded drives are characterized by two main features: there is an electric motor inside a pod and the total unit is azimuthing. In 1990, the auxiliary vessel 'Seiti' was the first ship to be equipped with a pod drive. Within only a decade, pod drives became the dominant choice of propulsion for certain ship types. The hydrodynamic unit efficiency of pod drives is approximately 5% lower than that of an identical conventional propeller with a rudder as a unit. In many cases, in addition, some efficiency is lost for pod units, because the pod propeller cannot have optimal diameter due to the torque limitation of the pod motor. A small gain in propeller efficiency can be expected for twin pod arrangements because the inflow to the propeller is more uniform (absence of shafts and shaft brackets). This leads to better design conditions for the propeller and therefore to higher propeller efficiencies.

  Mewis (2001) gives advantages and disadvantages of pod drives:

  + More cargo space because the engine can be located more freely
  + Better maneuverability
  + Lower noise level
  + Low speeds are possible
  + Suited as booster drive in order to increase the speed
  + Less work expense in ship manufacturing
  + (Potentially) lower power requirement for twin screw ships
  – Higher capital costs
  – Diesel electric system required (power loss)
  – Increased power requirement for single screw arrangements

– Limitation in power (up to 32 MW in 2003)
– Limitation in speed (up to 30 knots in 2003).

The following order of suitability depending on ship type is given:

Very well suited: cruise liner, RoPax ferry, icebreaker
Well suited: supply vessel, bulker, tanker (single and twin screw)
Hardly suited: container vessel (single screw) < 3000 TEU
Not well suited: container vessel (twin screw)
Not feasible: container vessel (single screw) > 3000 TEU.

- Waterjets. Waterjets as alternative propulsive systems for fast ships, or ships operating in extremely shallow water, are discussed by, e.g., Allison (1993), Kruppa (1994), and Terswiga (1996). For high ship speed, restricted propulsor diameters and cavitation-free operation, conventional propellers reach their limits. Since these problems are not generally new, pumps were introduced as a propulsion system already in the early 1920s. These ancient pump systems already included in principle the same components as a modern waterjet: water inlets with inboard tubing system, the pump inducing energy to the water and finally the nozzle, which deals with the propulsive power. With the early waterjet systems, the steering of the vessel was performed separately by conventional rudders, whereas in modern systems the steering and reversing systems are integrated in the jets.

  Special attention has to be paid to the shape of the waterjet inlets to avoid excessive additional resistance, cavitation, and noise and on the other hand to ensure sufficient flow of the pump.

  Waterjet propulsion has become a popular propulsor choice for fast ships. The Royal Institution of Naval Architects has in addition hosted dedicated conferences on waterjet propulsion in 1994 and 1998 and the ITTC has a subcommittee reviewing the continuing progress on waterjets.

  The application of waterjets ranges from fast monohull car/passenger ferries to catamarans, motor yachts, speed boats, hydrofoils, and surface effect ships. Waterjets feature the following advantages and disadvantages:

  + Higher efficiency at speeds above 35 knots ($\eta > 0.8$)
  + No appendage resistance for shafts, brackets, rudders
  + Operability in shallow water
  + More flexibility in locating main engines
  + Much smaller risk of cavitation
  + Excellent maneuvering
  + More thrust with smaller impeller diameter
  + Lower noise development (-7 to -10 dB (A9))
  + No need for reversing gear (except when gas turbines are used as prime mover)
  – Drastically lower efficiency at lower speeds
  – Larger weight of the waterjet unit incl. the added water mass within the pump

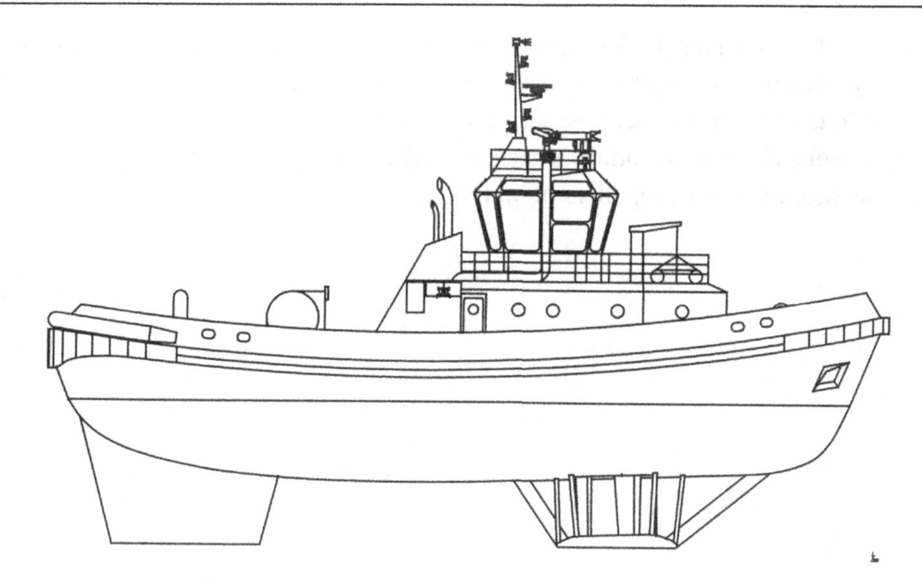

Figure 2.14:
Voith–Schneider propeller installed in a Voith water tractor tug. *Source: Voith Turbo*

– Power required to lift water from inlet to the nozzle level, especially for nozzle above water line (hydrofoils)
– Higher cost for pumps and inlets
– Larger space requirements inside hull.

Model tests with waterjets are mostly performed at high model speeds and therefore the available measuring time becomes relatively short so that the exact self-propulsion point is difficult to match. Therefore it is common practice that different runs will be performed for each speed with various impeller rpms. This procedure delivers a series of cross curves from which the actual self-propulsion point can be found at the condition that the residuary force is equal to the corresponding friction deduction.

- Surface-piercing propellers. Surface-piercing propellers operate only partially submerged, typically with 30–50% of the propeller being surfaced. The propeller blades are designed to operate such that the pressure face of the blade remains fully wetted and the suction side is fully ventilated or dry. Surface-piercing propellers are used on fast craft, typically racing boats and some fast naval vessels.
- Voith–Schneider propellers (cycloidal propellers). The Voith–Schneider propeller (VSP) is a cycloidal drive that combines propulsion and maneuvering. The VSP consists of a circular plate with an array of typically five vertical, foil-shaped blades. The plate is attached to the bottom of the vessel (Fig. 2.14). It rotates around a vertical axis. Each blade in turn can rotate around a vertical axis. An internal gear changes the angle of attack of the blades in sync with the rotation of the plate, so that each blade can provide thrust in the same direction. The VSP

makes vessels highly maneuverable, as the thrust can be adjusted in magnitude and direction arbitrarily without changing the engine's rpm, almost instantaneously. VSPs are usually arranged in tandem. It is widely used on tugs and ferries.

- Paddle-wheels. In the past, paddle-wheels played a large role in river boats, but have been largely replaced now by propellers or waterjets.

CHAPTER 3

Resistance and Propulsion

Chapter Outline

Practical Ship Hydrodynamics. DOI: 10.1016/B978-0-08-097150-6.10003-X

3.1. Resistance and Propulsion Concepts

3.1.1. Interaction Between Ship and Propeller

Any propulsion system interacts with the ship hull. The flow field is changed by the hull. The propulsion system changes, in turn, the flow field at the ship hull. However, traditionally naval architects have considered propeller and ship separately and introduced special efficiencies and factors to account for the effects of interaction. While this decomposition is seen by many as an important aid in structuring the complex problems of ship hydrodynamics, it also hinders a system approach in design and can confuse as much as it can help. Since it is still the backbone of our experimental procedures and ingrained in generations of naval architects, the most important concepts and quantities are covered here. The hope is, however, that in the future CFD will allow a more comprehensive optimization of the ship interacting with the propeller as a whole system.

The general definition 'power = force · speed' yields the effective power

$$P_E = R_T \cdot V_s \tag{3.1}$$

where R_T is the total calm-water resistance of the ship excluding resistance of appendages related to the propulsive organs. Sometimes the rudder is also excluded and treated as part of the propulsion system. (This gives a glimpse of the conceptual confusion likely to follow from different conventions concerning the decomposition. Remember that in the end the installed power is to be minimized. Then 'accounting' conventions for individual factors do not matter. What is lost in one factor will be gained in another.) V_s is the ship speed. P_E is the power we would have to use to tow the ship without a propulsive system.

Following the same general definition of power, we can also define a power formed by the propeller thrust and the speed of advance of the propeller, the so-called thrust power:

$$P_T = T \cdot V_A \tag{3.2}$$

The thrust T measured in a propulsion test is higher than the resistance R_T measured in a resistance test (without propeller). So the propeller induces an additional resistance:

1. The propeller increases the flow velocities in the aftbody of the ship which increases frictional resistance.
2. The propeller decreases the pressure in the aftbody, thus increasing the inviscid resistance.

The second mechanism dominates for usual propeller arrangements. The thrust deduction fraction t couples thrust and resistance:

$$t = 1 - \frac{R_T}{T} \quad \text{or} \quad T(1-t) = R_T \tag{3.3}$$

where t is usually assumed to be the same for model and ship, although the friction component introduces a certain scale effect. Empirical formulae for t are plagued by large margins of uncertainty.

The propeller inflow, i.e. the speed of advance of the propeller V_A, is generally slower than the ship speed due to the ship's wake. The wake is usually decomposed into three components:

- *Friction wake.* Due to viscosity, the flow velocity relative to the ship hull is slowed down in the boundary layer, leading in regions of high curvature (especially in the aftbody) to flow separation.
- *Potential wake.* In an ideal fluid without viscosity and free surface, the flow velocity at the stern resembles the flow velocity at the bow, featuring lower velocities with a stagnation point.
- *Wave wake.* The steady wave system of the ship changes the flow locally as a result of the orbital velocity under the waves. A wave crest above the propeller increases the wake fraction, a wave trough decreases it.

For usual single-screw ships, the frictional wake dominates. Wave wake is only significant for $F_n > 0.3$. The measured wake fraction in model tests is larger than in full scale as boundary layer and flow separation are relatively larger in model scale. Traditionally, correction formulae try to consider this overprediction, but the influence of separation can only be estimated and this introduces a significant error margin. In validation studies, CFD has shown good agreement with model test measurements. It is widely assumed that computing power and turbulence modeling had improved by 2010 to the point where also full-scale computations were expected to be accurate even though they could not be validated explicitly. Despite errors in predicting the wake, the errors in predicting the required power remain small, as the energy loss due to the wake is partially recovered by the propeller. However, the errors in predicting the wake propagate completely when computing optimum propeller rpm and pitch.

The wake behind the ship without propeller is called the nominal wake. The propeller action accelerates the flow field by typically 5–20%. The wake behind the ship with operating propeller is called the effective wake. The wake distribution is either measured by laser-Doppler velocimetry or computed by CFD. CFD also predicts the integral of the wake over the propeller plane, the wake fraction w, well. The wake fraction is defined as:

$$w = 1 - \frac{V_A}{V_s} \qquad (3.4)$$

Empirical formulae to estimate w in simple design approaches consider only a few main parameters, but actually the shape of the ship influences the wake considerably. Other important parameters like propeller diameter and propeller clearance are also not explicitly represented in these simple design formulae.

The ratio of the effective power to the thrust power is called the hull efficiency:

$$\eta_H = \frac{P_E}{P_T} = \frac{R_T \cdot V_s}{T \cdot V_A} = \frac{1-t}{1-w} \tag{3.5}$$

So the hull efficiency can be expressed solely by thrust deduction factor t and wake fraction w. η_H can be less or greater than 1. It is thus not really an efficiency, which by definition cannot be greater than 100%.

The power delivered at the propeller can be expressed by the torque and the rpm:

$$P_D = 2\pi \cdot n \cdot Q \tag{3.6}$$

This power is less than the 'brake power' directly at the ship engine P_B due to losses in shaft and bearings. These losses are comprehensively expressed in the shafting efficiency η_S: $P_D = \eta_S \cdot P_B$. The ship hydrodynamicist is not concerned with P_B and can consider P_D as the input power in all further considerations of optimizing the ship hydrodynamics. We use here a simplified definition for the shafting efficiency. Usually marine engineers decompose η_S into a shafting efficiency that accounts for the losses in the shafting only and an additional mechanical efficiency. For the ship hydrodynamicist it suffices to know that the power losses between engine and delivered power are typically 1.5–2%.

The losses from delivered power P_D to thrust power P_T are expressed in the (propeller) efficiency behind ship η_B: $P_T = \eta_B \cdot P_D$.

The open-water characteristics of the propeller are relatively easy to measure and compute. The open-water efficiency η_0 of the propeller is, however, different to η_B. Theoretically, the relative rotative efficiency η_R accounts for the differences between the open-water test and the inhomogeneous three-dimensional propeller inflow encountered in propulsion conditions: $\eta_B = \eta_R \cdot \eta_{B0}$. In reality, the propeller efficiency behind the ship cannot be measured and all effects not included in the hull efficiency, i.e. wake and thrust deduction fraction, are included in η_R. η_R again is not truly an efficiency. Typical values for single-screw ships range from 1.02 to 1.06.

The various powers and efficiencies can be expressed as follows:

$$P_B > P_D > P_T > P_E \tag{3.7}$$

$$P_E = \eta_H \cdot P_T = \eta_H \cdot \eta_B \cdot P_D = \eta_H \cdot \eta_0 \cdot \eta_R \cdot P_D = \eta_H \cdot \eta_0 \cdot \eta_R \cdot \eta_S \cdot P_B = \eta_D \cdot \eta_S \cdot P_B \tag{3.8}$$

The propulsive efficiency η_D collectively expresses the hydrodynamic efficiencies: $\eta_H \cdot \eta_0 \cdot \eta_R$.

3.1.2. Decomposition of Resistance

As the resistance of a full-scale ship cannot be measured directly, our knowledge about the resistance of ships comes from model tests. The measured calm-water resistance is usually decomposed into various components, although all these components interact and most of them cannot be measured individually. The concept of resistance decomposition is thought to help in designing the hull form as the designer can focus on how to influence individual resistance components. Larsson and Baba (1996) give a comprehensive overview of modern methods of resistance decomposition (Fig. 3.1).

The total calm-water resistance of a new ship hull can be decomposed into:

- *Friction resistance.* Due to viscosity, directly at the ship hull water particles 'cling' to the surface and move with ship speed. A short distance away from the ship, the water particles already have the velocity of an outer, quasi-inviscid flow. The region between the ship surface and the outer flow forms the boundary layer. In the aftbody of a container ship with $R_n \approx 10^9$, the boundary layer thickness may be 1 m. The rapid velocity changes in the normal direction in the boundary layer induce high shear stresses. The integral of the shear stresses over the wetted surface yields the friction resistance.
- *Viscous pressure resistance.* A deeply submerged model of a ship will have no wave resistance, but its resistance will be higher than just the frictional resistance. The form of the ship induces a local flow field with velocities that are sometimes higher and sometimes lower than the average velocity. The average of the resulting shear stresses is then higher. Also, energy losses in the boundary layer, vortices and flow separation prevent an increase to stagnation pressure in the aftbody as predicted in ideal fluid theory. Full ship forms have a higher viscous pressure resistance than slender ship forms.

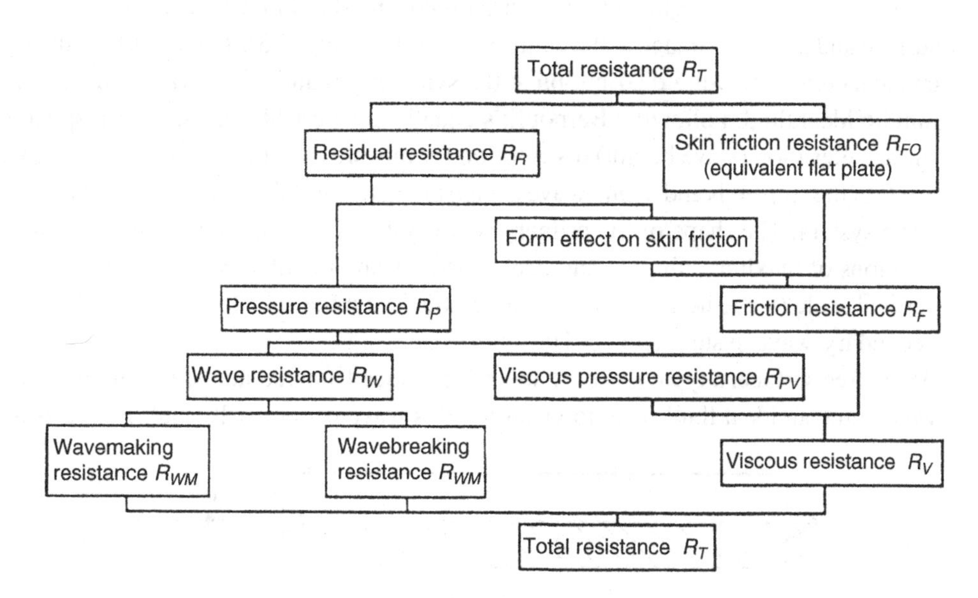

Figure 3.1:
Resistance decomposition

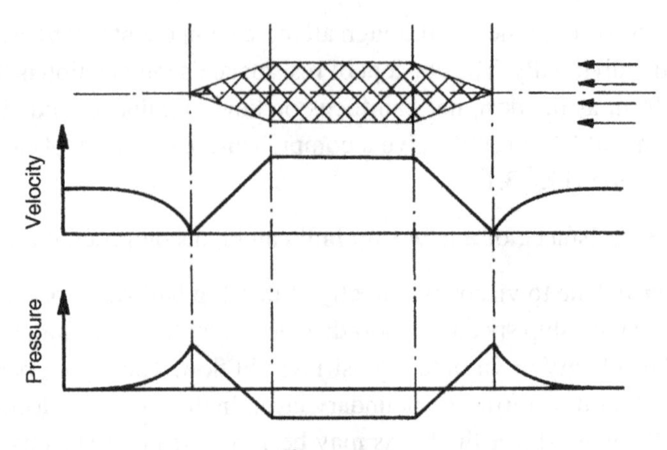

Figure 3.2:
'Primary' wave system

- *Wave resistance.* The ship creates a typical wave system which contributes to the total resistance. In the literature, the wave system is often (rather artificially) decomposed into a primary and a secondary wave system:
  1. Primary wave system (Fig. 3.2)
     In an ideal fluid with no viscosity, a deeply submerged body would have zero resistance (D'Alembert's paradoxon). The flow would be slower at both ends of the body and faster in the middle. Correspondingly at each end, the pressure will be higher than average, reaching at one point stagnation pressure, and the pressure in the middle will be lower than average. Now imagine a body consisting of the ship hull below the calm-water surface and its mirror image at the calm-water surface (Fig. 3.3). This double body would create a certain pressure distribution at the symmetry plane (calm-water surface) in an infinite ideal fluid. Following Bernoulli's equation, we could express a corresponding surface elevation (wave height) distribution for this pressure distribution, yielding wave crests at the ship ends and a long wave trough along the middle. This is called the primary wave system. The shape of the primary wave system is speed independent, e.g. the locations of maxima, minima, and zero crossings are not affected by the speed. The vertical scale (wave height) depends quadratically on the speed.
  2. Secondary wave system (Fig. 3.4)
     At the free surface, a typical wave pattern is produced and radiated downstream. Even if we assume an ideal fluid with no viscosity, this wave pattern will result in a resistance.

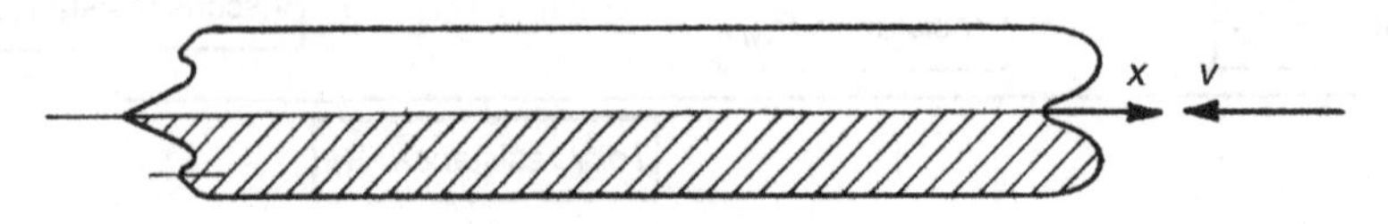

Figure 3.3:
Double-body flow

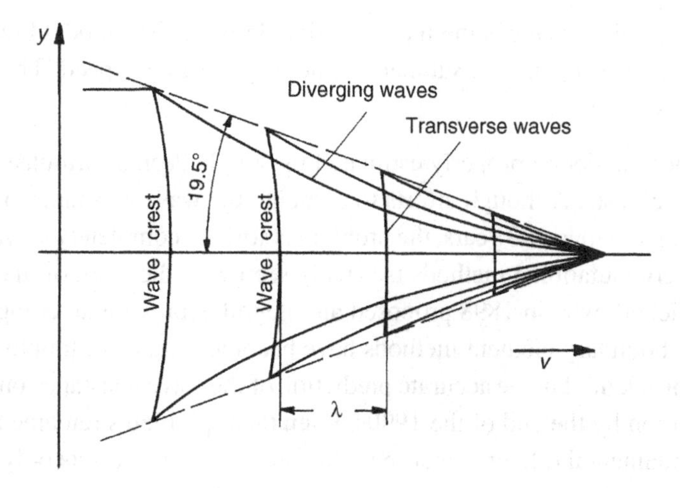

Figure 3.4:
'Secondary' wave system

The wave pattern consists of transverse and divergent waves. In deep water, the wave pattern is limited to a wedge-shaped region with a half-angle of 19.5°. This angle is independent of the actual shape of the ship. On shallow water, the half-angle widens to 90° (for depth Froude number $F_{nh} = 1.0$) and then becomes more and more narrow for supercritical speeds above $F_{nh} = 1$. The ship produces various wave patterns which interfere with each other. The main wave patterns are created where strong changes in the geometry near the water surface occur, i.e. at the bulbous bow, the bow, the forward shoulder, the aft shoulder, and the stern. The wave length λ depends quadratically on the ship speed. Unfavorable Froude numbers with mutual reinforcement between major wave systems, e.g. bow and stern waves, should be avoided. This makes, e.g., $F_n = 0.4$ an unfavorable Froude number. The interference effects result in a wave resistance curve with humps and hollows. If the wave resistance coefficient is considered, i.e. the wave resistance is made non-dimensional by an expression involving the square of the speed, the humps and hollows become very pronounced.

In reality, the problem is more complex:

- The steepness of waves is limited. The pressure in the 'primary wave system' changes rapidly at the ship ends enforcing unrealistically steep waves. In reality, waves break here and change the subsequent 'secondary wave pattern'. At Froude numbers around 0.25 usually considerable wave-breaking starts, making this Froude number often unfavorable, although many textbooks recommend it as favorable based on the above interference argument for the 'secondary wave pattern'.
- The free surface also results in a dynamic trim and sinkage. This also changes the wave pattern. Even if the double-body flow around the dynamically trimmed and sunk ship is computed, this is not really the ship geometry acting on the fluid, as the actually wetted

surface (wave profile) changes the hull. The double-body flow model breaks down completely if a transom stern is submerged, but dry at design speed. This is the case for many modern ship hulls.

The wave resistance cannot be properly estimated by simple design formulae. It is usually determined in model tests. Although efforts to compute the wave resistance by theoretical methods date back more than 100 years, the problem is still not completely solved satisfactorily. The beginning of computational methods is usually seen with the work of the Australian mathematician Michell, who in 1898 proposed an integral expression to compute the wave resistance. Today, boundary element methods have become a standard tool to compute the 'wave resistance problem', but the accurate prediction of the wave resistance only came close to a satisfactory solution by the end of the 1990s. Even then, problems remained with breaking waves and the fundamental dilemma that in reality ship resistance exists only as a whole quantity. Its separation into components is merely a hypothesis to facilitate analysis, but the theoretically cleanly divided resistance components interact and require a comprehensive approach for a completely satisfactory treatment. Free-surface RANSE has made a great deal of progress possible in this respect, despite some remaining problems in capturing accurately breaking waves and spray formation at the ship's bow (Peric and Bertram 2011).

Computational methods for the analysis of the wave resistance will be discussed in detail in Section 3.5.1.

3.2. Experimental Approach

3.2.1. Towing Tanks and Experimental Set-Up

Despite the ever-increasing importance of numerical methods for ship hydrodynamics, model tests in towing tanks are still seen as an essential part in the design of a ship to validate the power requirements in calm water, which form a fundamental part of each contract between shipowner and shipyard.

We owe the modern methodology of predicting a ship's resistance to William Froude, who presented his approach in 1874 to the predecessor of the Royal Institution of Naval Architects in England. His hypothesis was that the ship resistance is divisible into frictional and wavemaking resistance, with the wavemaking resistance following his 'law of comparison' (Froude similarity). This ingenious concept allowed Froude to show, for the first time, how the resistance of a full-scale ship may be determined by testing scale models. His success motivated building the first model basin in 1879 in Torquay, England. Soon further model basins followed in Europe and the USA.

Tests are usually performed in towing tanks, where the water is still and the model is towed by a carriage. (Alternatively, tests can also be performed in circulating tanks, where the model is still and the water moves.) The carriage in a towing tank keeps its speed with high precision. The model is kept on course by special wires at the ship ends. Usually, models are free to trim and sink. After the initial acceleration, some time has to pass before a stationary state is reached. Then the remaining measuring time is determined by the remaining towing tank distance and the deceleration time of the carriage. Therefore, towing tanks are usually several hundred meters long to allow sufficient measuring time (in most cases).

The model size is determined by a number of boundary conditions:

- The model should be as large as possible to minimize viscous scale effects, especially concerning laminar/turbulent flow and flow separation.
- The model should be small enough to avoid strength problems (both internal strength of the model and loads on the test carriage).
- The model should be small enough such that the corresponding test speed can be achieved by the carriage.
- The model should be small enough to avoid noticeable effects of restricted water in the test basin.

This leads to a bandwidth of acceptable model sizes. Typically models for resistance and propulsion tests have a size $4\,\text{m} \leq L_m \leq 10\,\text{m}$. Model scales range between $15 \leq \lambda \leq 45$. In practice, often the selected stock propeller decides the exact model scale.

Tests are performed keeping Froude similarity, i.e. Froude numbers of model and full scale are the same. The Reynolds numbers differ typically by two orders of magnitude. The scale effect (error of not keeping the Reynolds similarity) is then partially compensated by empirical corrections.

Models operate at considerably lower Reynolds numbers. (Typically for models $R_n \approx 10^7$ and for full-scale ships $R_n \approx 10^9$.) This means that in the model the transition from laminar to turbulent flow occurs relatively further aft. As a consequence, the resistance would be more difficult to scale. Therefore, the model is equipped with artificial turbulence stimulators (sand strip, studs, or trip wire) in the forebody. One assumes that the transition from laminar to turbulent flow occurs at a length corresponding to $R_n = 0.5 \cdot 10^6$ from the stem. In practice, often the turbulence stimulators are located somewhat further aft. Then the reduced resistance due to the longer laminar flow compensates (at least partially) the additional resistance of the turbulence stimulators.

The models are made of special paraffin wax or special tropical wood that hardly changes volume and shape with time or temperature. Wax models are cheaper, but less robust. Wooden models receive a smooth finish of paint. Yellow is the preferred color for regular models as this color contrasts nicely with the (blackish) water, which is important for visual observations, e.g. of the wave profile. For icebreakers, often for similar purposes, red is the preferred color as it appears to be a good compromise for contrasts of water and ice.

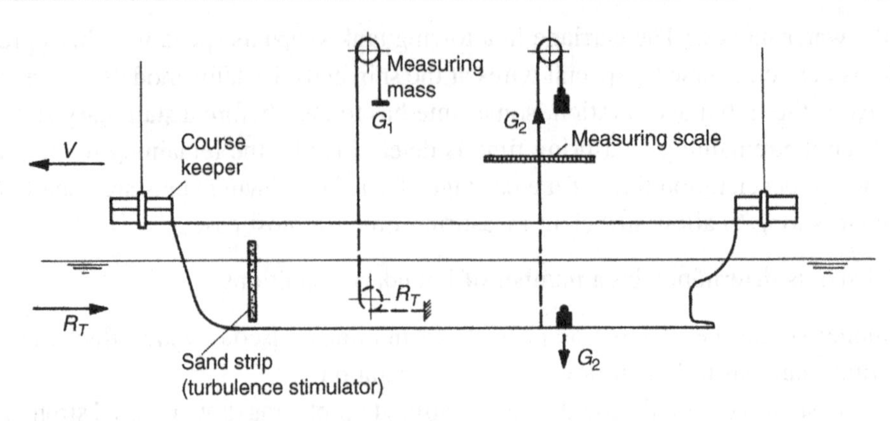

Figure 3.5:
Experimental set-up for resistance test

3.2.2. Resistance Test

Resistance tests determine the resistance of the ship without propeller (and often also without other appendages; sometimes resistance tests are performed for both the 'naked' hull and the hull with appendages). Propulsion tests are performed with an operating propeller and other relevant appendages. A problem is that the forces on appendages are largely driven by viscosity effects with small to negligible gravity effects. As Reynolds similarity is violated, the forces cannot be scaled easily to full scale. For ships with large and unusual appendages, the margins of errors in prediction are thus much larger than for usual hulls, where experience helps in making appropriate corrections.

The model is towed by weights and wires (Fig. 3.5). The main towing force comes from the main weight G_1. The weight G_2 is used for fine-tuning:

$$R_T = G_1 \pm G_2 \sin \alpha \tag{3.9}$$

The sign is positive if the vertical wire moves aft. The angle α is determined indirectly by measuring the distance on the length scale. Alternatively, modern experimental techniques also use strain gauges as these do not tend to oscillate as the wire-weight systems do.

The model test gives the resistance (and power) for towing tank conditions:

- (usually) sufficiently deep water;
- no seaway;
- no wind;
- fresh water at room temperature.

This model resistance has to be converted for a prediction of the full-scale ship. To do this conversion several methods are outlined in the following chapters, namely:

- Method ITTC 1957;
- Method of Hughes–Prohaska;
- Method ITTC 1978;
- Geosim method of Telfer.

The most important of these methods in practice is the method ITTC 1978. Resistance tests are also used to measure the nominal wake, i.e. the wake of the ship without propeller. Measurements of the nominal wake are usually limited to the propeller plane. The local velocities were traditionally measured by pitot tubes. Laser-Doppler velocimetry also allows non-intrusive measurements of the flow field. The results are usually displayed as contour lines of the longitudinal component of the velocity (Fig. 3.6). These data play an important role in the design of a propeller. For optimizing the propeller pitch as a function of the radial distance from the hub, the wake fraction is computed as a function of this radial distance by integrating the wake in the circumferential direction:

$$w(r) = \frac{1}{2\pi} \int_{0}^{2\pi} w(r,\phi)\, \mathrm{d}\phi \tag{3.10}$$

The wake field is also used in evaluating propeller-induced vibrations.

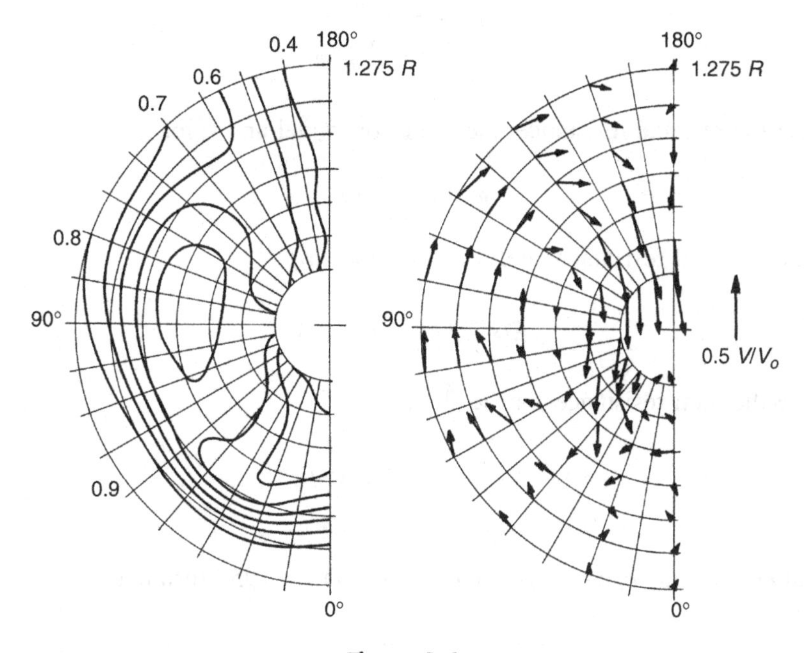

Figure 3.6:
Results of wake measurement

3.2.3. Method ITTC 1957

The resistance of the hull is decomposed as:

$$R_T = R_F + R_R \tag{3.11}$$

R_F is the frictional resistance, R_R the residual resistance. Usually the resistance forces are expressed as non-dimensional coefficients of the form:

$$c_i = \frac{R_i}{\frac{1}{2}\rho V_s^2 S} \tag{3.12}$$

S is the wetted surface in calm water, V_s the ship speed. The resistance coefficient of the ship is then determined as:

$$c_{Ts} = c_{Fs} + c_R + c_A = c_{Fs} + (c_{Tm} - c_{Fm}) + c_A \tag{3.13}$$

The index s again denotes values for the full-scale ship, the index m values for the model. c_R is assumed to be independent of model scale, i.e. c_R is the same for model and full scale. The model test serves primarily to determine c_R. The procedure is as follows:

1. Determine the total resistance coefficient in the model test:

$$c_{Tm} = \frac{R_{Tm}}{\frac{1}{2}\rho_m V_m^2 S_m} \tag{3.14}$$

2. Determine the residual resistance, the same for model and ship:

$$c_R = c_{Tm} - c_{Fm} \tag{3.15}$$

3. Determine the total resistance coefficient for the ship:

$$c_{Ts} = c_R + c_{Fs} + c_A \tag{3.16}$$

4. Determine the total resistance for the ship:

$$R_{Ts} = c_{Ts} \cdot \frac{1}{2}\rho_s V_s^2 S_s \tag{3.17}$$

The frictional coefficients c_F are determined by the ITTC 1957 formula:

$$c_F = \frac{0.075}{(\log_{10} R_n - 2)^2} \tag{3.18}$$

Table 3.1: Recommended values for c_A

L_{pp} (m)	c_A
50–150	0.00035–0.0004
150–210	0.0002
210–260	0.0001
260–300	0
300–350	−0.0001
>350	−0.00025

This formula already contains a global form effect, increasing the value of c_F by 12% compared to the value for flat plates (Hughes formula).

Historically c_A was a roughness allowance coefficient which considered that the model was smooth while the full-scale ship was rough, especially when ship hulls were still riveted. However, with the advent of welded ships c_A sometimes became negative for fast and big ships. Therefore, c_A is more appropriately termed the correlation coefficient. c_A encompasses collectively all corrections, including roughness allowance, but also particularities of the measuring device of the model basin, errors in the model–ship correlation line, and the method. Model basins use c_A not as a constant, but as a function of the ship size, based on experience. The correlation coefficient makes predictions from various model basins difficult to compare and may in fact be abused to derive overly optimistic speed predictions to please customers.

Formulae for c_A differ between various model basins and shipyards. Examples are Table 3.1 and:

$$c_A = 0.35 \cdot 10^{-3} - 2 \cdot L_{pp} \cdot 10^{-6} \tag{3.19}$$

3.2.4. *Method of Hughes–Prohaska*

This approach decomposes the total resistance (coefficient) as follows:

$$c_T = (1 + k) \cdot c_{F0} + c_w \tag{3.20}$$

Both form factor $(1 + k)$ and wave resistance coefficient c_w are assumed to be the same for model and full scale, i.e. independent of R_n. The model test serves primarily to determine the wave resistance coefficient. The procedure is as follows:

1. Determine the total resistance coefficient in the model test as for the ITTC 1957 method:

$$c_{Tm} = \frac{R_{Tm}}{\frac{1}{2}\rho_m V_m^2 S_m} \tag{3.21}$$

2. Determine the wave resistance coefficient, the same for model and ship:

$$c_W = c_{Tm} - c_{F0m} \cdot (1 + k) \tag{3.22}$$

3. Determine the total resistance coefficient for the ship:

$$c_{Ts} = c_W + c_{F0s} \cdot (1 + k) + c_A \tag{3.23}$$

4. Determine the total resistance for the ship:

$$R_{Ts} = c_{Ts} \cdot \frac{1}{2} \rho_s V_s^2 S_s \tag{3.24}$$

The frictional coefficients c_{F0} for flat plates are determined by Hughes' formula:

$$c_F = \frac{0.067}{(\log_{10} R_n - 2)^2} \tag{3.25}$$

The correlation coefficient c_A differs fundamentally from the correlation coefficient for the ITTC 1957 method. Here c_A does not have to compensate for scaling errors of the viscous pressure resistance. ITTC recommends universally $c_A = 0.0004$.

The Hughes–Prohaska method is a form factor method. The form factor $(1 + k)$ is assumed to be independent of F_n and R_n and the same for model and ship. The form factor is determined by assuming:

$$\frac{c_T}{c_{F0}} = (1 + k) + \alpha \frac{F_n^4}{c_{F0}} \tag{3.26}$$

Model test results for several Froude numbers (e.g. between 0.12 and 0.24) serve to determine α in a regression analysis (Fig. 3.7).

3.2.5. Method of ITTC 1978

This approach is a modification of the Hughes–Prohaska method. It is generally more accurate and also considers the air resistance. The total resistance (coefficient) is again written in a form factor approach:

$$c_{Ts} = (1 + k)c_{Fs} + c_W + c_A + c_{AA} \tag{3.27}$$

c_w is the wave resistance coefficient, assumed to be the same for model and ship, i.e. independent of R_n. c_{Fs} is the frictional coefficient, following the ITTC 1957 formula. c_A is the correlation coefficient, which depends on the hull roughness:

$$c_A \cdot 10^3 = 105 \cdot \sqrt[3]{\frac{k_s}{L_{oss}}} - 0.64 \tag{3.28}$$

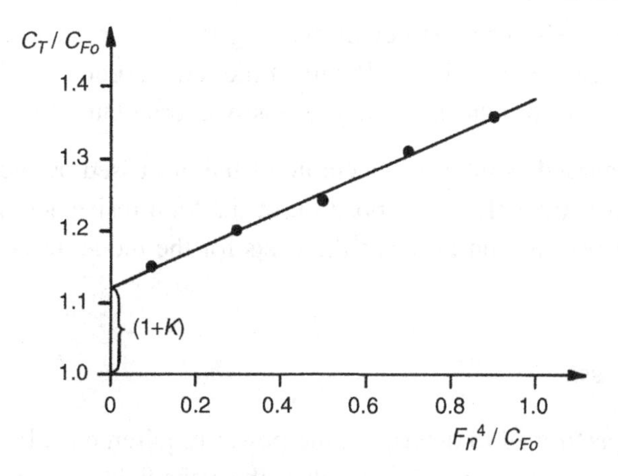

Figure 3.7:
Extrapolation of form factor

where k_s is the roughness of the hull and L_{oss} is the wetted length of the full-scale ship. For new ships, a typical value is $k_s/L_{oss} = 10^{-6}$, i.e. $c_A = 0.00041$.

c_{AA} considers globally the air resistance as follows:

$$c_{AA} = 0.001 \cdot \frac{A_T}{S} \tag{3.29}$$

where A_T is the frontal area of the ship above the waterline and S the wetted surface.

The model test serves primarily to determine the wave resistance coefficient. The procedure is similar to the procedure for Hughes–Prohaska, but the frictional coefficient is determined following the ITTC 1957 formula instead of Hughes' formula. The form factor is also determined slightly differently:

$$\frac{c_T}{c_F} = (1 + k) + \alpha \cdot \frac{F_n^n}{c_F} \tag{3.30}$$

Both n and α are determined in a regression analysis.

3.2.6. Geosim Method of Telfer

Telfer proposed in 1927 to perform model tests with families of models which are *geo*metrically *sim*ilar, but have different model scale. This means that tests are performed at the same Froude number, but different Reynolds numbers. The curve for the total resistance as a function of the Reynolds number is then used to extrapolate to the full-scale Reynolds number.

Telfer plotted the total resistance coefficient over log $R_n^{-1/3}$. For each model, a curve of the resistance is obtained as a function of F_n. Points of the same Froude number for various model scales are connected by a straight line which is easily extrapolated to full scale.

Telfer's method is regarded as the most accurate of the discussed prediction methods and avoids theoretically questionable decomposition of the total resistance. However, it is used only occasionally for research purposes as the costs for the model tests are too high for practical purposes.

3.2.7. Propulsion Test

Propulsion tests are performed to determine the power requirements, but also to supply wake and thrust deduction, and other input data (such as the wake field in the propeller plane) for the propeller design. The ship model is then equipped with a nearly optimum propeller selected from a large stock of propellers, the so-called stock propeller. The actual optimum propeller can only be designed after the propulsion test. The model is equipped with a propulsive drive, typically a small electro-motor (Fig. 3.8).

The tests are again performed for Froude similarity. The total resistance coefficient is then higher than for the full-scale ship, since the frictional resistance coefficient decreases with increasing Reynolds number. This effect is compensated by applying a 'friction deduction' force. This compensating force is determined as follows (see Section 3.2.5):

$$F_D = \frac{1}{2}\rho \cdot V_m^2 \cdot S_m \cdot ((1+k)(c_{Fm} - c_{Fs}) - c_A - c_{AA}) \tag{3.31}$$

The propeller then has to produce a thrust that has to compensate the total resistance R_T minus the compensating force F_D. The propulsion test is conducted with constant speed. The rpm of the propeller is adjusted such that the model is in self-propelled equilibrium. Usually the speed of the towing tank carriage is kept constant and the rpm of the propeller varied until an equilibrium is

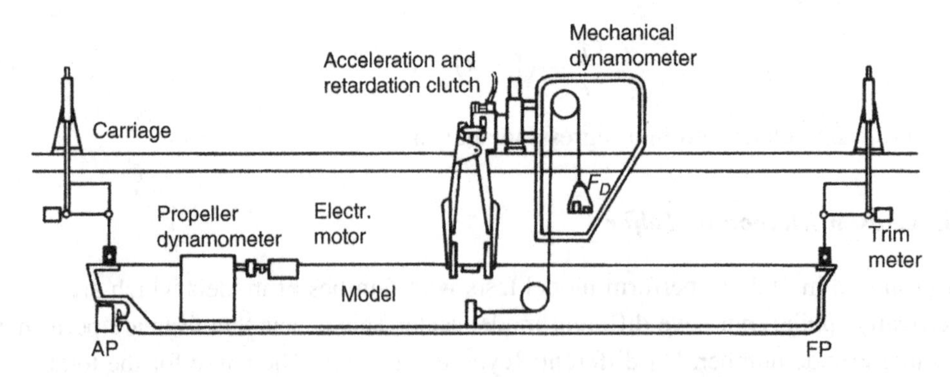

Figure 3.8:
Experimental set-up for propulsion test

reached. A propeller dynamometer then measures thrust and torque of the propeller as a function of speed. In addition, dynamical trim and sinkage of the model are recorded. The measured values can be transformed from model scale to full scale by the similarity laws: speed $V_s = \sqrt{\lambda} \cdot V_m$, rpm $n_s = n_m/\sqrt{\lambda}$, thrust $T_s = T_m \cdot (\rho_s/\rho_m) \cdot \lambda^3$, torque $Q_s = Q_m \cdot (\rho_s/\rho_m) \cdot \lambda^4$. A problem is that the propeller inflow is not geometrically similar for model and full scale due to the different Reynolds number. Thus the wake fraction is also different. Also, the propeller rpm should be corrected to be appropriate for the higher Reynolds number of the full-scale ship.

The scale effects on the wake fraction are attempted to be compensated by the empirical formula:

$$w_s = w_m \cdot \frac{c_{Fs}}{c_{Fm}} + (t + 0.04) \cdot \left(1 - \frac{c_{Fs}}{c_{Fm}}\right) \tag{3.32}$$

t is the thrust deduction coefficient and is assumed to be the same for model and full scale.

The evaluation of the propulsion test requires the resistance characteristics and the open-water characteristics of the stock propeller. There are two approaches:

1. *'Thrust identity' approach.* The propeller produces the same thrust in a wake field of wake fraction w as in open water with speed $V_s(1 - w)$ for the same rpm, fluid properties, etc.
2. *'Torque identity' approach.* The propeller produces the same torque in a wake field of wake fraction w as in open water with speed $V_s(1 - w)$ for the same rpm, fluid properties, etc.

The ITTC standard is the 'thrust identity' approach. It will be covered in more detail in the next chapter on the ITTC 1978 performance prediction method.

The results of propulsion tests are usually given in diagrams, as shown in Fig. 3.9. Delivered power and propeller rpm are plotted over speed. The results of the propulsion test prediction are validated in the sea trial of the ship introducing necessary corrections for wind, seaway, and shallow water. The diagrams contain not only the full-load design condition at trial speed, but also ballast conditions and service speed conditions. Service conditions feature higher resistance, reflecting the reality of the ship after some years of service: increased hull roughness due to fouling and corrosion, added resistance in seaway and wind.

3.2.8. ITTC 1978 Performance Prediction Method

The ITTC 1978 performance prediction method (IPPM78) has become a widely accepted procedure to evaluate model tests. It combines various aspects of resistance, propulsion, and open-water tests. These are comprehensively reviewed here. Further details may be found in Section 3.2.5, Section 3.2.7, and Section 2.5, Chapter 2. The IPPM78 assumes that the following tests have been performed yielding the corresponding results:

resistance test $R_{Tm} = f(V_m)$
open-water test $T_m = f(V_{Am}, n_m)$

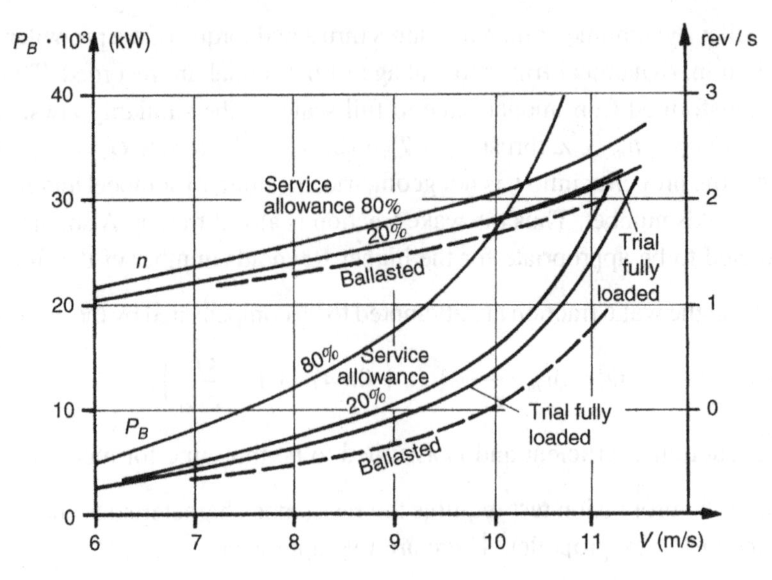

Figure 3.9:
Result of propulsion test

$$Q_m = f(V_{Am}, n_m)$$
$$\text{propulsion test } T_m = f(V_m, n_m)$$
$$Q_m = f(V_m, n_m)$$

R_T is the total resistance, V the ship speed, V_A the average inflow speed to the propeller, n the propeller rpm, K_T the propeller thrust coefficient, and K_Q the propeller torque coefficient. Generally, m denotes model, s full scale.

The resistance is evaluated using the ITTC 1978 method (for single-screw ships) described in Section 3.2.5:

1. Determine the total resistance coefficient in the model test:

$$c_{Tm} = \frac{R_{Tm}}{\frac{1}{2}\rho_m V_m^2 S_m} \tag{3.33}$$

2. Determine the frictional resistance coefficient for the model following ITTC 1957:

$$c_{Fm} = \frac{0.075}{(\log_{10} R_{nm} - 2)^2} \tag{3.34}$$

The Reynolds number of the model is $R_{nm} = V_m L_{osm}/\nu_m$, where L_{os} is the wetted length of the model. L_{os} is the length of the overall wetted surface, i.e. usually the length from the tip of the bulbous bow to the trailing edge of the rudder.

3. Determine the wave resistance coefficient, same for model and ship:

$$c_w = c_{Tm} - (1+k)c_{Fm} \tag{3.35}$$

The determination of the form factor $(1 + k)$ is described below.

4. Determine the total resistance coefficient for the ship:

$$c_{Ts} = c_w - (1+k)c_{Fs} + c_A + c_{AA} \tag{3.36}$$

c_{Fs} is the frictional resistance coefficient following ITTC 1957, but for the full-scale ship. c_A is a correlation coefficient (roughness allowance). c_{AA} considers the air resistance:

$$c_A = \left(105\sqrt[3]{\frac{k_s}{L_{oss}}} - 0.64\right) \cdot 10^{-3} \tag{3.37}$$

k_s is the roughness ($= 1.5 \cdot 10^{-4}$ m) and L_{oss} the wetted length of the ship.

$$c_{AA} = 0.001\frac{A_T}{S_s} \tag{3.38}$$

A_T is the frontal area of the ship above the water, S_s the wetted surface.

5. Determine the total resistance for the ship:

$$R_{Ts} = c_{Ts} \cdot \frac{1}{2}\rho_s V_s^2 S_s \tag{3.39}$$

The form factor is determined in a least square fit of α and n in the function:

$$\frac{c_{Tm}}{c_{Fm}} = (1+k) + \alpha \cdot \frac{F_n^n}{c_{Fm}} \tag{3.40}$$

The open-water test gives the thrust coefficient K_T and the torque coefficient K_Q as functions of the advance number J:

$$K_{Tm} = \frac{T_m}{\rho_m n_m^2 D_m^4} \qquad K_{Qm} = \frac{Q_m}{\rho_m n_m^2 D_m^5} \qquad J = \frac{V_{Am}}{n_m D_m} \tag{3.41}$$

D_m is the propeller diameter. The model propeller characteristics are transformed to full scale (Reynolds number correction) as follows:

$$K_{Ts} = K_{Tm} + 0.3 \cdot Z \cdot \frac{c}{D_s} \cdot \frac{P_s}{D_s} \cdot \Delta C_D \tag{3.42}$$

$$K_{Qs} = K_{Qm} - 0.25 \cdot Z \cdot \frac{c}{D_s} \cdot \Delta C_D \tag{3.43}$$

Z is the number of propeller blades, P_s/D_s the pitch—diameter ratio, D_s the propeller diameter in full scale, and c the chord length at radius $0.7D$.

$$\Delta C_D = C_{Dm} - C_{Ds} \tag{3.44}$$

This is the change in the profile resistance coefficient of the propeller blades. These are computed as:

$$C_{Dm} = 2\left(1 + 2\frac{t_m}{c_m}\right)\left(\frac{0.044}{R_{nco}^{1/6}} - \frac{5}{R_{nco}^{2/3}}\right) \tag{3.45}$$

t is the maximum blade thickness, c the maximum chord length. The Reynolds number $R_{nco} = V_{co}c_m/\nu_m$ at $0.7D_m$, i.e. $V_{co} = \sqrt{V_{Am}^2 + (0.7\pi n_m D_m)^2}$.

k_p is the propeller blade roughness, taken as $3 \cdot 10^{-5}$ if not otherwise known.

The evaluation of the propulsion test requires the resistance and open-water characteristics. The open-water characteristics are denoted here by the index *fv*. The results of the propulsion test are denoted by *pv*:

$$C_{Ds} = 2\left(1 + 2\frac{t_s}{c_s}\right)\left(1.89 + 1.62 \log_{10}\frac{c_s}{k_p}\right)^{-2.5} \tag{3.46}$$

Thrust identity is assumed, i.e. $K_{Tm,pv} = K_{Tm,fv}$. Then the open-water diagram can be used to determine the advance number J_m. This in turn yields the wake fraction of the model:

$$w_m = 1 - \frac{J_m D_m n_m}{V_m} \tag{3.47}$$

The thrust deduction fraction is:

$$t = 1 + \frac{F_D - R_{Tm}}{T_m} \tag{3.48}$$

F_D is the force compensating for the difference in resistance similarity between model and full-scale ship:

$$F_D = \frac{1}{2}\rho \cdot V_m^2 \cdot S \cdot ((1 + k)(c_{Fm} - c_{Fs}) - c_A - c_{AA}) \tag{3.49}$$

With known J_m the torque coefficient $K_{Qm,fv}$ can also be determined. The propeller efficiency behind the ship is then:

$$\eta_{bm} = \frac{K_{Tm,pv}}{K_{Qm,pv}} \cdot \frac{J_m}{2\pi} \tag{3.50}$$

The open-water efficiency is:

$$\eta_{0m} = \frac{K_{Tm,fv}}{K_{Qm,fv}} \cdot \frac{J_m}{2\pi} \tag{3.51}$$

This determines the relative rotative efficiency:

$$\eta_R = \frac{\eta_{bm}}{\eta_{0m}} = \frac{K_{Qm,fv}}{K_{Qm,pv}} \cdot \frac{K_{Tm,fv}}{K_{Tm,fv}} \tag{3.52}$$

While t and η_R are assumed to be the same for ship and model, the wake fraction w has to be corrected:

$$w_s = w_m \frac{c_{Fs}}{c_{Fm}} + (t + 0.04)\left(1 - \frac{c_{Fs}}{c_{Fm}}\right) \tag{3.53}$$

A curve for the parameter K_T/J^2 as function of J is introduced in the open-water diagram for the full-scale ship. The design point is defined by:

$$\left(\frac{K_T}{J^2}\right)_s = \frac{T_s}{\rho_s \cdot D_s^2 \cdot V_{As}^2} = \frac{S_s}{2D_s^2} \cdot \frac{c_{Ts}}{(1-t)(1-w_s)^2} \tag{3.54}$$

The curve for K_T/J^2 can then be used to determine the corresponding J_s. This in turn determines the torque coefficient of the propeller behind the ship $K_{Qs} = f(J_s)$ and the open-water propeller efficiency $\eta_{0s} = f(J_s)$. The propeller rpm of the full-scale propeller is then:

$$n_s = \frac{(1-w_s) \cdot V_s}{J_s \cdot D_s} \tag{3.55}$$

The propeller torque in full scale is then:

$$Q_s = \frac{K_{Qs}}{\eta_R} \rho_s \cdot n_s^2 \cdot D_s^2 \tag{3.56}$$

The propeller thrust of the full-scale ship is:

$$T_s = \left(\frac{K_T}{J^2}\right)_s \cdot J_s^2 \cdot \rho_s \cdot n_s^2 \cdot D_s^4 \tag{3.57}$$

The delivered power is then:

$$P_{Ds} = Q_s \cdot 2\pi \cdot n_s \tag{3.58}$$

The total propulsion efficiency is then:

$$\eta_{Ds} = \eta_0 \cdot \eta_R \cdot n_{Hs} \tag{3.59}$$

3.3. Additional Resistance Under Service Conditions

The model test conditions differ in certain important points from trial and service conditions for the real ship. These include effects of:

- appendages;
- shallow water;
- wind;
- roughness;
- seaway.

Empirical corrections (based on physically more or less correct assumptions) are then used to estimate these effects and to correlate measured values from one state (model or trial) to another (service). The individual additional resistance components will be briefly discussed in the following.

- Appendages
  Model tests can be performed with geometrically properly scaled appendages. However, the flow around appendages is predominantly governed by viscous forces and would require Reynolds similarity. Subsequently, the measured forces on the appendages for Froude similarity are not properly scaled up to the real ship. Appendages may be tested separately and often the resistance of the appendages is scaled separately and added in a prediction for the full-scale ship. Unfortunately, this procedure does not account for interaction between hull and appendages and also introduces considerable error margins. Fortunately, most ships have only a few appendages and errors in estimating their resistance can be accepted. For unconventional ships with many and complex appendages, the difficulties in estimating the resistance of the appendages properly leads to a larger margin of uncertainty for the global full-scale prediction.
  Schneekluth and Bertram (1998) compiled some data from shipbuilding experience:
  - Properly arranged bilge keels contribute only 1–2% to the total resistance of ships. However, trim and ship motions in seastates increase the resistance more than for ships without bilge keels. Thus, in evaluation of model tests, a much higher increase of resistance should be made for ships in ballast condition.
  - Bow thrusters, if properly designed and located, do not significantly increase resistance. Transverse thrusters in the aftbody may increase resistance by 1–6%.
  - Shaft brackets and bossings increase resistance by 5–12%. For twin-screw ships with long propeller shafts, the resistance increase may be more than 20%.
  - Rudders increase resistance little (~1%) if in the neutral position and improve propulsion. But moderate rudder angles may increase resistance already by 2–6%.
- Shallow water
  Shallow water increases friction resistance and usually also wave resistance. Near the critical depth Froude number $F_{nh} = 1$, the resistance is strongly increased. Figure 3.10

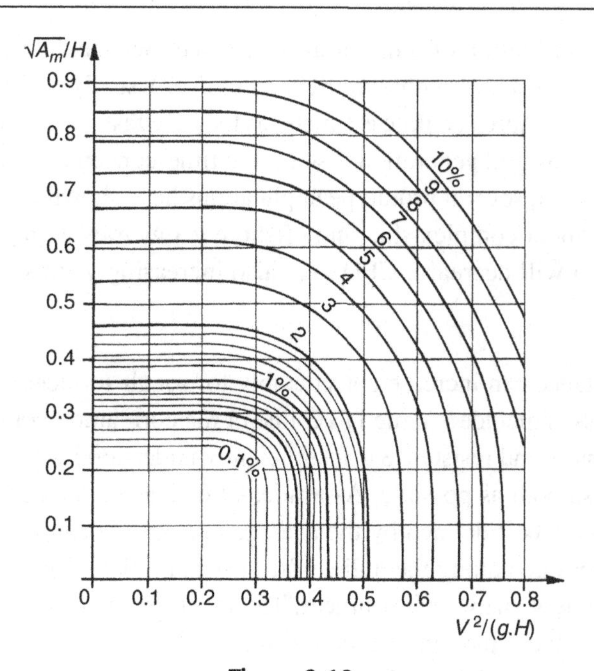

Figure 3.10:
Percentage loss of speed in shallow water (Lackenby 1963); A_m = midship section area, H = water depth, $V^2/(gH) = F_{nh}^2$

allows estimating the speed loss for weak shallow-water influence. The figure follows Schlichting's hypothesis that the wave resistance is the same if the wave lengths of the transversal waves are the same. Similar, but more sophisticated, diagrams are still popular in practice. For strong shallow-water influence, a simple correction is impossible as wave-breaking, squat and deformation of the free surface introduce complex physical interactions. In this case, only model tests or CFD may help.

In numerical simulations (CFD), the inclusion of shallow water is relatively simple. Boundary element methods based on Rankine elements use mirror images of the elements with respect to the water bottom. The image elements have the same strength as the original elements. This automatically yields zero normal velocity on the water bottom due to symmetry. The analytical inclusion of the bottom in Green function methods is more difficult, but also feasible. Field methods discretize the fluid domain to the water bottom and enforce a suitable boundary condition there. Shallow-water flows often feature stronger non-linearities than deep-water flows, making them in turn more difficult to solve numerically. CFD is much better suited to predict squat (dynamic trim and sinkage) in restricted waters than empirical formulae or even model tests.

- Wind

  Wind resistance is important for ships with large lateral areas above the water level, e.g. container ships and car ferries. Fast and unconventional ships, e.g. air-cushioned vehicles, also require the contribution of wind or air resistance. Schneekluth and Bertram (1998) give simple design estimates with empirical formulae. Usually wind tunnel tests are the preferred choice for a more accurate estimate, as they are fast and cheap to perform. CFD is not yet competitive, as grid generation is so far too time-consuming and expensive for most applications. However, several prototype applications have shown the capability of CFD to compute air flow about complex ship and offshore geometries with good results. As costs for grid generation will decrease, CFD may also increasingly substitute for wind tunnel tests.

- Roughness
  The friction resistance can increase considerably for rough surfaces. For new ships, the effect of roughness is included in the ITTC line or the correlation constant. The problem of correlating roughness and resistance is insufficiently understood. Model tests try to produce a hull surface as smooth as possible. As a rule, CFD does not consider roughness at all. Coating roughness is best defined by the average maximum peak-to-trough in a 50 mm sample length, measured along the hull surface (Swain 2010). This value can be measured using a hull roughness analyzer instrument. The mean roughness is obtained by making typically 10–15 of these measurements in one pass. It is recommended that at least 100 locations distributed around the hull are measured and the values combined to get an average hull roughness (AHR) value. The AHR can then be used to correct the resistance coefficient, following ITTC (1990):

$$\Delta c_f = \left(44\left[\sqrt[3]{\frac{AHR}{L}} - 10R_n^{-1/3}\right] + 0.125\right) \cdot 10^{-3} \tag{3.60}$$

This formula is valid for coatings with roughness values up to 225 microns. According to this formula, a rough hull surface (without fouling) may increase the frictional resistance by up to 5%. Fouling can increase the resistance by much more. Swain (2010) gives an example for the required power increase for a frigate at 15 knots, using the Naval Ships' Technical Manual (NSTM) rating:

NSTM rating	Description	Increase in required power
0	Hydraulically smooth surface	0%
0	Typical as applied antifouling coating	2%
10–20	Deteriorated coating or light slime	11%
30	Heavy slime	21%
40–60	Small calcareous fouling or weed	35%
70–80	Medium calcareous fouling	54%
90–100	Heavy calcareous fouling	86%

Fouling is estimated to be a serious problem. Munk and Kane (2009) found (based on statistical data of 10 years) that some 20% of the world's fleet are in poor condition with added resistance due to fouling in excess of 50% of the total resistance.

- Seaway
  The added resistance of a ship in a seaway is generally determined by computational methods and will be discussed in more detail in the chapters treating ship seakeeping. Such predictions for a certain region or route depend on the accuracy of seastate statistics, which usually introduce a larger error than the actual computational simulation. Ship size is generally more important than ship shape. Schneekluth and Bertram (1998) give simple design estimates for the speed loss due to added resistance in waves.

3.4. Fast Ships

3.4.1. Fast Monohulls

Most fast monohulls operate at Froude numbers $0.3 < F_n < 1.7$. There is a considerable overlap in operational speed ranges for various fast ship hull forms, but care should be taken in selecting the appropriate hull form. For example, planing hulls operated at $F_n < 0.6$ require more power than round-bilge non-planing hulls of the same displacement.

The most common representatives of fast monohulls are:

- *Displacement ships.* Typical examples are corvettes, frigates, and working boats. The hulls are characterized by straight V-shaped sections in the forebody, slender waterlines, round bilge with decreasing radius going to the transom stern and centerline skeg. They are frequently fitted with an integrated trim wedge. The LCB (longitudinal center of buoyancy) positions usually lie between 2% and 3% aft of $L_{pp}/2$ for larger ships. Displacement ships operate up to $F_n = 0.4$–0.6, i.e. they approach only the beginning of the planing condition. Advantages of this hull form are good seakeeping behavior, good course-keeping ability, and – if the vessel operates above the resistance hump – relatively low dynamic trim at top speed. The steep run of the power curve at higher speeds, caused by the fact that little hydrodynamic lift is produced, is the main disadvantage and determines the operational limits of this type.
- *Semi-displacement ships.* Typical examples are patrol boats, special navy craft, pleasure yachts, pilot boats, etc. Semi-displacement ships achieve higher speeds than displacement ships due to increased dynamical lift and corresponding reduction in resistance. Vessels can reach the planing condition with speeds of up to $F_n \approx 1$. The course-changing and course-keeping behavior is similar to that of pure displacement ships. The seakeeping is in general good. At high speeds, roll-induced transverse instability can arise under certain circumstances.

- *Planing hulls.* Typical examples are fast patrol boats, racing boats, search and rescue boats, and fast small passenger ferries. Planing hull designs should normally be used for high-speed vessels only. The stations have straight sections and knuckle lines (with a bilge knuckle running from the stem over the entire length to the transom), relatively large deadrise angles in the forebody decreasing further aft to approximately $L/2$ and continuing at nearly constant angles of not less than 10° to the transom. Early planing hull designs with warped deadrise are not common today. The forward part of the longitudinal knuckle is designed to work as a spray rail. Trim wedges with adjustable tabs are often installed to control the dynamic trim. These become less effective for $F_n > 1$ as there is generally a reduction in dynamic trim in that speed range. The typical advantages of this hull form develop at speeds $F_n > 1$. The seakeeping qualities of these vessels are not as good as for displacement and semi-displacement hulls. This disadvantage can be partially compensated by selecting relatively high L/B ($L/B \approx 7-8$) and deadrise angles $\tau > 10°$ in the aft part. The high-speed stability problem of semi-displacement hulls may also occur with planing hulls.

The power requirements of fast ships can be estimated following Bertram and Mesbahi (2004), who derived formulae based on graphs given by Fritsch and Bertram (2002). The resistance of high-speed vessels is primarily a function of the vessel's displacement, wetted length and surface, speed and additionally breadth for planing hulls. Therefore significant parameters are the slenderness $L/\nabla^{1/3}$ and the specific resistance R_T/∇. The total resistance R_T is decomposed as usual:

$$R_T = R_F + R_R \tag{3.61}$$

$$R_F = c_F \cdot \frac{\rho}{2} \cdot V^2 \cdot S \tag{3.62}$$

$$R_R = R_W + R_{APP} + R_{AA} + R_{PARAS} \tag{3.63}$$

The wetted surface S is defined at rest except for planing hulls as described in more detail below; c_F follows ITTC'57 with Reynolds number based on L_{wl}. The appendage resistance R_{APP}, the air and wind resistance R_{AA}, and the parasitic resistance R_{PARAS} (resistance of hull openings such as underwater exhaust gas exits, scoops, zinc anodes, etc.) can be estimated globally with 3–5% R_F for a projected vessel, but the determination of R_R (which includes wave, wave-making, spray and viscous pressure or separation resistance) is more difficult. It is common practice to take data from one of the systematical series, e.g. Bailey (1976) or Blount and Clement (1963). However, these prediction methods are time-consuming and semi-empirical formulae are more helpful for design engineers.

R_{AA} can be calculated following empirical formulae given in Section 3.6.

Considering the propulsive efficiencies yields the necessary engine power P_B from the effective power P_E:

$$P_B = \frac{P_E}{\eta_D \cdot \eta_M} \tag{3.64}$$

$\eta_M = 95\%$ is a typical mechanical efficiency of gearbox and shaft bearings. The propulsive efficiency is $\eta_D = \eta_H \cdot \eta_R \cdot \eta_0$. Since $\eta_H \approx 1$ and $\eta_R \approx 1$ for these hull forms, the main influence is the propeller efficiency η_0. Modern propeller designs and waterjet propulsion systems can reach values of more than 70% under good operational conditions.

The power may be predicted for conventional fast ships as follows:

- *Planing hulls.* Different test series are available for the necessary reliable power prediction in the early design phase. The most useful is the DTMB Series 62 (Blount and Clement 1963). More recently, HSVA formulae offer a simple estimate (Fritsch and Bertram 2002):

$$P_B = 0.7354 \cdot \left(\frac{\Delta \cdot V}{765.2} + \frac{B_C^2 \cdot V^3}{1051.1} \right) \tag{3.65}$$

  B_C [m] is the mean of the maximum beam at chines and the chine beam at the transom. V [kn] is the speed and P_B [kW] the brake power. B_C can be estimated in the design stage by:

$$B_C = 0.215 \cdot \Delta^{0.275} \tag{3.66}$$

  Δ [t] is the displacement mass.
- *Semi-displacement hulls.* The procedure for estimating resistance and power is very similar as for planing hulls. The NPL High Speed Round Bilge Displacement Hull Series (Bailey 1976) is available to aid the selection of main dimensions, lines design, resistance, and power prediction. This series also deals with examples for practical application. A simple estimate following HSVA is (Bertram and Mesbahi 2004):

$$R_T = C_{T\nabla} \cdot \frac{\rho}{2} \cdot V^2 \cdot \nabla^{2/3} \tag{3.67}$$

  $C_{T\nabla}$ is a function of the Froude number, given by the following relations:

  For $0.002 < C_{T\nabla} < 0.005$, $0.4 < F_n < 1.2$:

$$\begin{aligned} C_{T\nabla} = c_0 + c_1 \cdot \text{sig}\,[b_0 &+ b_1 \cdot \text{sig}(a_{10} + a_{11} \cdot x_1 + a_{12} \cdot x_2) \\ &+ b_2 \cdot \text{sig}(a_{20} + a_{21} \cdot x_1 + a_{22} \cdot x_2) + b_3 \cdot \text{sig}(a_{30} + a_{31} \cdot x_1 + a_{32} \cdot x_2) \\ &+ b_4 \cdot \text{sig}(a_{40} + a_{41} \cdot x_1 + a_{42} \cdot x_2) + b_5 \cdot \text{sig}(a_{50} + a_{51} \cdot x_1 + a_{52} \cdot x_2)] \end{aligned} \tag{3.68}$$

$x_1 = 1.125 \cdot F_n - 0.4 \quad x_2 = 0.299 \cdot C_{T\nabla} - 0.5481$

$a_{10} = 0.07700 \quad a_{11} = -5.28009 \quad a_{12} = 1.58541$

$a_{20} = -4.15170 \quad a_{21} = -4.40526 \quad a_{22} = 1.49122$

$a_{30} = 0.49661 \quad a_{31} = 2.49719 \quad a_{32} = -5.00270$

$a_{40} = -12.78673 \quad a_{41} = 13.63660 \quad a_{42} = -0.00639$

$a_{50} = 1.38547 \quad a_{51} = -29.74828 \quad a_{52} = -0.04912$

$b_0 = -1.04511 \quad b_1 = 5.51238 \quad b_2 = -0.98215$

$b_3 = -2.13594 \quad b_4 = 4.56969 \quad b_5 = -6.46482$

$c_0 = 0.02688 \quad c_1 = 0.0522$

For $0.0025 < C_{T\nabla} < 0.007$, $0.2 < F_n < 0.45$:

$$\begin{aligned} C_{T\nabla} = c_0 + c_1 \cdot \text{sig}\,[b_0 &+ b_1 \cdot \text{sig}(a_{10} + a_{11} \cdot x_1 + a_{12} \cdot x_2) \\ &+ b_2 \cdot \text{sig}(a_{20} + a_{21} \cdot x_1 + a_{22} \cdot x_2) \\ &+ b_3 \cdot \text{sig}(a_{30} + a_{31} \cdot x_1 + a_{32} \cdot x_2) \\ &+ b_4 \cdot \text{sig}(a_{40} + a_{41} \cdot x_1 + a_{42} \cdot x_2)] \end{aligned} \tag{3.69}$$

$x_1 = 3 \cdot F_n - 0.55 \quad x_2 = 0.1992 \cdot C_{T\nabla} - 0.44642$

$a_{10} = 2.47120 \quad a_{11} = -4.70440 \quad a_{12} = -0.71328$

$a_{20} = 2.80191 \quad a_{21} = -5.08604 \quad a_{22} = -0.80876$

$a_{30} = 0.53110 \quad a_{31} = -2.42700 \quad a_{32} = -0.61778$

$a_{40} = 0.15070 \quad a_{41} = 0.85700 \quad a_{42} = 0.72333$

$b_0 = 0.89195 \quad b_1 = -1.74315 \quad b_2 = -1.91516$

$b_3 = -0.80806 \quad b_4 = 0.60328$

$c_0 = 0.019134 \quad c_1 = 0.05333$

For frigates and corvettes, for $0.0016 < C_{T\nabla} < 0.0029$, $0.25 < F_n < 0.8$:

$$\begin{aligned} C_{T\nabla} = c_0 + c_1 \cdot \text{sig}\,[&b_0 + b_1 \cdot \text{sig}(a_{10} + a_{11} \cdot x_1 + a_{12} \cdot x_2) \\ &+ b_2 \cdot \text{sig}(a_{20} + a_{21} \cdot x_1 + a_{22} \cdot x_2) + b_3 \cdot \text{sig}(a_{30} + a_{31} \cdot x_1 + a_{32} \cdot x_2) \\ &+ b_4 \cdot \text{sig}(a_{40} + a_{41} \cdot x_1 + a_{42} \cdot x_2) + b_5 \cdot \text{sig}(a_{50} + a_{51} \cdot x_1 + a_{52} \cdot x_2)] \end{aligned} \tag{3.70}$$

$x_1 = 1.636 \cdot F_n - 0.359 \quad x_2 = 0.75541 \cdot C_{T\nabla} - 1.1959$

$a_{10} = -5.38402 \quad a_{11} = 16.26584 \quad a_{12} = -0.6375$

$a_{20} = -2.84961 \quad a_{21} = 9.28172 \quad a_{22} = -1.9176$

$a_{30} = -3.62339 \quad a_{31} = -4.52883 \quad a_{32} = 6.16248$

$a_{40} = 1.91471 \quad a_{41} = -3.14575 \quad a_{42} = 2.17611$

$a_{50} = 0.93425 \quad a_{51} = -11.05182 \quad a_{52} = 1.45755$

$b_0 = -1.1453 \quad b_1 = 3.40313 \quad b_2 = -3.32619$

$b_3 = 1.88172 \quad b_4 = 1.22925 \quad b_5 = -3.5398$

$c_0 = 0.01331 \quad c_1 = 0.04177$

Since the value found for the effective power is valid for the bare hull only, allowances for R_{APP} and R_{AA} must be added. R_{APP} can be calculated directly, e.g. Bailey (1976), or estimated from statistical data:

Two propellers:

$$R_{APP}/R_T\,[\%] = c_0 + c_1 \cdot \text{sig}\left[b_0 + b_1 \cdot \text{sig}(a_{10} + a_{11} \cdot x_1) + b_2 \cdot \text{sig}(a_{20} + a_{21} \cdot x_1)\right] \quad (3.71)$$

$$x_1 = 0.6544 \cdot F_n + 0.0338$$
$$a_{10} = -7.3461 \quad a_{11} = 14.1181 \quad a_{20} = -3.5455 \quad a_{21} = 13.3944$$
$$b_0 = 2.9959 \quad b_1 = 4.0700 \quad b_2 = -6.8369$$
$$c_0 = 7.0235 \quad c_1 = 8.7183$$

Three propellers:

$$R_{APP}/R_T\,[\%] = c_0 + c_1 \cdot \text{sig}\left[b_0 + b_1 \cdot \text{sig}(a_{10} + a_{11} \cdot x_1) + b_2 \cdot \text{sig}(a_{20} + a_{21} \cdot x_1)\right] \quad (3.72)$$

$$x_1 = 0.6453 \cdot F_n + 0.0477$$
$$a_{10} = -4.231 \quad a_{11} = 15.0686 \quad a_{20} = -7.375 \quad a_{21} = 14.0019$$
$$b_0 = 2.7373 \quad b_1 = -6.4811 \quad b_2 = 4.0462$$
$$c_0 = 10.7197 \quad c_1 = 12.5462$$

Four propellers:

$$R_{APP}/R_T\,[\%] = c_0 + c_1 \cdot \text{sig}\left[b_0 + b_1 \cdot \text{sig}(a_{10} + a_{11} \cdot x_1) + b_2 \cdot \text{sig}(a_{20} + a_{21} \cdot x_1)\right] \quad (3.73)$$

$$x_1 = 0.6859 \cdot F_n - 0.007964$$
$$a_{10} = 3.7972 \quad a_{11} = -16.4323 \quad a_{20} = 5.9647 \quad a_{21} = -11.98701$$
$$b_0 = 0.3437 \quad b_1 = 6.2153 \quad b_2 = -3.7455$$
$$c_0 = 14.2334 \quad c_1 = 16.4206$$

These formulae do not include interference effects from the individual parts of the appendages.

Appendages play a special role for fast ships. Many fast displacement, semi-displacement, and also planing hulls are characterized by moderate to severe spray generation. The spray comes from the bow wave rising up the hull with speed. This is particularly caused by the relatively blunt waterlines and hard buttock forward when $L/\nabla^{1/3}$ is unfavorably small or the beam too large. Severe spray generation has a number of disadvantages:

- The increase of frictional (due to larger wetted surface) and wave-making resistance.
- Wetness of deck and superstructures, unfavorable for yachts and unacceptable for gas turbine-powered ships (due to their demand for very dry and salt-free combustion air).
- Increased radar signature (for navy craft).

Spray generation can be taken into account when designing the hull before entering the construction phase. Sometimes hull changes are not possible. Then spray rails can often be an effective and relatively cheap measure to reduce spray generation. Spray rails can also improve the performance of existing fast ships. Typical spray rail arrangements either use an additional triangular profile or integrate a two-step knuckle line into the form. These run from the stem to about amidships. In both cases a horizontal deflection area with a sharp edge must be created. Spray rails also influence the dynamic lift on the forebody, thus improving the resistance also indirectly.

The resistance of a fast ship is fundamentally linked with the dynamic trim. Recommended optimum trim angles for fast vessels in modern practice according to HSVA are (Bertram and Mesbahi 2004):

Displacement and semi-displacement hulls

$$\begin{aligned}\theta = 0.7 \cdot \{c_0 + c_1 \cdot \text{sig}\,[b_0 &+ b_1 \cdot \text{sig}(a_{10} + a_{11} \cdot x_1 + a_{12} \cdot x_2) \\ &+ b_2 \cdot \text{sig}(a_{20} + a_{21} \cdot x_1 + a_{22} \cdot x_2) \\ &+ b_3 \cdot \text{sig}(a_{30} + a_{31} \cdot x_1 + a_{32} \cdot x_2) \\ &+ b_4 \cdot \text{sig}(a_{40} + a_{41} \cdot x_1 + a_{42} \cdot x_2)]\}\end{aligned} \tag{3.74}$$

$$\begin{aligned}&x_1 = 0.9 \cdot \left(\nabla^{2/3}/B \cdot T\right) - 1.975 \quad x_2 = 0.71 \cdot F_n - 0.0795 \\ &a_{10} = -3.75198 \quad a_{11} = -1.69432 \quad a_{12} = 9.49288 \\ &a_{20} = -2.49216 \quad a_{21} = 3.86243 \quad a_{22} = -1.65272 \\ &a_{30} = 3.87188 \quad a_{31} = 0.61239 \quad a_{32} = -17.00609 \\ &a_{40} = -2.68088 \quad a_{41} = -3.55418 \quad a_{42} = 3.42624 \\ &b_0 = 1.63558 \quad b_1 = -2.18713 \quad b_2 = 2.15603 \\ &b_3 = -4.84437 \quad b_4 = -1.51677 \\ &c_0 = -0.17276 \quad c_1 = 2.364\end{aligned} \tag{3.75}$$

B is the width, T the draft, ∇ the volumetric displacement.

Planing hulls

$$\begin{aligned}\theta = 0.7 \cdot \{c_0 + c_1 \cdot \text{sig}\,[b_0 + b_1 \cdot \text{sig}(a_{10} + a_{11} \cdot x_1 + a_{12} \cdot x_2) \\ + b_2 \cdot \text{sig}(a_{20} + a_{21} \cdot x_1 + a_{22} \cdot x_2)]\}\end{aligned}$$

$$\begin{aligned}&x_1 = 0.6 \cdot (\nabla^{2/3}/B \cdot T) - 0.85 \quad x_2 = 0.6624 \cdot F_n - 0.01936 \\ &a_{10} = -2.66906 \quad a_{11} = 0.12856 \quad a_{12} = 8.06127 \\ &a_{20} = 6.30112 \quad a_{21} = -3.37513 \quad a_{22} = -3.66594 \\ &b_0 = -0.83869 \quad b_1 = 4.16294 \quad b_2 = -2.71566 \\ &c_0 = -0.39046 \quad c_1 = 6.535\end{aligned}$$

Fixed trim wedges or moveable trim flaps can be used to optimize the dynamic trim for a given speed and slenderness. Trim wedges should normally be considered during the design phase, but they are also acceptable for improving craft already in service. Trim wedges are most effective at speeds in the resistance hump range at $F_n \approx 0.4$–0.5. They have almost no effect for $F_n > 1.2$. Reductions in total resistance of more than 10% are possible in the resistance hump range. The most effective trim wedge for a certain craft and operational range is best found in model tests.

Fixed or adjustable interceptors offer an alternative to control the dynamic trim of a vessel. An interceptor is basically a vertical extension of the transom beyond the shell plating. Forward of the interceptor plate the flow is decelerated and the local pressure is increased which generates a lift force to the vessel's stern. The effect is identical to that of a conventional stern wedge. However, the height of the interceptor needs only to be 50% of that of a wedge for the same effect on the dynamic trim and resistance. This is an advantage at lower speed due to the smaller immersed transom area.

Appendages strongly influence resistance and propulsive efficiency of fast ships ($R_{APP} = 6-15\%$ R_T). Recommendations are:

- Avoid over-sizing the shaft brackets, bossings, and rudder profiles.
- V-bracket designs may have approximately 5–7% higher R_{APP} than I-bracket designs.
- If V-brackets are obligatory for whatever reason the inner and outer legs should be aligned with the flow to minimize resistance and wake disturbance (vibration, cavitation). Optimization of the brackets may employ CFD or model tests (three-dimensional wake measurements).
- For twin-screw vessels, power consumption may differ by 3–5%, changing the sense of propeller rotation, depending on the aftbody lines. The propulsive coefficient η_D is also influenced by the degree of shaft inclination ε, expressed by an additional efficiency η_ε, (Hadler 1966):

$$\eta_\varepsilon = 1 - 0.00187 \cdot \varepsilon^{1.5} \tag{3.76}$$

  The decreasing tendency at increasing shaft angles ε indicates that the shaft arrangement should be considered carefully in the design. The phenomenon is due to the inhomogeneous flow to the propeller blades which reduces the propeller efficiency. Also cavitation may be increased to a certain degree.
- For twin-rudder arrangements, an inward inclination of the rudders' trailing edges by 2–3° can increase the propulsive efficiency by up to 3%.
- Strut barrels should be kept as small as possible and their noses should be rounded or have parabolic shapes.
- Bilge keels should generally be aligned with the flow at the bilge. The line of flow may be determined in paint tests or CFD.
- If non-retractable stabilizer fins are projected, the angle of attack with least resistance can be determined in model tests (with different adjusted fin angles) or employing CFD.

3.4.2. Catamarans

One of the advantages of catamarans vs. monohulls is the up to 70% larger deck area. On the other hand, catamarans have typically 20% more weight and 30–40% larger wetted surface. Catamarans usually require 20–80% (the higher values near $F_n = 0.5$) more power than monohulls due to higher frictional resistance and higher wave resistance. Catamarans feature high transverse stability, but roll periods are similar to monohulls due to high moments of inertia. Catamaran designs come at low, medium and high speeds. Thus catamaran hull forms range from pure displacement up to real planing hulls.

Displacement catamarans usually operate near the hydrodynamically unfavorable hump speed ($F_n \approx 0.5$). The design is then usually driven by the demand for a large and stable working platform, high transverse stability and shallow draft where speed is not so important, e.g. for buoy layers, sight-seeing boats, etc. There is no typical hull form for displacement catamarans. Round bilge, hard chine, and combinations of both are used. Asymmetric hull forms are common to reduce the wave interference effects between the hulls. For catamarans with low design speed, a relatively large $L/\nabla^{1/3}$ should be selected to minimize the resistance. The majority of displacement catamarans are driven by fully immersed conventional propellers. Due to the frequent shallow draft requirements for catamarans the clearance for the propellers becomes rather small. Then arrangements of tunnels and propeller nozzles are usual.

Semi-displacement catamarans operate at higher speeds, frequently at the beginning of the planing condition at $F_n \approx 1$ or slightly above. Again, no typical hull characteristic is observed; both round-bilge and hard-chine sections are common. For rough seas (like the North Sea), round-bilge sections are more advantageous with respect to ride comfort. Most wave-piercer catamarans also have round-bilge sections. Semi-displacement catamarans may have propeller drives or waterjet propulsion.

Planing catamarans operate at speeds up to 50 knots or more and F_n up to 2.0 and higher. Typical knuckled planing hull forms dominate. Symmetric and asymmetric hull forms show only marginal performance differences. For high speeds, waterjets offer better efficiencies than conventional propellers with lower cavitation risk. Thus for planing catamarans, waterjets are the most favorable propulsion system. Surface-piercing propellers are also an option which has been employed by some racing boats and navy craft.

Foils may reduce resistance and improve seakeeping. Foil-assisted catamarans (FACs) have forward and aft foils, supporting part of the total weight. The bow is usually lifted clear of the water, but the stern remains partially immersed, which is necessary for waterjet operation and stability. Increasing the foil area decreases the resistance. For modern FACs, the foils are equipped with efficient ride control systems which usually adjust a movable flap on the forward foil and in more advanced systems also on the rear foils. Controllable flaps are recommended for several reasons. The risk of broaching in quartering or side waves can be reduced, especially when operating with foils in maximum lift condition. Controllable flaps also help to tune dynamical trim and foil adjustment for maximum lift and minimum resistance. For FACs, wetted length and surface of the model change very much with speed.

The highest stresses for fast catamarans are slamming impacts on the fore part of the wetdeck. The most common anti-slamming device (ASD) is a deep-V part in the forward wetdeck above the calm waterline, as in wave-piercing catamarans. The wave energy in slamming events remains unchanged by ASDs, but is smeared over a longer period, thus reducing peaks. Arranging longitudinal rails and steps on the bottom of the wetdeck also reduces the slamming impacts as air–water cushions are formed between the longitudinal rails. Alternatively, longitudinal stiffeners with holes have been proposed.

The resistance for catamarans can be estimated following HSVA (Fritsch and Bertram 2002):

$$C_{T\nabla} = \frac{0.2}{L/\nabla^{1/3}} + \frac{2.05}{[1+25(F_n-0.45)^2]\cdot(L/\nabla^{1/3})^2} \quad \text{round bilge} \tag{3.77}$$

$$C_{T\nabla} = \frac{0.25}{[1+25(F_n-0.45)^2]\cdot(L/\nabla^{1/3})} + \frac{2.5}{[1+25(F_n-0.45)^2]\cdot(L/\nabla^{1/3})^2} \quad \text{hard chine} \tag{3.78}$$

R_{APP} and R_{AA} must be added separately.

3.4.3. Problems for Fast and Unconventional Ships

Model testing has a long tradition for the prediction and optimization of ship performance of conventional ships. The scaling laws are well established and the procedures correlate model and ship with a high level of accuracy. The same scaling laws generally apply to high-speed craft, but two fundamental problems may arise:

1. Physical quantities may have major effects on the results which cannot be deduced from classical model tests. The physical quantities in this context are: surface tension (spray), viscous forces and moments, aerodynamic forces, cavitation.
2. Limitations of the test facilities do not allow an optimum scale. The most important limitations are generally water depth and carriage speed.

Fast and unconventional ships are often 'hybrid' ships, i.e. they produce the necessary buoyancy by more than one of the three possible options: buoyancy, dynamic lift (foils or planing), aerostatic lift (air cushion). For the propulsion of fast ships, subcavitating, cavitating, and ventilated propellers as well as waterjets with flush or pitot-type inlets are used. Due to viscous effects and cavitation, correlation to full-scale ships causes additional problems.

Generally we cannot expect the same level of accuracy for a power prediction as for conventional ships. The towing tank should provide an error estimate for each individual case. Another problem arises from the fact that the resistance curves for fast ships are often quite flat near the design point as are the curves of available thrust for many propulsors. For example, errors in predicted resistance or available thrust of 1% would result in an error of the attainable speed of also about 1%, while for conventional cargo ships the error in speed would often be only 1/3%, i.e. the speed prediction is more accurate than the power prediction.

The main problems for model testing are discussed individually:

- Model tank restrictions
  The physics of high-speed ships are usually highly non-linear. The positions of the ship in resistance (without propeller) and propulsion (with propeller) conditions differ strongly. Viscosity and free-surface effects, including spray and overturning waves, play significant roles, making both experimental and numerical predictions difficult.

  Valid predictions from tank tests for the resistance of the full-scale ship in unrestricted water are only possible if the tank is sufficiently large, as compared to the model to allow similarity in flow. Blockage, i.e. the ratio of the submerged cross-section of the model to the tank cross-section, will generally be very low for models of high-speed ships. However, shallow-water effects depend mainly on the model speed and the tank water depth. The depth Froude number F_{nh} should not be greater than 0.8 to be free of significant shallow-water effects.

  The frictional resistance is usually computed from the frictional resistance of a flat plate of similar length as the length of the wetted underwater body of the model. This wetted length at test speed differs considerably from the wetted length at zero speed for planing or semi-planing hull forms. In addition, the correlation requires that the boundary layer is fully turbulent. Even when turbulence stimulators are used, a minimum Reynolds number has to be reached. We can be sure to have a turbulent boundary layer for $R_n > 5 \cdot 10^6$. This gives a lower limit to the speeds that can be investigated depending on the model length used.

  Figure 3.11 illustrates, using a towing tank with water depth $H = 6$ m and a water temperature 15°C, how an envelope of possible test speeds evolves from these two restrictions. A practical

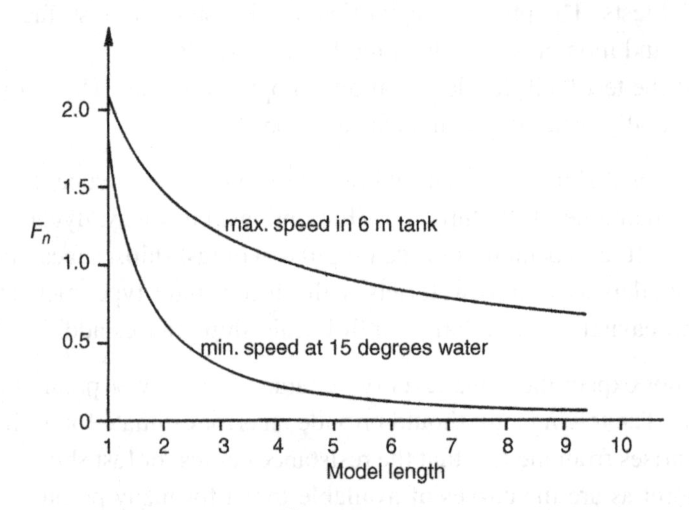

Figure 3.11:
Possible speed range to be safely investigated in a 6m deep towing tank at 15°C water temperature

limitation may be the maximum carriage speed. However, at HSVA the usable maximum carriage speed exceeds the maximum speed to avoid shallow-water effects.

- Planing hulls
  In the planing condition a significant share of the resistance is frictional and there is some aerodynamic resistance. At the design speed, the residual resistance, i.e. the resistance component determined from model tests, may only be 25–30% of the total resistance. In model scale, this part is even smaller. Therefore the measurements of the model resistance must be very accurate. Resistance of planing hulls strongly depends on the trim of the vessel. A careful test set-up is needed to ensure that the model is towed in the correct direction. The most important problem, however, is the accurate determination of the wetted surface and the wetted length which is needed to compute the frictional resistance for both the model and the ship. Side photographs, while popular, are not adequate. Preferably underwater photographs should be used. In many cases, the accurate measurement of trim and sinkage may be adequate in combination with hydrostatic computation of wetted surface and length. As the flotation line of such vessels strongly depends on speed, proper arrangement of turbulence stimulation is needed as well. Depending on the propulsion system, planing vessels will have appendages like rudders and shafts. For typical twin-screw ships with shafts, one pair of I-brackets and one pair of V-brackets, the appendage resistance could account for 10% of the total resistance of the ship. As viscous resistance is a major part of the appendage resistance and as the Reynolds number of the appendages will be small for the model in any case or the appendage may be within the boundary layer of the vessel, only a crude correlation of the appendage resistance is possible: the resistance of the appendage is determined in model scale by comparing the resistance of the model with and without appendages. Then an empirical correction for transferring the appendage resistance to the full-scale ship is applied. In many cases, it may be sufficient to perform accurate measurements without any appendages on the model and then use empirical estimates for the appendage resistance.
- Craft with hydrofoils
  Hydrofoils may be used to lift the hull out of the water to reduce resistance. Besides classical hydrofoil boats which are lifted completely out of the water and are fully supported by foil lift, hybrid hydrofoil boats may be used which are partially supported by buoyancy and partially by foil lift, e.g. catamarans with foils between the two hulls. When performing and evaluating resistance and propulsion tests for such vessels, the following problems have to be kept in mind:
  - The Reynolds number of the foils and struts will always be very low. Therefore the boundary layer on the foil may become partially laminar. This will influence the lift and the frictional resistance of the foils in a way requiring special correlation procedures to compensate at least partially for these scaling errors. The uncertainty level is still estimated as high as 5%, which is definitely higher than for conventional craft.

- Cavitation may occur on the full-scale hydrofoil. This will influence the lift and drag of the foils. Significant cavitation will certainly occur if the foil loading exceeds 10^5 N/m^2. With configurations not fully optimized for cavitation avoidance, significant cavitation is expected for foil loadings already in excess of $5 \cdot 10^4$ N/m^2. Another important parameter is the vessel's speed. Beyond 40 knots, cavitation has to be expected on joints to struts, flaps, foil tips, and other critical parts. At speeds beyond 60 knots, cavitation on the largest part of the foil has to be expected. When model testing these configurations in model tanks, no cavitation will occur. Therefore similarity of forces cannot be expected. To overcome this problem, resistance and propulsion tests could be performed in a free-surface cavitation tunnel. However, due to the usually small cross-sections of these tunnels, shallow-water effects may then be unavoidable. Therefore HSVA recommends the following procedure:

  1. Perform tests in the towing tank using non-cavitating foils from stock, varying angle of attack, and measure the total resistance and the resistance of the foils.
  2. Test the foils (including struts) in a cavitation tunnel varying angle of attack, observe cavitation and measure forces.
  3. Combine the results of both tests by determining the angle of attack for similar lift of foils and summing the resistance components.

  In the preliminary design phase, the tests in the cavitation tunnel may be substituted by corresponding flow computations. Alternatively, full-scale RANSE computations can be used.

- Surface effect ships (SES)
  SES combine aerostatic lift and buoyancy. The wave resistance curve of SES exhibits humps and hollows as in conventional ships. The magnitude of the humps and hollows in wave resistance depends strongly on the cushion L/B ratio. Wave-making of the submerged hulls and the cushion can simply be scaled according to Froude similarity as long as the tank depth is sufficient to avoid shallow-water effects. Otherwise a correction based on the potential flow due to a moving pressure patch is applied. Due to the significant influence of trim, this method has some disadvantages. To determine the wetted surface, observations inside the cushion are required with a video camera. The frictional resistance of the seals cannot be separated out of the total resistance. The pressure distribution between seals and cushion has to be controlled and the air flow must be determined. Also, the model aerodynamic resistance in the condition under the carriage has to be determined and used for separating the wave resistance. Generally separate wind tunnel tests are recommended to determine the significant aerodynamic resistance of such ships.
- Propulsion with propellers
  - *Conventional propellers.* Most of the problems concerning the scaling of resistance also appear in the propulsion test, as they determine the propeller loading. The use of a thrust deduction fraction is formally correct, but the change in resistance is partially due to a change of trim with operating propellers. For hydrofoils, the problem is that cavitation

is not present at model scale. Therefore, for cases with propeller loading where significant cavitation is expected, additional cavitation tests are used to determine the thrust loss due to cavitation. Z-drives which may even be equipped with contra-rotating propellers are expensive to model and to equip with accurate measuring devices. Therefore propulsion tests with such units are rarely performed. Instead the results of resistance and open-water tests of such units in a proper scale are numerically combined.

- *Cavitating propellers.* Certain high-speed propellers are designed to operate with a controlled extent of cavitation on the suction side of the blades. They are called super-cavitating or partially cavitating (Newton–Rader) propellers. These propulsors cannot be tested in a normal towing tank. Here again either resistance tests or propulsion tests with non-cavitating stock propellers are performed and combined with open-water tests in a cavitation tunnel.
- *Surface-piercing propellers.* Surface-piercing or ventilated propellers operate directly at the free surface. Thus the suction side is ventilated and therefore the collapse of cavitation bubbles on the blade surface is avoided. Due to the operation at the free surface, Froude similarity has to be maintained in model tests. On the other hand, thrust and torque, but more important also the side and vertical forces, strongly depend on the cavitation number. The vertical force may amount to up to 40% of the thrust and therefore will strongly influence the resistance of planing vessels or SES, ships where this type of propeller is typically employed.

- Waterjet propulsion
  A common means of propulsion for high-speed ships is the waterjet. Through an inlet in the bottom of the craft, water enters into a bent duct to the pump, where the pressure level is raised. Finally the water is accelerated and discharged in a nozzle through the transom. Power measurements on a model of the complete system cannot be properly correlated to full scale. Only the inlet and the nozzle are built to scale and an arbitrary model pump with sufficient capacity is used. The evaluation of waterjet experiments is difficult and usually involves several special procedures involving a combination of computations, e.g. the velocity profile on the inlet by boundary layer or RANSE computations, and measured properties, e.g. pressures in the nozzle. The properties of the pump are determined either in separate tests of a larger pump model, taken from experience with other pumps, or supplied by the pump manufacturer. A special committee of the ITTC was formed to cover waterjet propulsion and latest recommendation and literature references may be found in the ITTC proceedings.

3.5. CFD Approaches for Steady Flow

3.5.1. Wave Resistance Computations

The wave resistance problem considers the steady motion of a ship in initially smooth water assuming an ideal fluid, i.e. especially neglecting all viscous effects. The ship will create waves at the freely deformable water surface. The computations involve far more information than the mere resistance, which is of minor importance in many applications and usually computed quite inaccurately. But the expression 'wave resistance problem' is easier than 'steady, inviscid straight-ahead course problem', and thus more popular.

The work of the Australian mathematician J. H. Michell in 1898 is often seen as the birth of modern theoretical methods for ship wave resistance predictions. While Michell's theory cannot be classified as computational fluid dynamics in the modern sense, it was a milestone at the time and is still inspiring mathematicians today. Michell expressed the wave resistance of a thin wall-sided ship as:

$$R_w = \frac{4}{\pi} \rho V^2 \nu_2 \int_1^\infty \frac{\lambda^2}{\sqrt{\lambda^2 - 1}} |A(\lambda)|^2 \, d\lambda \tag{3.79}$$

with:

$$A(\lambda) = -i\nu\lambda \int_S e^{\nu\lambda^2 z + i\nu\lambda x} f(x, z) \, dz \, dx \tag{3.80}$$

V is the ship speed, ρ water density, $\nu = g/V^2$, g gravity acceleration, $f(x,z)$ half-width of ship, x longitudinal coordinate (positive forward), z vertical coordinate (from calm waterline, positive upwards), and S ship surface below the calm waterline. The expression gives realistic results for very thin bodies (width/length ratio very small) for arbitrary Froude number, and for slender ships (width/length ratio and depth/length ratio very small) for high Froude numbers. Michell's theory (including all subsequent refinements) is in essence unacceptable for real ship geometries and ship speeds. However, on occasions it is still useful. An example may be the prediction of the wave resistance of a submarine near the free surface with a streamlined snorkel piercing the free surface. While CFD can discretize the main submarine, it will neglect all appendages of much smaller scale. Then Michell's theory can be applied to analyze the additional influence of the snorkel, which will have a very large Froude number based on the chord length of its profile cross-section. Söding (1995) gives a FORTRAN routine to compute Michell's integral.

The classical methods (thin ship theories, slender-body theories) introduce simplifications which imply limitations regarding the ship's geometry. Real ship geometries are generally not thin or slender enough. The differences between computational and experimental results are consequently unacceptable. Practical applications in industry are based largely on boundary element methods. These remain the most important design tools for naval architects despite the increasing application of viscous flow tools.

Classical methods using so-called Kelvin or Havelock sources fulfill automatically a crude approximation of the dynamical and kinematical free-surface conditions. Kelvin sources are complicated and require great care in their numerical evaluation. Rankine sources on the other hand are quite simple. Wave resistance codes represent the flow as a superposition of Rankine sources and sometimes also dipoles or vortices. The potential of a Rankine point source is a factor divided by the distance between the point source and the considered point in the fluid domain. The factor is called the source strength. The derivative of the potential in an arbitrary spatial direction gives the velocity in this direction. This mathematical operation is simple to perform for Rankine sources.

Boundary element methods discretize surfaces into a finite number of elements and a corresponding number of collocation points. A desired (linear) condition is fulfilled exactly at these collocation points by proper adjustment of the initially unknown source strengths. One hopes/claims that between these points the boundary condition is fulfilled at least in good approximation. Laplace's equation and the decay condition (far away the ship does not disturb the flow) are automatically fulfilled. Mirror images of the panels at the bottom of the fluid domain walls may enforce a no-penetration condition there for shallow-water cases. Repeated use of mirror images at vertical canal walls can enforce in similar fashion the side-wall condition. For numerical reasons, this is preferable to a treatment of the side walls as collocation points similar as for the ship hull.

In the wave resistance problem, we consider a ship moving with constant speed V in water of constant depth and width. For inviscid and irrotational flow, this problem is equivalent to a ship being fixed in an inflow of constant speed. The following simplifications are generally assumed:

- Water is incompressible, irrotational, and inviscid.
- Surface tension is negligible.
- There are no breaking waves.
- The hull has no knuckles which cross streamlines.
- Appendages and propellers are not included in the model. (The inclusion of a propeller makes little sense as long as viscous effects are not also included.)

The governing field equation is Laplace's equation. A unique description of the problem requires further conditions on all boundaries of the modeled fluid domain:

- Hull condition: water does not penetrate the ship's surface.
- Transom stern condition: for ships with a transom stern, we generally assume that the flow separates and the transom stern is dry. Atmospheric pressure is then enforced at the edge of the transom stern. The condition is usually linearized assuming that the water flows only in the longitudinal direction. This can only approximately reflect the real conditions at the stern, but apparently works well as long as the transom stern is moderately small, as for most container ships. For fast ships which have a very large transom stern, several researchers report problems. For submerged transom sterns at low speed, the potential flow model is inapplicable and only field methods are capable of an appropriate analysis.
- Kinematic condition: water does not penetrate the water surface.

- Dynamic condition: there is atmospheric pressure at the water surface. Beneath an air cushion, this condition modifies to the air cushion pressure. The inclusion of an air cushion in wave resistance computations has been reported in various applications. However, these computations require the user to specify the distribution of the pressure, especially the gradual decline of the pressure at the ends of the cushion. In reality, this is a difficult task as the dynamics of the air cushion and the flexible skirts make the problem more complicated. Subsequently, the computations must be expected to be less accurate than for conventional displacement hulls.
- Radiation condition: waves created by the ship do not propagate ahead. (This condition is not valid for shallow-water cases when the flow becomes unsteady and soliton waves are pulsed ahead. For subcritical speeds with depth Froude number $F_{nh} < 1$, this poses no problem.)
- Decay condition: the flow is undisturbed far away from the ship.
- Open-boundary condition: waves generated by the ship pass unreflected any artificial boundary of the computational domain.
- Equilibrium: the ship is in equilibrium, i.e. trim and sinkage are changed in such a way that the dynamical vertical force and the trim moment are counteracted.
- Bottom condition (shallow-water case): no water flows through the sea bottom.
- Side-wall condition (canal case): no water flows through the side walls.
- Kutta condition (for catamaran/SWATH): at the stern/end of the strut the flow separates. The Kutta condition describes a phenomenon associated with viscous effects. Potential flow methods use special techniques to ensure that the flow separates. However, the point of separation has to be determined externally 'by higher insight'. For geometries with sharp aftbodies (foils), this is quite simple. For twin-hull ships, the disturbance of the flow by one demi-hull induces a slightly non-uniform inflow at the other demi-hull. This resembles the flow around a foil at a very small angle of incidence. A simplified Kutta condition usually suffices to ensure a realistic flow pattern at the stern: zero transverse flow is enforced. This is sometimes called the 'Joukowski condition'.

The decay condition substitutes the open-boundary condition if the boundary of the computational domain lies at infinity. The decay condition also substitutes the bottom and side-wall conditions if bottom and side wall are at infinity, which is the usual case.

Hull, transom stern, and Kutta conditions are usually enforced numerically at collocation points. Also, a combination of kinematic and dynamic conditions is numerically fulfilled at collocation points. Combining dynamic and kinematic boundary conditions eliminates the unknown wave elevation, but yields a non-linear equation to be fulfilled at the a priori unknown free-surface elevation.

Classical methods linearize the differences between the actual flow and uniform flow to simplify the non-linear boundary condition to a linear condition fulfilled at the calm-water surface. This condition is called the Kelvin condition. For practical applications, this crude approximation is unsuitable.

Dawson proposed in 1977 to use the potential of a double-body flow and the undisturbed water surface as a better approximation. Double-body linearization was popular until the early 1990s. The original boundary condition of Dawson was inconsistent. This inconsistency was copied by most subsequent publications following Dawson's approach. Sometimes this inconsistency is accepted deliberately to avoid evaluation of higher derivatives, but in most cases and possibly also in the original it was simply an oversight. Dawson's approach requires the evaluation of terms on the free surface along streamlines of the double-body flow. This required either more or less elaborate schemes for streamline tracking or some 'courage' in simply applying Dawson's approach on smooth grid lines on the free surface which were algebraically generated.

The first consistently linearized free-surface condition for arbitrary approximations of the base flow and the free-surface elevation was developed in Hamburg by Söding. This condition is rather complicated involving up to a third of the derivatives of the potential, but it can be simply repeated in an iterative process which is usually started with uniform flow and no waves.

Fully non-linear methods were first developed in Sweden and Germany in the late 1980s. The success of these methods quickly motivated various other research groups to copy the techniques and apply the methods commercially. The best-known codes used in commercial applications include SHIPFLOW-XPAN, SHALLO, v-SHALLO, RAPID, SWIFT, and FSWAVE/VSAERO. The development is very near the limit of what potential flow codes can achieve. The state of the art is well documented in two PhD theses, Raven (1996) and Janson (1996). Despite occasional other claims, all 'fully non-linear' codes have similar capabilities when used by their designers or somebody well trained in using the specific code. Everybody loves his own child best, but objectively the differences are small. All 'fully non-linear' codes in commercial use share similar shortcomings when it comes to handling breaking waves, semi-planing or planing boats or extreme non-linearities. Free-surface RANSE methods are the appropriate tools in these cases where wave resistance codes are no longer applicable.

Waves propagate only downstream (except for rare shallow-water cases). This radiation condition has to be enforced by numerical techniques. Most methods employ special finite difference (FD) operators to compute second derivatives of the potential in the free-surface condition. Dawson proposed a four-point FD operator for second derivatives along streamlines. Beside the considered collocation point, the FD operator uses the next three points upstream. Dawson's method automatically requires grids oriented along streamlines of the double-body flow approximate solution. Dawson determined his operator by trial and error for a two-dimensional flow with a simple Kelvin condition. His criteria were that the wave length should correspond to the analytically predicted wave length and the wave amplitude should remain constant some distance behind the disturbance causing the waves.

Dawson approximated the derivative of any function H with respect to ℓ at the point i numerically by:

$$H_{\ell i} \approx CA_i H_i + CB_i H_{i-1} + CC_i H_{i-2} + CD_i H_{i-3} \tag{3.81}$$

$H_{\ell i}$ is the derivative with respect to ℓ at point P_i. H_i to H_{i-3} are the values of the function H at points P_i to P_{i-3}, all lying on the same streamline of the double-body flow upstream of P_i. The coefficients CA_i to CD_i are determined from the arc lengths L_j ($j = 1$ to $i-3$) of the streamline between point P_i and point P_j:

$$L_j = \int_{P_i}^{P_j} d\ell \quad \text{on the streamline} \tag{3.82}$$

$$CA_i = -(CB_i + CC_i + CD_i) \tag{3.83}$$

$$CB_i = L_{i-2}^2 L_{i-3}^2 (L_{i-3} - L_{i-2})(L_{i-3} + L_{i-2})/D_i \tag{3.84}$$

$$CC_i = -L_{i-1}^2 L_{i-3}^2 (L_{i-3} - L_{i-1})(L_{i-3} + L_{i-1})/D_i \tag{3.85}$$

$$CD_i = L_{i-1}^2 L_{i-2}^2 (L_{i-2} - L_{i-1})(L_{i-2} + L_{i-1})/D_i \tag{3.86}$$

$$D_i = L_{i-1} L_{i-2} L_{i-3} (L_{i-3} - L_{i-1})(L_{i-2} - L_{i-1})(L_{i-3} - L_{i-2}) \cdot (L_{i-3} + L_{i-2} + L_{i-1}) \tag{3.87}$$

This four-point FD operator dampens the waves to some extent and gives usual discretizations (about ten elements per wave length) wave lengths which are about 5% too short. Strong point-to-point oscillations of the source strength occur for very fine grids. Various FD operators have been subsequently investigated to overcome these disadvantages. Of all these, only the spline interpolation developed at MIT was really convincing as it overcomes all the problems of Dawson (Nakos 1990, Nakos and Sclavounos 1990).

An alternative approach to FD operators involves 'staggered grids' as developed in Hamburg. This technique adds an extra row of source points (or panels) at the downstream end of the computational domain and an extra row of collocation points at the upstream end (Fig. 3.12).

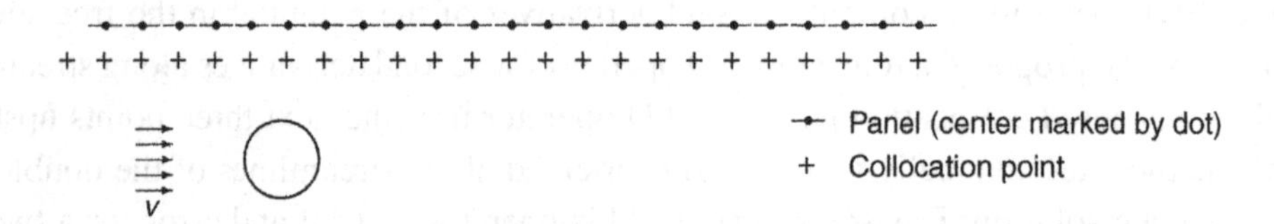

Figure 3.12:
'Shifting' technique (in 2d)

For equidistant grids this can also be interpreted as shifting or staggering the grid of collocation points vs. the grid of source elements. This technique shows absolutely no numerical damping or distortion of the wave length, but requires all derivatives in the formulation to be evaluated numerically.

Only part of the water surface can be discretized. This introduces an artificial boundary of the computational domain. Disturbances created at this artificial boundary can destroy the whole solution. Methods based on FD operators use simple two-point operators at the downstream end of the grid which strongly dampen waves. At the upstream end of the grid, where waves should not appear, various conditions can be used, e.g. the longitudinal component of the disturbance velocity is zero. Nakos (1990) has to ensure in his MIT method (SWAN code) based on spline interpolation that waves do not reach the side boundary. This leads to relatively broad computational domains. Time-domain versions of the SWAN code use a 'numerical beach'. For the wave resistance problem, the time-domain approach seems unnecessarily expensive and is rarely used in practice. Norwegian researchers tried to reduce the computational domain by matching the panel solution for the near-field to a thin-ship-theory solution in the far field. However, this approach saved only little computational time at the expense of a considerably more complicated code and was subsequently abandoned. The 'staggered grid' technique is again an elegant alternative. Without further special treatment, waves leave the computational domain without reflection.

Most methods integrate the pressure on the ship's surface to determine the forces (especially the resistance) and moments. 'Fully non-linear' methods integrate over the actually wetted surface while older methods often take the CWL as the upper boundary for the integration. An alternative to pressure integration is the analysis of the wave energy behind the ship (wave cut analysis). The wave resistance coefficients should theoretically tend to zero for low speeds. Pressure integration usually gives resistance coefficients which remain finite for small Froude numbers. However, wave cut analysis requires larger grids behind the ship, leading to increased computational time and storage. Most developers of wave resistance codes have at some point tried to incorporate wave cut analysis to determine the wave resistance more accurately. So far the evidence has not been compelling enough to abandon the direct pressure integration.

Most panel methods give as a direct result the source strengths of the panels. A subsequent computation determines the velocities at the individual points. Bernoulli's equation then gives pressures and wave elevations (again at individual points). Integration of pressures and wave heights finally yields the desired forces and moments, which in turn are used to determine dynamical trim and sinkage ('squat').

Fully non-linear state-of-the-art codes fulfill iteratively an equilibrium condition (dynamical trim and sinkage) and both kinematic and dynamic conditions on the actually deformed free surface. The differences in results between 'fully non-linear' and linear or 'somewhat non-linear' computations are considerable (typically 25%), but the agreement of computed and measured resistances is not consistently better in 'fully non-linear' methods. This may in part be due to the computational procedure or inherent assumptions in computing a wave resistance from experimental data (usually using a form factor method), but also due to computational errors in determining the resistance, which are of similar magnitude as the actual resistance. One reason for the unsatisfactory accuracy of the numerical procedures lies in the numerical sensitivity of the pressure integration. The pressure integration basically involves subtracting forces of the same magnitude which largely cancel. The relative error is strongly propagated in such a case. Initial errors stem from the discretization. For example, integration of the hydrostatic pressure for the ship at rest should give zero longitudinal force, but usual discretizations show forces that may lie within the same order of magnitude as the wave resistance. Still, there is consensus that panel methods capture the pressure distribution at the bow quite accurately. The vertical force is not affected by the numerical sensitivity. Predictions for the dynamical sinkage usually differ by less than 5% for a large bandwidth of Froude numbers. Trim moment is not predicted as well due to viscous effects and numerical sensitivity. This tendency is amplified by shallow water.

Panel methods are still the most important CFD instrument for form improvement of ships. They are widely used by ship designers. For at least a decade, they have also been used in formal hull optimization in industrial applications. The fundamental limitation of panel methods lies in the neglect of viscosity (aftbody and appendages) and breaking waves. The intersection between water surface and ship will remain a problem zone for panel methods, because the problem is ill-posed here within a potential flow model. The immediate vicinity of the bow of a ship always features breaking waves and spray, which cannot be included by panel methods. Ad hoc solutions are subject to research, but no convincing solution has been published yet. In industry practice, these limitations are overcome by using free-surface RANSE methods rather than boundary element methods, when breaking waves must be captured. Free-surface RANSE methods can simulate flows with complicated free-surface geometries (breaking waves, splashes), allowing the analyses of problems beyond the realm of BEM applications.

3.5.2. Viscous Flow Computations

RANSE solvers are state of the art for viscous ship flows. A computational prediction of the total calm-water resistance using RANSE solvers to replace model tests would be desirable. So far the accuracy of the RANSE predictions is largely perceived as insufficient, but this is expected to change within the next decade. Nevertheless, RANSE solvers are widely applied to analyze:

- the flow around aftbodies of ships;
- the flow around appendages.

The first research applications for RANSE solutions with wave-making for ships appeared in the late 1980s. By the late 1990s various research groups also presented results for ships free to trim and sink. Ten years later, free-surface RANSE computations were used regularly in many industry projects.

The basic techniques of RANSE codes have been discussed in Section 1.5. Various applications to ship design and research applications are found in the literature. Representative for the development of the state of the art for ship design applications are the proceedings of the Numerical Towing Tank Symposium (NuTTS) and surveys by leading companies in the field such as Flowtech (Larsson 1997, Larsson et al. 1998), or HSVA (Bertram and Jensen 1994, Bertram 1998a). The state of the art in research is documented in validation workshops like the Tokyo 1994 and 2005 workshops, the Gothenburg 2000 and 2010 workshops. RANSE computations require considerable skill and experience in grid generation and should therefore as a rule be executed by experts usually found in special consulting companies.

3.6. Simple Design Approaches

In early design stages, the power requirements have to be estimated to judge the weight and volume requirements of the main engine and fuel. As this has to be done repeatedly in design loops, model tests are not suitable solutions for reasons of time and costs. Instead, simple, largely empirical methods are employed which only require a few global design parameters. These methods are discussed in more detail by Schneekluth and Bertram (1998).

The main approaches are:

- Estimate from parent ship, e.g. by admiralty or similar formulae
  The estimate from a parent ship may give good estimates if the parent ship is close enough (in geometrical properties and speed parameters) to the design ship. The 'admiralty formula' is still used today, but only for a very rough estimate:

$$P_B = \frac{\Delta^{2/3} \cdot V^3}{C} \tag{3.88}$$

  The admiralty constant C is assumed to be constant for similar ships with similar Froude numbers, i.e. ships that have almost the same C_B, $C_B\ \nabla/L^3$, F_n, ∇, etc. Typical values for C [in $t^{2/3} \cdot kn^3 = kW$] are:

multi-purpose vessel	400–600
bulker and tanker	600–750
reefer	550–700
feeder ship	350–500
warship	150

  These values give an order of magnitude only. The constant C should be determined individually for basis ships used in design.

Generalized admiralty formulae are of the form:

$$P_B = \frac{\Delta^m \cdot V^n}{C} \tag{3.89}$$

where m and n are determined from regression analysis of databases.

Völker (1974) gives a modified admiralty formula for cargo ships with smaller scatter for C:

$$P_B = \frac{\Delta^{0.567} \cdot V^{3.6}}{C \cdot \eta_D} \tag{3.90}$$

η_D in this formula may be estimated by empirical formulae. Strictly speaking, the exponent of V should be a function of speed range and ship hull form. The admiralty formula is thus only useful if a ship of the same type, size, and speed range is selected to determine C. It is possible to increase the accuracy of the Völker formula by adjusting it to specific ship types.

The admiralty formula is very coarse and not recommended (unless a very close similar ship is used to determine C), but an estimate based on a form factor approach is popular in practice. Here, it is usually assumed that the parameter c_w/F_n^4 and the form factor remain constant in the conversion from parent ship to design ship. Such a more or less sophisticated plus/minus conversion from a parent ship is currently the preferred choice for a quick estimate.

Tugs are special ships which differ in many ways from regular cargo ships (Allan 2004). The main design specification concerns maneuverability and ability to assist escort vessels in maneuvering. This requires a somewhat different approach in design. Bertram and Bentin (2001) use neural nets to express the required power P_B [kW] of harbor tugs as function of design speed V [kn], bollard pull b_p [t] and length between perpendiculars L_{pp} [m]:

$$\begin{aligned} P_B = 1060 + 3354 \cdot \mathrm{sig}(1.23 - 6.44 \cdot \mathrm{sig}(0.08652 \cdot L_{pp} - 0.3171 \cdot b_p - 3.84 \cdot V \\ + 60.4709) + 2.97 \cdot \mathrm{sig}(0.8539 \cdot L_{pp} + 0.2307 \cdot b_p - 0.484 \cdot V \\ - 23.07) - 5.98 \cdot \mathrm{sig}(0.2596 \cdot L_{pp} + 0.0856 \cdot b_p + 0.51 \cdot V \\ - 17.577) + 2.61 \cdot \mathrm{sig}(0.2857 \cdot L_{pp} + 0.7132 \cdot b_p + 0.476 \cdot V \\ - 25.7645)) \end{aligned} \tag{3.91}$$

- Systematic series (e.g. Taylor–Gertler, Series-60, SSPA) or regression analysis of many ships (e.g. Lap–Keller, Holtrop–Mennen, Hollenbach)
  All of the systematic series and most of the regression analysis approaches are outdated. They often underestimate the actual resistance of modern ship hulls. It may come as a surprise that older ships were apparently better in terms of resistance. There are several explanations:
  - suitability for container stowage plays a larger role in modern ships;
  - modern ships often have a higher propulsive efficiency compensating partially for the higher resistance;

- more severe safety regulations, e.g. concerning stability, pose additional constraints on the hydrodynamic optimization.

  It is fairly difficult to estimate accurately the residual resistance or the wave resistance. The method of choice today would be a CFD code, but for a quick estimate one may accept larger margins of errors and resort to classical estimates. Schneekluth and Bertram (1998) list some of the older methods for resistance predictions including the Lap–Keller method. These methods are historical and should no longer be applied in modern ship design. More 'modern' methods which are often found imbedded in ship design systems are (Table 3.2):
- 'Taylor–Gertler' (for slender ships) (Gertler 1954)
- 'Guldhammer–Harvald' (Guldhammer and Harvald 1974)
- 'Holtrop–Mennen' (Holtrop and Mennen 1978, 1982, Holtrop 1977, 1978, 1984)
- 'Hollenbach' (Hollenbach 1998, 1999)
- 'NPL' (for fast ships) (Bailey 1976).

The older methods like 'Taylor–Gertler' do not consider the bulbous bow. The effect of a bulbous bow may then be approximately introduced by increasing the length in the calculation by two-thirds of the bulb length.

Oortmerssen (1971) presents a simple method to estimate the residual resistance of tugs and trawlers based on regression analysis of model basin data. The range of parameters for which the coefficients of the basic expressions are valid, are as follows: $8\text{ m} < L_{WL} < 80\text{ m}$; $5\text{ m}^3 < \nabla < 3000\text{ m}^3$; $3 < L/B < 6.2$; $1.9 < B/T < 4.0$; $0.50 < C_P < 0.73$; $0.70 < C_M < 0.97$; $-7\%\, L < lcb < 2.8\%\, L$ forward of $0.5\, L$; $0 < F_n < 0.5$; $10° < i_E < 46°$. i_E is the half angle of entrance of the design waterline. The residual resistance made non-dimensional by the displacement weight in his expression:

$$\frac{R_R}{\Delta \cdot g} = C_1 \cdot \exp\left(-\frac{mF_n^2}{9}\right) + C_2 \cdot \exp\left(-\frac{m}{F_n^2}\right) + C_3 \cdot \exp(-mF_n^2) \cdot \sin\left(\frac{1}{F_n^2}\right) + C_4 \cdot \exp(-mF_n^2) \cdot \cos\left(\frac{1}{F_n^2}\right) \qquad (3.92)$$

$$C_1 \cdot 10^3 = 79.32134 - 0.09287\, lcb - 0.00209\, lcb^2 - 246.45896\, C_P + 187.13664\, C_P^2 - 1.42893\, L/B + 0.11898\, (L/B)^2 + 0.15727\, C_{IE} - 0.00064\, C_{IE}^2 - 2.52862\, B/T + 0.50619\, (B/T)^2 + 1.62851\, C_M \qquad (3.93)$$

$$C_2 \cdot 10^3 = 6714.88397 + 19.83\, lcb + 2.66997\, lcb^2 - 19662.024\, C_P + 14099.9\, C_P^2 + 137.33613\, L/B - 13.36938\, (L/B)^2 - 4.49852\, C_{IE} + 0.021\, C_{IE}^2 + 216.44923\, B/T - 35.07602\, (B/T)^2 - 128.72535\, C_M \qquad (3.94)$$

$$C_3 \cdot 10^3 = -908.44371 + 2.52704\, lcb - 0.35794\, lcb^2 + 755.1866\, C_P - 48.93952\, C_P^2 + 9.86873\, L/B - 0.77652\, (L/B)2 + 3.79020\, C_{IE} - 0.01879\, C_{IE}^2 - 9.24399\, B/T + 1.28571\, (B/T)^2 + 250.6491\, C_M \qquad (3.95)$$

Table 3.2: Resistance prediction methods

Resistance procedure 'Taylor–Gertler'

Basis for procedure: Systematic model tests with a model warship
Target value: C_R
Input values: L_{wl}; $F_{n,wl} = V/\sqrt{g \cdot L_{wl}}$; $C_{P,wl} = \nabla/L_{wl}^3$; B/T; S

Remarks:

1. Influence of bulb not taken into account
2. The procedure generally underestimates by 5–10%
3. Area of application: fast cargo ships, warships
4. Constant or dependent variable values: $C_M = 0.925 =$ constant, $lcb = 0.5\ L_{wl}$

Resistance procedure 'Guldhammer–Harvald'

Basis for procedure: Evaluation of well-known resistance calculation procedures (Taylor, Lap, Series 60, Gothenburg, BSRA, etc.)
Target value: C_R and C_F
Input values: L_{wl}; $F_{n,wl}$; B/T; lcb, section shape; A_{BT} (bulb); S; $C_{P,wl}$; $L_{wl}/\nabla^{1/3}$

Remarks:

1. Influence of bulb taken into account
2. Reference to length in WL
3. Area of application: universal, tankers
4. The correction for the center of buoyancy appears (from area to area) overestimated
5. The procedure underestimates resistance for ships with small L/B

Resistance procedure 'Holtrop–Mennen'

Basis for procedure: Evaluation of database of the Dutch Model Basin MARIN
Target value: C_T
Input values: F_n; L_{pp}; L_{wl}; B; T; ∇; lcb; C_{WP}; S; section shape; trim; ...

Remarks:

1. Resistance decomposition like ITTC'78
2. Considers bulbous bow and transom stern
3. Covers wide range of ships
4. Many parameters; some may have to be estimated in early design

Resistance procedure 'Hollenbach'

Basis for procedure: Evaluation of database of Vienna ship model basin
Target value: R_T
Input values: F_n; L_{pp}; L_{wl}; B; T; C_B; D_P; trim; number of appendages

Remarks:

1. Considers twin-screw ships
2. Relatively modern database
3. Applicable to modern cargo ship
4. Several typing mistakes between various publications
5. Gives also 'minimum' and 'maximum' resistance curves

Resistance procedure 'NPL Series'

Basis for procedure: Systematic model tests with high-speed, round-bilge displacement forms
Target value: $R_R/(\nabla \cdot g)$

Table 3.2: Continued

Input values: $F_{n\nabla}$; L/B; $L/\nabla^{1/3}$ *Remarks:*   1. For fast displacement ships; (originally) graphical method   2. Simple HSVA formulae recommended instead as easy to program and based on more modern ship designs
Resistance procedure 'Van Oortmerssen'
Basis for procedure: Evaluation of database of Dutch model basin MARIN for small ships   *Target value:* $R_R/(\nabla \cdot g)$ *Input values:* F_n; lcb, L/B; B/T; C_M; C_{IE}; C_P; $L/\nabla^{1/3}$ *Remarks:*   1. For tugs and trawlers up to $F_n = 0.5$ 2. $C_{IE} = i_E \cdot L/B$, where i_E is the half angle of entrance of the design waterline   3. Easy to program

$$\begin{aligned} C^4 \cdot 10^3 &= 3012.14549 + 2.71437\ lcb + 0.25521\ lcb^2 - 9198.8084\ C_P \\ &+6886.60416\ C_P^2 - 159.92694\ L/B + 16.23621\ (L/B)^2 - 0.82014\ C_{IE} \\ &+0.00225\ C_{IE}^2 + 236.3797\ B/T - 44.1782\ (B/T)^2 + 207.2558\ C_M \end{aligned} \tag{3.96}$$

$$m = 0.14347\ C_P^{-2.1976} \tag{3.97}$$

$C_{IE} = i_E \cdot L/B$, where i_E is taken in degrees. $L = (L_{PP}+L_{WL})/2$ in Van Oortmerssen's formula.

MacPherson (1993) provides some background and guidance to designers for simple computer-based prediction methods, and these are recommended for further studies.

Some of the old estimation methods are still popular as they are easy to program. Thus they are embedded in naval architectural CAD systems or more recently in design expert systems. However, they are fundamentally limited to global predictions, as they represent the hull shape by few global parameters.

The following compiles assorted simple design formulae, mostly taken from Schneekluth and Bertram (1998):

- Propulsive efficiency η_D
  Typical values are: $\eta_D \approx 0.6-0.7$ for cargo ships
  $\eta_D \approx 0.4-0.6$ for tugs
  Danckwardt (1969) gives the following estimate (Henschke 1965):

$$\eta_D = 0.836 - 0.000165 \cdot n \cdot \nabla^{1/6} \tag{3.98}$$

n is the propeller rpm and ∇ [m^3] the displacement volume. All ships checked were within $\pm10\%$ of this estimate; half of the ships were within $\pm2.5\%$.
Keller (1973) gives:

$$\eta_D = 0.885 - 0.00012 \cdot n \cdot \sqrt{L_{pp}} \quad (3.99)$$

HSVA gave, for twin-screw ships, in 1957:

$$\eta_D = 0.69 - 12000 \cdot \left(0.041 - \frac{V_s}{n \cdot D_p}\right)^3 \pm 0.02 \quad (3.100)$$

Ship speed V_s [in kn], propeller diameter D_p [in m], $0.016 \leq V_s/(n \cdot D_p) \leq 0.04$.

- Hull efficiency η_H
  The hull efficiency can be estimated indirectly by estimating thrust deduction fraction t and wake fraction w separately or directly. For small ships with rake of keel, Helm (1980) gives an empirical formula:

$$\eta_H = 0.895 - \frac{0.0065 \cdot L}{\nabla^{1/3}} - 0.005 \cdot \frac{B}{T} - 0.033 \cdot C_P + 0.2 \cdot C_M + 0.01 \cdot lcb \quad (3.101)$$

  lcb here is the longitudinal center of buoyancy taken from $L_{pp}/2$ [in $\%L_{pp}$]. The basis for this formula covers $3.5 \leq L/\nabla^{1/3} \leq 5.5$, $0.53 \leq C_P \leq 0.71$, $2.25 \leq B/T \leq 4.50$, $0.60 \leq C_M \leq 0.89$, rake of keel $40\%T$, $D_P = 0.75T$. T is taken amidships.
  Usually, it is preferable to estimate t and w separately and then deduct η_H from there.
- Thrust deduction fraction t
  For single-screw ships:

$$t = 0.5 \cdot C_P - 0.12; \quad \text{Heckscher for cargo ships} \quad (3.102)$$

$$t = 0.77 \cdot C_P - 0.30; \quad \text{Heckscher for trawlers} \quad (3.103)$$

$$t = 0.5 \cdot C_B - 0.15; \quad \text{Danckwardt for cargo ships} \quad (3.104)$$

$$t = w \cdot (1.57 - 2.3 \cdot C_B/C_{WP} + 1.5 \cdot C_B); \quad \text{SSPA for cargo ships} \quad (3.105)$$

$$t = 0.001979 \cdot \frac{L}{B(1 - C_P)} + 1.0585 \cdot \frac{B}{L} - 0.00524 - 0.1418 \cdot \frac{D_P^2}{BT}; \quad \text{Holtrop and Mennen} \quad (3.106)$$

  For twin-screw ships:

$$t = 0.5 \cdot C_P - 0.18; \quad \text{Heckscher for cargo ships} \quad (3.107)$$

$$t = 0.52 \cdot C_B - 0.18; \quad \text{Danckwardt for cargo ships} \quad (3.108)$$

$$t = w \cdot (1.67 - 2.3 \cdot C_B/C_{WP} + 1.5 \cdot C_B); \quad \text{SSPA for cargo ships} \quad (3.109)$$

$$t = 0.325 \cdot C_B - 0.1885 \cdot \frac{D_P}{\sqrt{BT}}; \quad \text{Holtrop and Mennen} \tag{3.110}$$

Alte and Baur (1986) give an empirical coupling between t and the wake fraction w:

$$(1 - t) = (1 - w)^{0.4-0.8} \tag{3.111}$$

In general, in the early design stage it cannot be determined which t will give the best hull efficiency η_H. t can be estimated only roughly in the design stage and all of the above formulae have a much larger uncertainty margin than those for w given below. t thus represents the largest uncertainty factor in the power prognosis.

- Wake fraction w

For single-screw ships:

$$w = 0.5 \cdot C_P \cdot \frac{1.6}{1 + D_P/T} \cdot \frac{16}{10 + L/B}; \quad \text{Schneekluth for ships with stern bulb} \tag{3.112}$$

$$w = 0.75 \cdot C_B - 0.24; \quad \text{Krüger} \tag{3.113}$$

$$w = 0.7 \cdot C_P - 0.18; \quad \text{Heckscher for cargo ships} \tag{3.114}$$

$$w = 0.77 \cdot C_P - 0.28; \quad \text{Heckscher for trawlers} \tag{3.115}$$

$$w = 0.25 + 2.5(C_B - 0.6)^2; \quad \text{Troost for cargo ships} \tag{3.116}$$

$$w = 0.5 \cdot C_B; \quad \text{Troost for coastal feeders} \tag{3.117}$$

$$w = C_B/3 + 0.01; \quad \text{Caldwell for tugs with } 0.47 < C_B < 0.56 \tag{3.118}$$

$$w = 0.165 \cdot C_B \cdot \frac{\nabla^{1/3}}{D_P} - 0.1 \cdot (F_n - 0.2); \quad \text{Papmehl} \tag{3.119}$$

$$w = \frac{3}{1 - (C_P/C_{WP})^2} \cdot \frac{B}{L} \cdot \frac{E}{T} \cdot \left[1 - \frac{1.5 \cdot D + (\varepsilon + r)}{B}\right]; \quad \text{Telfer for cargo ships} \tag{3.120}$$

ε is the skew angle in radians, r is the rake angle in radians, and E is height of the shaft center over keel.

For twin-screw ships:

$$w = 0.81 \cdot C_B - 0.34; \quad \text{Krüger} \tag{3.121}$$

$$w = 0.7 \cdot C_P - 0.3; \quad \text{Heckscher for cargo ships} \tag{3.122}$$

$$w = C_B/3 - 0.03; \quad \text{Caldwell for tugs with } 0.47 < C_B < 0.56 \tag{3.123}$$

Holtrop and Mennen (1978) and Holtrop (1984) give further more complicated formulae for w for single-screw and twin-screw ships, which can be integrated in a power prognosis program.

All the above formulae consider only a few main parameters, but the shape of the ship, especially the aftbody, influences the wake considerably. Other important parameters are propeller diameter and propeller clearance, which are not explicitly represented in the above formulae. For bulk carriers with $C_B \approx 0{:}85$, $w < 0.3$ have been obtained by form optimization. The above formulae can thus predict too high w values for full ships.

- Relative rotative efficiency η_R
  The relative rotative efficiency is driven by many different effects. This makes it difficult to express η_R as a function of just a few parameters.
  Holtrop and Mennen (1978) and Holtrop (1984) give:

$$\eta_R = 0.9922 - 0.05908 \cdot A_E/A_0 + 0.07424 \cdot (C_P - 0.0225 \cdot lcb) \quad \text{for single-screw ships} \tag{3.124}$$

$$\eta_R = 0.9737 - 0.111 \cdot (C_P - 0.0225 \cdot lcb) - 0.06325 \cdot P/D_P \quad \text{for twin-screw ships} \tag{3.125}$$

lcb here is the longitudinal center of buoyancy taken from $L_{wl}/2$ [in $\%L_{wl}$]
A_E/A_0 is the blade area ratio of the propeller
P/D_P is the pitch-to-diameter ratio of the propeller.
Helm (1980) gives for small ships:

$$\eta_R = 0.826 + 0.01 \frac{L}{\nabla^{1/3}} + 0.02 \frac{B}{T} + 0.1 \cdot C_M \tag{3.126}$$

The basis for this formula is the same as for Helm's formula for η_H.
Alte and Baur (1986) recommend, as a simple estimate, $\eta_R = 1.00$ for single-screw ships, $\eta_R = 0.98$ for twin-screw ships.
Jensen (1994) gives $\eta_R = 1.02\text{–}1.06$ for single-screw ships, depending also on details of the experimental and correlation procedure.

- Wetted surface S
  Non-dimensional resistance coefficients require the wetted surface S, usually taken at calm-water conditions. CAD systems compute S accurately, but if only the main dimensions are known, one may resort to estimates:

$$S = \nabla^{1/3} \cdot (3.4 \cdot \nabla^{1/3} + 0.5 \cdot L_{wl}) \quad \text{Lap (1954) for cargo ships and ferries} \tag{3.127}$$

$$S = L \cdot (1.8 \cdot T + C_B \cdot B) \quad \text{Schneekluth for warships} \tag{3.128}$$

$$S = \frac{\nabla}{B} \cdot \left[\frac{1.7}{C_B - 0.2 \cdot (C_B - 0.65)} + \frac{B}{T}\right] \quad \text{Danckwardt for cargo ships and ferries} \tag{3.129}$$

$$S = \frac{\nabla}{B} \cdot \left[\frac{1.7}{C_B} + \frac{B}{T} \cdot (0.92 + \frac{0.092}{C_B}) \right] \quad \text{Danckwardt for trawlers} \tag{3.130}$$

$$S = L \cdot (2T + B) \cdot C_M^{0.5} \cdot (0.453 + 0.4425 \cdot C_B - 0.2862 \cdot C_M - 0.003467 \cdot B/T + 0.3696 \cdot C_{WP}) \quad \text{Holtrop–Mennen for cargo ships} \tag{3.131}$$

- Viscous pressure resistance coefficient C_{PV}

$$C_{PV} \cdot 10^3 = (26 \cdot C_\nabla + 0.16) + \left(\frac{B}{T} - \frac{13 - 10^3 \cdot C_\nabla}{6} \right) \cdot (C_P + 58 \cdot C_\nabla - 0.408) \cdot (0.535 - 35 \cdot C_\nabla) \tag{3.132}$$

The formula was derived by Schneekluth from the Taylor experiments (dating back to 1910 and 1954), based on $B/T = 2.25{-}4.5$, $C_P = 0.48{-}0.8$, $C_\nabla = \nabla/L^3 = 0.001{-}0.007$.

- Form factor $k = R_{PV}/R_F$

$$k = 18.7 \cdot (C_B \cdot B/L)^2 \quad \text{Granville (1956)} \tag{3.133}$$

$$k = 14 \cdot (\nabla/L^3) \cdot (B/T) \quad \text{Alte and Baur (1986)} \tag{3.134}$$

$$k = -0.095 + 25.6 \cdot \frac{C_B}{(L/B)^2 \cdot \sqrt{B/T}} \quad \text{Watanabe (1986)} \tag{3.135}$$

- Appendage resistance R_{APP}
  Simple semi-empirical formula for appendages are:
  Exposed shafting, stern tubes and bossings:

$$R_{APP} = \frac{1}{2} \rho \, V^2 \, Ld \, (1.1 \sin^3 \varepsilon + \pi \, C_F) \tag{3.136}$$

Here C_F is calculated with a Reynolds number based on the diameter d. L is the length of shaft and ε its inclination relative to the keel.
Struts and rudders:

$$R_{APP} = \rho \, V^2 \, S \, C_F \, (1 + 2(t/c) + 60(t/c)^4) \tag{3.137}$$

Here C_F is calculated with a Reynolds number based on chord length c. t is the thickness of the strut and S the projected surface (one side) of the strut.
Bilge keels:

$$R_{APP} = \rho \, V^2 \, S_B \, C_F \tag{3.138}$$

Here C_F is calculated with a Reynolds number based on the bilge keel length.
Transom wedges:

$$R_{APP} = 0.0001196\, c\, \delta\, \rho\, V^2\, S(\tau + \delta) \quad (3.139)$$

Here δ is the wedge angle, τ the trim angle of the vessel (positive stern down), S the wetted surface, and c the chord length of the wedge.

- Wind resistance R_{AA}
  Wind resistance is important for ships with large lateral areas above the water level, e.g. container ships and car ferries. Fast and unconventional ships, e.g. air-cushioned vehicles, also require inclusion of the contribution of wind or air resistance. Jensen (1994) gives a very simple estimate for the wind resistance of cargo ships:

$$R_{AA} = C_{AA} \cdot \frac{\rho_{air}}{2} \cdot (V + V_{wind})^2 \cdot A_F \quad (3.140)$$

For cargo ships, Jensen (1994) gives $C_{AA} = 0.8\text{–}1.0$. $\rho_{air} = 1.25\ \text{kg/m}^3$ the density of air, V_{wind} is the absolute value of wind speed and A_F is the frontal projected area of the ship above sea level.
The wind resistance may be estimated with more accuracy following Blendermann (1993, 1996):

$$R_{AA} = \frac{\rho_{air}}{2} \cdot u^2 \cdot A_L \cdot CD_l \cdot \frac{\cos\varepsilon}{1 - \frac{\delta}{2}\left(1 - \frac{CD_l}{CD_t}\right)\sin^2 2\varepsilon} \quad (3.141)$$

Here u is the apparent wind velocity, A_L the lateral-plane area, ε the apparent wind angle ($\varepsilon = 0°$ in head wind), and δ the cross-force parameter. CD_t and CD_l are the non-dimensional drag coefficients in beam wind and head wind, respectively. It is convenient to give the longitudinal drag with respect to the frontal projected area A_F:

$$CD_{l,AF} = CD_l \cdot \frac{A_L}{A_F} \quad (3.142)$$

Table 3.3 gives typical values for CD_t, $CD_{l,AF}$ and δ. The maximum wind resistance usually occurs for $0° < \varepsilon < 20°$. The above formulae and the values in the table are for uniform or nearly uniform flow, e.g. above the ocean. The wind speed at a height of 10 m above sea level u_{10} is usually taken as reference speed. Wind speed in Beaufort (Beaufort number BN) is converted to m/s by:

$$u_{10} = 0.836 \cdot BN^{1.5} \quad (3.143)$$

- Speed loss in wind and waves

Table 3.3: Coefficients to estimate wind resistance (Blendermann 1996)

	CD_t	$CD_{l,AF}$	δ
Car carrier	0.95	0.55	0.80
Cargo ship, container on deck, bridge aft	0.85	0.65/0.55	0.40
Containership, loaded	0.90	0.55	0.40
Destroyer	0.85	0.60	0.65
Diving support vessel	0.90	0.60	0.55
Drilling vessel	1.00	0.70–1.00	0.10
Ferry	0.90	0.45	0.80
Fishing vessel	0.95	0.70	0.40
LNG tanker	0.70	0.60	0.50
Offshore supply vessel	0.90	0.55	0.55
Passenger liner	0.90	0.40	0.80
Research vessel	0.85	0.55	0.60
Speedboat	0.90	0.55	0.60
Tanker, loaded	0.70	0.90	0.40
Tanker, in ballast	0.70	0.75	0.40
Tender	0.85	0.55	0.65

Townsin and Kwon (1983) give simple approximate formulae to estimate the speed loss due to added resistance in wind and waves:

$$\Delta V = C_\mu \cdot C_{\text{ship}} \cdot V\% \tag{3.144}$$

C_μ is a factor considering the predominant direction of wind and waves, depending on the Beaufort number BN:

$$C_\mu = 1.0 \quad \text{for } \mu = 0\text{–}30° \tag{3.145}$$

$$C_\mu = 1.7 - 0.03 \cdot (BN - 4)^2 \quad \text{for } \mu = 30\text{–}60° \tag{3.146}$$

$$C_\mu = 0.9 - 0.06 \cdot (BN - 6)^2 \quad \text{for } \mu = 60\text{–}150° \tag{3.147}$$

$$C_\mu = 0.4 - 0.03 \cdot (BN - 8)^2 \quad \text{for } \mu = 150\text{–}180° \tag{3.148}$$

C_{ship} is a factor considering the ship type:

$$C_{\text{ship}} = 0.5BN + BN^{6.5}/(2.7 \cdot \nabla^{2/3}) \quad \text{for tankers, laden} \tag{3.149}$$

$$C_{\text{ship}} = 0.7BN + BN^{6.5}/(2.7 \cdot \nabla^{2/3}) \quad \text{for tankers, ballast} \tag{3.150}$$

$$C_{\text{ship}} = 0.7BN + BN^{6.5}/(2.2 \cdot \nabla^{2/3}) \quad \text{for container ships} \tag{3.151}$$

∇ is the volume displacement in m^3. Tables 3.4 and 3.5 give relations between Beaufort number, wind speeds, and average wave heights.

Table 3.4: Wind strengths in Beaufort (Bft) (Henschke 1965)

Bft	Wind description	Wind speed
0	No wind	0.0–0.2 m/s
1	Gentle current of air	0.3–1.5 m/s
2	Gentle breeze	1.6–3.3 m/s
3	Light breeze	3.4–5.4 m/s
4	Moderate breeze	5.5–7.9 m/s
5	Fresh breeze	8.0–10.7 m/s
6	Strong wind	10.8–13.8 m/s
7	Stiff wind	13.9–17.1 m/s
8	Violent wind	17.2–20.7 m/s
9	Storm	20.8–24.4 m/s
10	Violent storm	24.5–28.3 m/s
11	Hurricane-like storm	28.5–32.7 m/s
12	Hurricane	>32.7 m/s

Table 3.5: Sea strengths for North Sea coupled to wind strengths (Henschke 1965)

			Approximate average	
Sea state	**Bft**	**Sea description**	**Wave height**	**Wave length**
0	0	Smooth sea	–	–
1	1	Calm, rippling sea	0–0.5 m	0–10 m
2	2–3	Gentle sea	0.5–0.75 m	10–12.5 m
3	4	Light sea	0.75–1.25 m	12.5–22.5 m
4	5	Moderate sea	1.25–2.0 m	22.5–37.5 m
5	6	Rough sea	2.0–3.5 m	37.5–60.0 m
6	7	Very rough sea	3.5–6.0 m	60.0–105.0 m
7	8–9	High sea	>6.0 m	>105.0 m
8	10	Very high sea	Up to 20 m	Up to 600 m
9	11–12	Extremely heavy sea	Up to 20 m	Up to 600 m

- Natural periods for ship motions
  For 'normal' ships, the natural frequencies in roll, heave, and pitch can be estimated by simple formulae.
  Natural roll period [s]:

$$T_{\text{roll}} = C \cdot \frac{B}{\sqrt{GM}} \tag{3.152}$$

with

$$C = 0.746 + 0.046\frac{B}{T} - 0.086 \cdot \frac{L}{100} \quad \text{following IMO} \tag{3.153}$$

$C = 0.7627–0.8229$ (typically 0.8) for cargo ships following Parsons (2004)

$C = 0.6924–1.0035$ generally following Parsons (2004)
B [m] is the width, L [m] the length in the waterline, T [m] the draft, GM [m] the metacentric height.
Natural pitch period [s]:

$$T_{\text{pitch}} = C \cdot \frac{L}{\sqrt{GM_L}} \tag{3.154}$$

$C = 0.4817–0.5218$ generally following Parsons (2004)
L [m] is the length in the waterline.

$$T_{\text{pitch}} = 1.776 \cdot \frac{\sqrt{TC_B(0.6 + 0.36 \cdot B/T)}}{C_{WP}} \quad \text{following Lamb (1969)} \tag{3.155}$$

B [m] is the width, T [m] the draft, C_B the block coefficient, and C_{WP} the waterplane coefficient.
Natural heave period [s]:

$$T_{\text{heave}} = 2.007 \cdot \sqrt{T\frac{C_B}{C_{WP}} \cdot \left(\frac{B}{3T} + 1.2\right)} \quad \text{following Lamb (1969)} \tag{3.156}$$

Variables as above for pitch period.

3.7. Fuel-Saving Options

3.7.1. Introduction

For decades, ships have been designed for much lower fuel costs. Increasing fuel prices and IMO regulations to curb CO_2 (carbon dioxide) emissions put pressure on ship owners to obtain more fuel-efficient ships. As a result, we have seen a renaissance of some concepts of the 1970s which were developed in response to the first oil crisis, as well as new proposals for fuel-saving devices. Many publications (including promotional material by companies) give unrealistically optimistic claims for fuel-saving potential of these devices. There are various reasons for false estimations:

- The published savings achieved with a particular device are normally for the best case. For example, formal hull optimization has improved the fuel efficiency of one vessel by 16% at design speed and draft. Subsequent literature then – correctly – states that up to 16% may be gained. This is quoted as '16% gains' in a subsequent survey or report and taken as typical value.

- Quoted savings are valid for initially bad designs, whereas hydrodynamically optimized designs would never reach that saving.
- Numbers valid for one certain ship type (say high-speed container vessels) are taken for other ships (e.g. bulk carriers), where they do not apply.
- Numbers are taken for design speed and draft. Frequently encountered off-design conditions are ignored. Utilization of a fuel-saving device is often incorrectly assumed to be 100% of the time at sea for a ship and 100% over fleets for global estimates.
- Saving potential refers to calm-water resistance, but is applied to total resistance or total fuel consumption (including the on-board energy consumption).
- For propulsion-improving devices, published savings are based on a comparison of power requirement measured before and after conversion. Measurements are not corrected for hull and propeller roughness (ship and propeller are often cleaned while the ship is refitted with a propulsion-improving device), sea state and loading condition. If measures are corrected for a 'neutral condition', the correction procedure in itself has an uncertainty of 2–3%.
- Saving potential is quoted based on model tests and questionable extrapolation to full scale. Model tests violate Reynolds similarity and hence boundary layers and flows at appendages in the boundary layer are not similar. Most quoted figures are based on publications (and model tests) of the 1970s and 1980s. There is usually no documentation on how figures were derived. In my personal experience, re-analyses and detailed full-scale measurements with today's technology always showed substantially lower figures.

3.7.2. Global Measures to Reduce Resistance

On the most global level, there are two (almost trivial) options following from the admiralty formula:

- Reduce ship size. The ship size (or displacement) is driven by the cargo weight, ballast, steel weight and equipment, and outfit weight. The fuel consumption scales with displacement to the power 2/3. As cargo weight usually is a fixed quantity and dominates the overall displacement, savings through minimizing steel weight and ballast are usually only small to moderate. However, other measures to reduce power requirements lead to smaller engines and associated periphery (power trains, cooling pumps, fuel tanks, etc.). This yields secondary savings in new designs due to smaller ship size.
- Reduce speed. Speed reduction is a very effective way to reduce fuel consumption and emission. The admiralty formula assumes a cubic relationship between power and speed. This is a widely used assumption for small speed changes, but actual speed curves exhibit local deviations from this rule of thumb. The rule applies for the bare-hull, calm-water condition. A 10% speed reduction (i.e. taking 90% of the reference speed) yields a 27% reduction in required power ($0.9^3 = 0.73$). In addition, slower design speeds allow higher propeller efficiency. This may add another 2% fuel savings for 10% design speed

reduction. Slower speed often also results in lower added resistance in seaways. As mentioned above, there are secondary savings to the smaller installed engine power. For new buildings, design for slower speed is thus a very effective lever to reduce fuel consumption. Necessary measures to keep delivery capacity constant (larger fleet size or larger cargo capacity) may increase fuel consumption in fleets by 6–8% (as port times are not affected by ship speed). The net reduction in fuel consumption is then 23–25% for 10% design speed reduction and constant delivery capacity. Several factors introduce penalties or constraints for lower speeds:

- Lower speed often attracts less cargo.
- Capital costs of cargo depend on transport time and cargo value. Slower transport increases capital costs on the cargo and reduces freight rates accordingly.
- Transitional costs for logistics pose barriers in intermodal transport chains. These costs occur once for adapting existing schedules, but can be considerable in large transport networks.
- Slower ships transport less and additional ships are needed to maintain a transport capacity. Correspondingly, crew costs increase.
- The auxiliary power needed for crew (hotel-load), navigation and (if applicable) cargo care is independent of speed. Correspondingly, the associated costs increase.
- Safety aspects pose lower limits for very low speeds. However, a 10% reduction in design speed is generally not critical in this respect.

Reduced speed for existing ships is called slow steaming. Slow steaming is less effective than designing for lower speed as there are no savings for better propeller efficiency and lower ship weight. Instead, hull, propeller, and engine operate in an off-design condition and thus at a lower efficiency. Slow steaming is adopted only when there is a slump in demand for shipping transport. Extended operation in off-design conditions leads to increased maintenance and down-time costs. In addition to technical obstacles, non-technical obstacles (like existing delivery contracts and logistics chains) hinder wider adoption.

In the following, we consider more detailed options for given speed and displacement. The attractiveness or sense of these measures depends generally on the composition of the total resistance of a ship, which differs significantly between various ship types. It is recommended to estimate at the beginning of a project the composition of the total resistance to facilitate a subsequent discussion on the effectiveness of fuel-saving measures.

Ships experience added resistance in seaways. This resistance is dominated in long waves by the ship motions, in short waves by wave reflection/diffraction. The motions can be influenced mainly by the length of the ship and to some extent by local shape details (flare of foreship or X-bow for example). The added resistance in seaways (and the saving potential for this item) is generally larger for smaller ships. For large ships, the reflection/diffraction can be reduced by different bow forms. Such proposals appear to be academic and not attractive in a holistic view.

Ideally, total power requirements should be minimized, considering also added resistance in waves in design (or even formal optimization). This has been proposed, but requires reliable prediction of the added resistance in waves. Added resistance in waves is difficult to measure and compute. Options to reduce added resistance in seaways by routing are discussed further below.

3.7.3. Hull Coatings and Air Lubrication

The frictional resistance is generally the largest part of the total resistance. The frictional resistance (for a given speed) is governed by wetted surface (main dimensions and trim) and surface roughness of the hull (average hull roughness of coating, added roughness due to fouling and coating degradation). Ships with severe fouling may require twice the power as with a smooth surface. Munk (2006) estimates that only one-third of the world fleet is in good coating condition with less than 20% added resistance compared to smooth surface condition.

Advanced hull coatings can reduce frictional resistance. An average hull roughness (AHR) of 65 μm is very good, 150 μm standard, and AHR > 200 μm sub-standard. As a rule of thumb, every 20 μm of hull roughness adds 1% to the required propulsion power (Townsin et al. 1980).

Low-surface-energy (LSE) coatings or foul release coatings create non-stick surfaces similar to those known in Teflon-coated pans, but best-practice LSE coatings reach barnacle adhesion strengths 10–20 times lower than Teflon. On the other hand, dynamic tests on moving ships have shown that well-attached barnacles (e.g. after longer stays in port) may require relatively high ship speeds to be released (Swain 2010). Some publications claim fuel savings in excess of 10% due to LSE coatings as compared to copper-based 'standard' coatings. These figures are misleading. Large improvements may be measured directly after coating, with the prerequisite hull cleaning, blasting and possibly also propeller cleaning. However, an appropriate assessment should consider the period between dry dockings. Here, a major supplier of marine coatings gives average savings of 4% for a supertanker, which can be seen as the upper limit for this ship type. All other ship types will have smaller savings, corresponding to the percentage that frictional resistance contributes to total resistance.

Coatings based on nanotechnologies have been on the market for several years and enjoy increasing popularity. It is difficult to judge claims concerning their fuel-saving potential, but a major supplier of marine coatings rated in 2010 their fuel-saving potential not higher than that of LSE coatings.

Surface-treated composites (STC) use embedded glass flakes to achieve a hard outer finish. This hard surface can be cleaned without damaging the coating. In principle, one coating would then suffice for the lifetime of a ship, but in practice local touch-ups may still be necessary. This approach is seen by many experts as very promising.

Air lubrication has attracted considerable media and industry attention. The basic idea is that a layer of air (on part of the hull) reduces the frictional resistance. The considerable technical effort is most attractive for large, slow ships with small draft. Air lubrication concepts can be classified into (Foeth 2008): air bubble concept (injection of air bubbles along the hull), air cavity concept (recesses underneath the hull are filled with air), and air film concept (using a larger film of air to cover the ship bottom).

There is no consensus on the saving potential with estimates ranging from −5% (i.e. increased fuel consumption) to 15% fuel savings. With no reliable, third-party evaluation, it remains to be seen whether this technology lives up to its claims.

3.7.4. Optimization of Hull and Appendages

Much can be gained in fuel efficiency in the proper selection of main dimensions and ship lines. Ship model basins should be consulted to assess the impact of main dimensions based on their experience and databases. For given main dimensions, wave resistance offers the largest design potential, as moderate changes may yield significant improvements. In most cases fast codes based on simplified potential flow models suffice (Abt and Harries 2007). For fuller hull shapes (tankers, bulkers), viscous flow computations are required, as viscous pressure resistance and hull–propeller interaction are significant. For limited computational resources, simplified approaches using resistance and wake fraction may be used, but proper simulations of the propulsion case at full scale are expected to become standard as computer hardware increases in power. Gains of formal optimization vary between 1.5% and 17%, with 4–5% as typical value.

The term appendages includes here negative appendages, i.e. recesses, e.g. for side thrusters or sea chests. Appendages contribute disproportionately to the resistance of a ship. Hydrodynamic analyses (model tests or CFD simulations) can determine proper local design and alignment of appendages.

Rudders offer an often underestimated potential for fuel savings. Improving the profile or changing to a highly efficient flap rudder allows reducing rudder size, thus weight and resistance. Due to the rotational component of the propeller, conventional straight rudders (at zero degree rudder angle) encounter oblique flow angles to one side at the upper part and to the other side in the lower part. This creates for most rudder profiles a slight additional thrust which recuperates part of the rotational losses of the propeller and improves propulsion. Some experts therefore recommend straight rudders. Others argue in favor of twisted rudders (e.g. Hollenbach and Friesch 2007). Dedicated CFD analyses are recommended to resolve these contradictions and to quantify expected savings in actual projects.

Ships are usually optimized for the trial or design speed in calm water, but later operated most of the time at lower speeds, even when they are not slow steaming. Fuel consumption is expected to be lower if a ship were to be designed for a more realistic mix of operational speeds, load conditions, and environmental conditions. The fuel savings gained are estimated to be 0.5–1.0% at the expense of a higher design effort.

3.7.5. Improved Propeller Designs

Modern CFD methods should lead to better propeller design, especially if design methods progress to reliable prediction of full-scale wake fields and hull–propeller interaction, considering speed and load case ranges instead of just a single operation point. Such improved propeller design procedures may be in place within the next 10 years. Potential savings of 1–4% were estimated by experts from various ship model basins in a confidential survey. The variability of propeller design and the high degree of interaction with the hull make it difficult to predict globally a fuel-saving potential.

Propellers with tip-modified blades form one special class of high-efficiency propellers. These propellers increase the efficiency without increasing diameter, similar to the tip fins often seen on aircraft wings. There are several variations on the theme (ITTC 1999, Carlton 2007):

- contracted and loaded tip (CLT) propellers with blade tips bent sharply towards the rudder (Perez Gomez and Gonzalez-Adalid 1997);
- Sparenberg–DeJong propellers with two-sided shifted end plates (Sparenberg and de Vries 1987);
- Kappel propellers with smoothly curved winglets (Andersen et al. 2002).

In interviews, propeller experts estimated 4–6% efficiency gains feasible for tankers and bulkers, but only negligible savings for ferries. Tip-modified propellers seem best suited for ships trading long-distance at a given speed.

3.7.6. Propulsion-Improving Devices (PIDs)

The propeller transforms the power delivered from the main engine via the shaft into a thrust power to propel the ship. Typically, only two-thirds of the delivered power is converted into thrust power. Various devices to improve propulsion – often by obtaining a more favorable flow in the aftbody – have been developed and installed since the early 1970s, motivated largely by the oil crisis (Blaurock 1990, Östergaard 1996). Some of the systems date back much further, but the oil crisis gave the incentive to research them more systematically and to install them on a larger scale. ITTC (1999) discussed extensively assorted propulsion-improving devices. Opinions on these devices differ widely, from negative effects (increasing fuel consumption) to more than 10% improvements. Model tests for these devices suffer from scaling errors, making any resulting quantification of savings for the full-scale ship questionable. Instead, CFD simulations for the full-scale ship are recommended to evaluate the effectiveness of a propulsion-improving device in design; the detailed insight in CFD simulations also allows a better comprehension of why a device is effective or not. While the absolute prediction accuracy of CFD is still questioned by many experts, the relative gain between two variants (with and without a duct, for example) is predicted with much higher accuracy.

Many devices have been proposed to recover rotational energy losses of the propeller. These can be categorized into pre-swirl (upstream of the propeller) and post-swirl (downstream of the propeller) devices. Devices can only recover losses partially; 30−50% of the losses are an upper limit of what a device may recuperate. Buhaug et al. (2009) give the following indicators for rotational losses:

- tanker/bulker: 3.4% at 10.9 knots, 3.9% at 15.6 knots
- container vessel: 3.9% at 15.5 knots, 5.3% at 21.2 knots
- multi-purpose vessel: 4.5% at 9.5 knots, 6.0% at 13.4 knots
- ro-pax vessel: 4.7% at 14.7 knots, 5.0% at 20.1 knots.

This would indicate (optimistic) upper limits for fuel-saving potential for devices targeted at rotational losses of 1.5−2% for tankers/bulkers, 2−2.5% for container vessels, 2−3% for MPVs and 2−2.5% for ro-pax vessels. Rudders behind the propeller already recover some of the rotational energy, reducing the fuel-saving potential further. Higher estimates found in various publications are then probably based on considering the propeller in open-water condition without rudder.

Pre-swirl devices are generally easier to integrate with the hull structure. Pre-swirl devices include the pre-swirl fin (proposed by SVA Potsdam) and pre-swirl stator blades. Asymmetric aftbodies (Schneekluth and Bertram 1998; Fig. 3.13) are a very robust way to generate swirl, but involve major changes in design. The added costs in ship construction are named frequently as an argument why asymmetric aftbodies are not considered as fuel-saving devices.

The Grim vane wheel (Fig. 3.14; Grim 1980, Schneekluth and Bertram 1998, Carlton 2007) is a freely rotating device installed behind the propeller (on the tail shaft or the rudder horn). The vane wheel is composed of a turbine section inside the propeller slipstream and a propeller section (vane tips) outside the propeller slipstream. The system appears suitable for a wide range of conventional cargo ships, but only few actual installations have been reported. Operators remain hesitant to use this device, as it appears mechanically delicate and involves considerable investment. There are concerns that collisions with wood or ice floes may damage the vane wheel. Improvements of 7−10% are reported (Breslin and Andersen 1994). The higher values are possible for higher propeller loading.

Rudder thrust fins are foils attached at the rudder. Both x-shaped thrust configurations and configurations with only two blades have been proposed. The blades are designed to generate thrust in the rotating propeller slipstream. Full hull forms (tanker, bulker) are expected to benefit more from such fins than slender ships (container vessels, ro-pax vessels). Fuel-saving potential of up to 9% has been claimed (Buhaug et al. 2009). However, no competent third-party proof for such claims is available and interviews with experts in several ship model basins resulted in rather pessimistic average fuel-saving potential estimates of 0.05%. Stator fins are another post-swirl device fixed on the rudder and intended for slender, high-speed ships like car carriers (Hoshino et al. 2004). No explicit claims on their fuel-saving potential have been published.

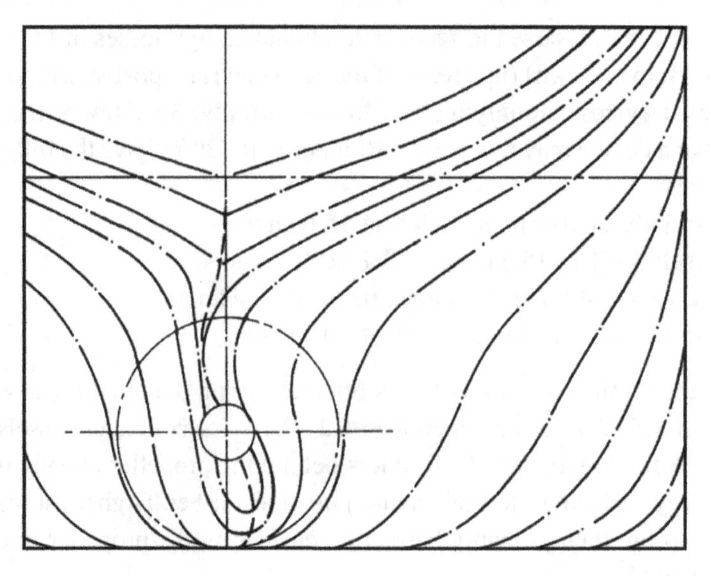

Figure 3.13:
Hull sections of asymmetric aftbody

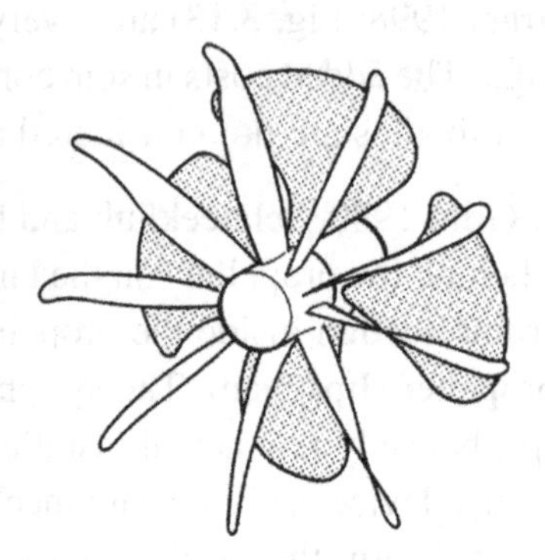

Figure 3.14:
Vane wheel

Contra-rotating propellers combine recuperation of rotational energy losses with better propeller loading (Van Manen and Sentic 1956, Schneekluth and Bertram 1998, Carlton 2007). Reported claims range from 6% to 20% in fuel efficiency improvement. However, contra-rotating propellers also have larger surface, more losses in bearing and recuperate rotational energy that otherwise would be recuperated by the rudder. Buhaug et al. (2009) give much lower estimates of 3–6% based on the estimates of rotational energy losses. This appears to be realistic. The mechanical complexity associated with frequent failure and down-time problems make the adoption of contra-rotating propellers unlikely. However, podded drives and

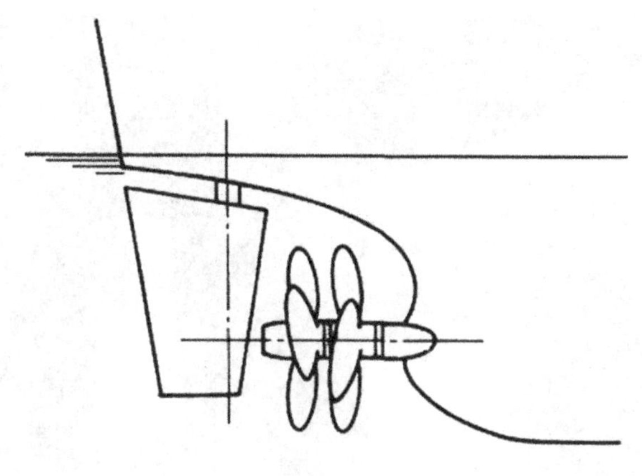

Figure 3.15:
Contra-rotating propeller

conventional propellers have been combined to hybrid CRP–POD propulsion. This option appears attractive for vessels that require redundant propulsion anyway, e.g. dangerous goods tankers.

Devices may be added to the propeller hub to suppress the hub vortex. Propeller boss cap fins (PBCF) were developed in Japan (ITTC 1999). The Hub Vortex Vane (HVV), a small vane propeller fixed to the tip of a cone-shaped boss cap, may have more blades than the propeller. There is no consensus about the effectiveness of the device that is popular due to its low costs, with estimates ranging from 0.1% to 7%.

3.7.7. Wake-Improving Devices

The propeller operates in an inhomogeneous wake behind the ship. The inhomogeneous wake induces pressure fluctuations on the propeller and the ship hull above the propeller, which in turn excite vibrations. The magnitude of these vibrations poses more or less restrictive constraints for the propeller design. A more homogeneous wake then translates into better propeller efficiency. Ideally, the hull lines (including discontinuities like appendages and inlets) should already be optimized in the design stage to have good hull–propeller interaction.

Wake-equalizing devices, such as Schneekluth nozzles, the Sumitomo Integrated Lammeren Duct (SILD) or the Hitachi Zosen nozzle (Carlton 2007), may improve propulsion in suboptimal designs, particularly for full hulls (tanker, bulker). Arguably the best-known wake-equalizing device is the Schneekluth nozzle (Fig. 3.16; Schneekluth 1986, Schneekluth and Bertram 1998). The Schneekluth nozzle is a ring-shaped flow vane with foil-type cross-section fitted to the hull in front of the upper propeller area. The Schneekluth nozzle is the propulsion-improving device with (by far) the most installations.

Figure 3.16:
Wake-equalizing duct

Independent analyses came to contradicting evaluations of the effectiveness of wake-equalizing devices (Ok 2005, Celik 2007). ITTC (1999) states cautiously: 'In conclusion, partial ducts may result in energy saving at full scale, but this was not, and probably cannot be proven by model tests.' Mewis (2009) combines a wake-equalizing duct with pre-swirl fins. The same general comments as for wake-equalizing devices apply. The effectiveness may depend on local flow details like the strength and position of the bilge vortex in the propeller plane, making the wake-equalizing devices effective in some cases and ineffective in others. The effectiveness should then be assessed on an individual case base by full-scale CFD simulations.

Grothues-Spork (1988) proposed spoilers – fitted before the propeller on both sides of the stern post – to straighten horizontally the boundary layer flow right before the propeller, thus creating direct thrust and improving the propeller efficiency. He used parts of a cylindrical surface such that they divert stronger near the hull and less further out. These fins are called Grothues spoilers (Fig. 3.17). Older literature, as quoted in Schneekluth and Bertram (1998), and Carlton (2007), gives power savings up to 9%, based on model tests. However, they are expected to increase fuel consumption rather than lead to any fuel

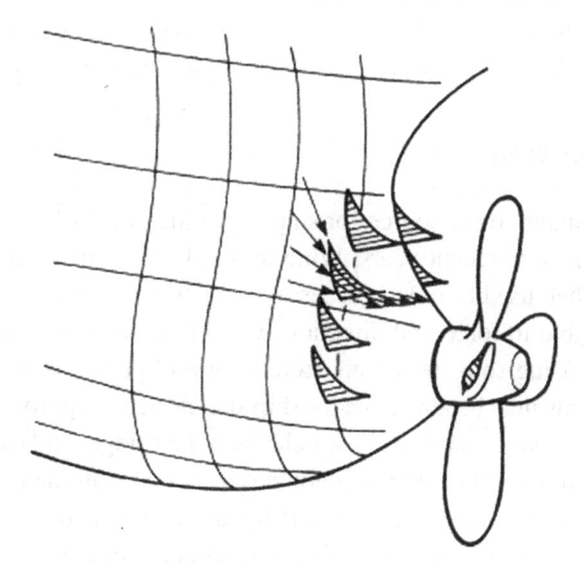

Figure 3.17:
Grothues spoilers

savings. Grothues spoilers and vortex generators have been employed to fix vibration problems in suboptimal designs.

Ducted propellers have been proposed as propulsion-improving devices (Buhaug et al. 2009). Tugs, offshore supply vessels, and fishing vessels frequently feature ducted propellers (Schneekluth and Bertram 1998). The Kort nozzle is an annular forward-extending duct around the propeller (Schneekluth and Bertram 1998). The nozzle ring has a cross-section shaped as a hydrofoil or similar section. The nozzle supplies the propeller with a larger water quantity (increasing ideal efficiency) and the foil shape serves to produce additional thrust. Kort nozzles feature the following advantages and disadvantages:

+ At high thrust-loading coefficients, better efficiency is obtainable. For tugs and pusher boats, efficiency improvements of around 20% are reported. Bollard pull can be raised by more than 30%.
+ The reduction of propeller efficiency in a seaway is lower for nozzle propellers than for non-ducted propellers.
+ Course stability is substantially improved by the nozzle.
− Course-changing ability during astern operation is somewhat impaired.
− Due to circulation in shallow water, the nozzle propeller tends to draw into itself shingle, stones, and ice floes.
− Due to the pressure drop in the nozzle, cavitation occurs earlier.

Only a small number of tankers were fitted with ducted propellers, back in the 1970s. Then ducts were no longer used for large ships, probably due to vibration and cavitation problems.

These problems could be overcome in view of present analysis capabilities (CFD and finite element analyses), leading to a renaissance of ducted propellers for large ships.

3.7.8. Wind-Assisted Ships

Wind was the predominant power source for ships until the late 19th century. Wind assistance has enjoyed a renaissance in recent years. Wind-assisted ships are mainly driven by engine propulsion. Sails are then used to reduce necessary power for a given service speed, provided that the wind is favorable in force and direction. Wind assistance becomes increasingly unattractive with increasing ship speed and decreasing fuel prices. Based on a fuel price of $500 per ton, the systems may be attractive for ship speeds below approximately 14–16 knots. Stability considerations, safety aspects (view field from the bridge) and cargo-handling aspects prevent wide use of sail assistance. For modern cargo vessels, automatic systems are the only viable option and the additional structural effort for mast support on ships with sails can be considerable. Sails for cargo vessels are typically high-performance rigid sails allowing automatic handling and giving propulsive forces even in apparent wind directions in the forward sector. Kites and Flettner rotors are generally more efficient than sails per surface area, but smaller in overall size. Optimum solutions depend on many parameters, most notably ship type, route, and speed.

Modern sails can be controlled automatically. They may be reefable cloth type (sail wings) or rigid profile type such as wing sails.

Kites have been brought to commercial maturity. Kites harness wind power at larger heights without the stability penalties of high masts. They move with much higher speeds than wind speed through the air, exploiting lifting forces similar to foils. By 2010, four ships were equipped with kites, 3 years after the first installation. Kites are claimed to be 25 times as effective (per given surface area) as regular sails. By May 2011, the largest available size was a 32 t pull (320 kN) kite. Kites with up to 130 t pull are envisioned. Savings of 10–35% are claimed for smaller ships on transatlantic routes.

Flettner rotors are another technology harnessing wind energy for ship propulsion. After 80 years of obscurity, they resurfaced in 2010 with the delivery of the E-Ship 1, a freighter equipped with Flettner rotors. These four cylinders, each 27 m tall and 4 m in diameter, are claimed to save 30% of the conventional fuel needed by the ship (10 000 tdw at 17.5 kn design speed, 7000 kW installed power). Flettner rotors create additional wind resistance for head winds and typically increase air draft (unless they are retractable, which requires additional system effort and complexity).

Solar power and wind power can be combined, using fixed sails equipped with solar panels. This option is employed successfully on the SolarSailor ferry operated in Sydney. The fuel-saving potential for large cargo vessels should be comparable to that of best-practice sails.

Sufficiently large units are yet to be developed and the technology, including high-performance solar panels, is still expensive.

There are few wind-assisted modern cargo ships. Kites are most mature with four installations (August 2010). The potential of other wind assistance options may be similar in magnitude. The saving potential differs largely between ship types, ship sizes, and trading routes. Therefore detailed studies are recommended on an individual case basis.

Reported fuel savings for wind assistance are probably in significant part due to reduced ship motions due to the dampening effect of sails. In many wind conditions, the sails cause more resistance and side drift than propulsion and are thus counter-productive.

3.7.9. Voyage Optimization

Trim optimization: for each draft and speed, there is a fuel-optimum trim. For ships with large transom sterns and bulbous bows, the power requirements for the best and worst trim may differ by more than 10%. Systematic model tests or CFD simulations are recommended to assess the best trim and the effect of different trim conditions. Decision support systems for fuel-optimum trim have been proven to result in considerable fuel savings for relatively low investment (Hansen and Freund 2010). For full hulls (tanker, bulker) the saving potential is smaller.

Weather routing (i.e. optimization of a ship's course and speed) may reduce the average added resistance in seaways. Buhaug et al. (2009) give 1–5%. The saving potential beyond what is already widely done may be less than 1% in practice. It depends among other factors on the routes which are traded (for example, Mediterranean or Atlantic).

Even engine load profile (rather than an even speed profile) offers considerable saving potential (Söding 1992). An even load profile during an entire voyage requires accurate ship models and accurate prediction at the beginning of the voyage of weather, currents, and possible other constraints during the voyage.

Adjusting an autopilot to more fuel-efficient setting has been claimed to save up to 2.5% fuel, but no reliable source is known. A significantly lower value appears to be more realistic for most large ships under professional management.

CHAPTER 4

Ship Seakeeping

Chapter Outline

4.1. Introduction

Seakeeping of ships is investigated with respect to the following issues:

- Maximum speed in a seaway: 'involuntary' speed reduction due to added resistance in waves and 'voluntary' speed reduction to avoid excessive motions, loads, etc.
- Route optimization (routing) to minimize, e.g., transport time, fuel consumption, or total cost.
- Structural design of the ship with respect to loads in seaways.
- Habitation comfort and safety of people on board: motion sickness, danger of accidental falls, man overboard.

Practical Ship Hydrodynamics. DOI: 10.1016/B978-0-08-097150-6.10004-1

- Ship safety: capsizing, large roll motions and accelerations, slamming, wave impact on superstructures or deck cargo, propeller racing resulting in excessive rpm for the engine.
- Operational limits for ships (e.g. for offshore supply vessels or helicopters landing on ships).

Tools to predict ship seakeeping are:

- Model tests.
- Full-scale measurements on ships at sea.
- Computations in the frequency domain: determination of the ship reactions to harmonic waves of different wave lengths and wave directions.
- Computations in the time domain (simulation in time): computation of the forces on the ship for given motions at one point in time; based on that information the computation of the motions at a following point in time, etc.
- Computations in the statistical domain: computation of statistically significant seakeeping values in natural (irregular) seaways, e.g. average frequency (occurrence per time) of events, such as exceeding certain limits for motions or loads in a given seaway or ocean region.

For many seakeeping issues, seakeeping is determined as follows:

1. Representation of the natural seaway as superposition of many regular (harmonic) waves (Fourier decomposition).
2. Computation (or sometimes measurement in model tests) of the ship reactions of interest in these harmonic waves.
3. Addition of the reactions in all these harmonic waves to a total reaction (superposition).

This procedure assumes (respectively requires) that the reaction of one wave on the ship is not changed by the simultaneous occurrence of another wave. This assumption is valid for small wave heights for almost all ship reactions with the exception of the added resistance.

This procedure is often applied also for seaways with large waves. However, in these cases it can only give rough estimates requiring proper corrections. One consequence of the assumed independence of the individual wave reactions is that all reactions of the ship are proportional to wave height. This is called linearization with respect to wave height.

The computations become considerably more expensive if this simplification is not made. Non-linear computations are usually necessary for the treatment of extreme motions (e.g. for capsizing investigations); here simulation in the time domain is the proper tool. However, for the determination of maximum loads it often suffices to apply corrections to initially linearly computed loads. The time-averaged added resistance is in good approximation proportional to the square of the wave height. Here the effect of harmonic waves of different lengths and direction can be superimposed as for the linear ship reactions.

To determine global properties (e.g. ship motions and accelerations) with sufficient accuracy, simpler methods suffice than for the determination of local properties (pressures, relative motions between water and ship).

Further recommended reading includes Faltinsen (1993, 2005) and Lewis (1990).

4.2. Experimental Approaches (Model and Full Scale)

Seakeeping model tests usually employ self-propelled models in narrow towing tanks or broad, rectangular seakeeping basins. The models are sometimes completely free, being kept on course by a rudder operated in remote control or by an autopilot. In other cases, some degrees of freedom are suppressed (e.g. by wires). If internal forces and moments are to be determined, the model is divided into a number of sections. The individual watertight sections are coupled to each other by gauges. These gauges consist of two rigid frames connected by rather stiff flat springs with strain gauges. Model motions are determined either directly or by measuring the accelerations and integrating them twice in time. Waves and relative motions of ships and waves are measured using two parallel wires penetrating the water surface. The change in the voltage between the wires is then correlated to the depth of submergence in water. The accuracy of ultrasonic devices is slightly worse. The model position in the tank can be determined from the angles between the ship and two or more cameras at the tank side. Either lights or reflectors on the ship give the necessary clear signal.

The waves are usually created by flaps driven by hydraulic cylinders. The flaps are inclined around a horizontal axis lying at the height of the tank bottom or even lower. Traditionally, these flaps were controlled mechanically by shaft mechanisms which created a (nearly) sinusoidal motion. Modern wave-makers are computer controlled following a prescribed time function. Sinusoidal flap motion creates harmonic waves. The superposition of many sinusoidal waves of different frequency creates irregular waves similar to natural wind seas. Some wave-makers use heightwise segmented flaps to simulate better the exponential decay of waves with water depth. Sometimes, but much less frequently, vertically moved bodies or air cushions are used to generate waves. These facilities create not only the desired wave, but also a near-field disturbance which decays with distance from the body or the air cushion. More harmful is the generation of higher harmonics (waves with an integer multiple of the basic wave frequency), but these higher harmonics can be easily filtered from the measured reactions if the reactions are linear. In computer-controlled wave-makers they can be largely eliminated by proper adjustment of the flap motions.

In towing tanks, waves are usually generated by one flap at one tank end spanning the complete tank width. The other tank end has a 'beach' to absorb the waves (ideally completely) so that no reflected waves influence the measurements and the water comes to rest as soon as possible after a test. If several, independently controlled flaps are used over the tank width, waves with propagation direction oblique to the tank longitudinal axis can be generated. These waves will then be reflected at the side walls of the tank. This is unproblematic if a superposition of many waves of different direction ('short-crested sea') is created as long as the distribution of the wave energy over the propagation direction is symmetrical to the tank longitudinal axis. In natural wind seas the energy distribution is similarly distributed around the average wind direction.

Rectangular wide seakeeping basins typically have a large number of wave-making flaps at two adjacent sides. An appropriate phase shift in the flap motions can then create oblique wave propagation. The other two sides of such a basin are then equipped with 'beaches' to absorb waves.

Seakeeping model tests are usually only performed for strongly non-linear seakeeping problems which are difficult to compute. Examples are roll motion and capsizing, slamming and water on deck. Linear seakeeping problems are only measured for research purposes to supply validation data for computational methods. In these cases many different frequencies can be measured at the same time. The measured data can then be decomposed (filtered) to obtain the reactions to the individual wave frequencies.

Seakeeping tests are expensive due to the long waiting periods between tests until the water has come to rest again. The waiting periods are especially long in conventional towing tanks. Also, the scope of the experiments is usually large as many parameters need to be varied, e.g. wave length, wave height, angle of encounter, ship speed, draught and trim, metacentric height, etc. Tests keep Froude similarity just as in resistance and propulsion tests. Gravity and inertia forces then correspond directly between model and full-scale ship. However, scale effects (errors due to the model scale) occur for forces which are due to viscosity, surface tension, compressibility of the water, or model elasticity. These effects are important, for example, for slamming pressure, water on deck, or sway, roll and yaw motions. However, in total, scale effects play a lesser role for seakeeping tests than for resistance and propulsion tests or maneuvering tests.

Seakeeping can also be measured on ships in normal operation or during special trial tests. Ship motions (with accelerometers and gyros) and sometimes also global and local loads (strain gauges), loss of speed, propeller rpm and torque are all measured. Recording the seaway is difficult in full-scale measurements. The options are:

1. No recording of actual seaway during trial; instead measurements of seaway over many years such that, for example, the expected maximum values during the lifetime of the ship can be extrapolated from the recorded distribution of long-term measured values (long-term measurement). The random variation of the actual seastate encountered by the ship introduces considerable inaccuracies for the predicted extreme values even if several years of measurements are available.

2. Computation of the seaway from the ship motions based on computed or model-test measured response amplitude operators for the motions. This allows only a rather rough estimate of the seaway. In following seas this method is hardly applicable. Nevertheless, averaging over, say, 10–100 half-hour measurements usually yields good estimates for the correlation of loads and seaway (medium-term measurement).
3. Parallel measurement of the seaway. Options are:
   - Using seastate measuring buoys (brought by the ship).
   - Performing the sea trials near a stationary seaway measuring installation.
   - Measuring the ship motions (by accelerometers) and the relative motion between water and ship (by pressure measurements at the hull or water level measurements using a special radar device); based on these data indirect determination of the absolute motion of the water surface is possible.
   - Measuring the wave spectrum (energy distribution over frequency and propagation direction) by evaluating radar signals reflected by the waves.
   - Computation or estimation of the seaway from the wind field before and during the experiments.
   - Estimation of significant wave height and period from 'experienced' seamen. This common practice is far too inaccurate: the correlation coefficient between measured (actual) and estimated wave period is typically $<50\%$! This holds also if the estimates are used to derive statistical distributions. For most extreme values of interest the errors in the estimates do not cancel, but are rather concentrated around the extreme values.

4.3. Waves and Seaway

4.3.1. Airy Waves (Harmonic Waves of Small Amplitude)

Wind-induced seaways can be approximated by the superposition of regular waves of small wave height (elementary waves, Airy waves). Each elementary wave has a sinusoidal profile with an infinite number of wave troughs and wave crests (Fig. 4.1). The wave troughs and crests are perpendicular to the direction of wave propagation. Such elementary waves are an important building block for all computational methods for linear seakeeping problems. Steep regular waves can be computed by, for example, Stokes' theory or panel methods. However, the superposition principle no longer applies to these waves. Therefore they play virtually no role at all in the prediction of ship seakeeping and are of rather academic interest for naval architects. Unfortunately, in using the superposition principle for elementary waves, all properties of the seaway which are non-linear with wave steepness (= wave height/wave length) are lost.

These are, for example, the broader wave troughs and steeper wave crests, the higher celerity of steeper waves which results in a tendency to form wave groups in natural wind seas: groups of waves with low wave height are followed by groups of waves with larger wave heights.

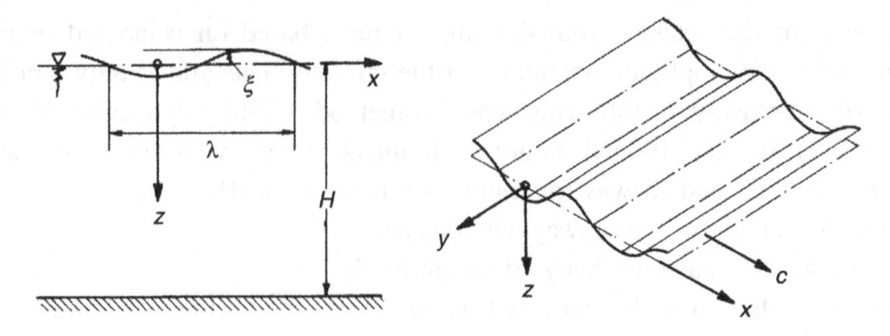

Figure 4.1:
Elementary waves

For ship seakeeping, the relevant waves are dominated by gravity effects. Surface tension, water compressibility and (for deep and moderately shallow water) viscosity can be neglected. Computations can then assume an ideal fluid (incompressible, inviscid) without surface tension. Consequently potential theory can be applied to describe the waves.

Generally, regular waves are described by a length parameter (wave length λ or wave number k) and a time parameter (wave period T or (circular) frequency ω). k and ω are defined as follows:

$$k = \frac{2\pi}{\lambda}; \quad \omega = \frac{2\pi}{T} \tag{4.1}$$

The celerity c denotes the speed of wave propagation, i.e. the speed of an individual wave crest or wave trough:

$$c = \frac{\lambda}{T} = \frac{\omega}{k} \tag{4.2}$$

For elementary waves, the following (dispersion) relation holds:

$$k = \frac{\omega^2}{g} \quad \text{on deep water} \tag{4.3}$$

$$k \tanh(kH) = \frac{\omega^2}{g} \quad \text{on finite depth} \tag{4.4}$$

$g = 9.81\ \text{m/s}^2$ and H is the water depth (Fig. 4.1).

The above equations can then be combined to give the following relations (for deep water):

$$c = \sqrt{\frac{g}{k}} = \frac{g}{\omega} = \sqrt{\frac{g\lambda}{2\pi}} = \frac{gT}{2\pi} \tag{4.5}$$

The potential ϕ of a wave traveling in the $+x$ direction is:

$$\phi = \text{Re}(-ic\widehat{h}e^{-kz}e^{i(\omega t - kx)}) \quad \text{for deep water} \tag{4.6}$$

$$\phi = \text{Re}\left(\frac{-ic\widehat{h}}{\sinh(kH)}\cosh(k(z-H))e^{i(\omega t - kx)}\right) \quad \text{for finite depth} \tag{4.7}$$

Re denotes the real part of a complex quantity; $i = \sqrt{-1}$; z as in Fig. 4.1; ^ denotes as usual a complex amplitude; $\widehat{h}$ = the complex amplitude of the wave. $h = |\widehat{h}|$ is the (real-valued) wave amplitude, i.e. half the wave height (from wave trough to wave crest). The real part of $\widehat{h}$ gives the distance of the wave trough from the calm-water level at time $t = 0$ at $x = 0$; the imaginary part gives the same value at ¼ period earlier. The deep-water formulae are applicable with errors of $< 0.5\%$ if the water depth is larger than half a wave length.

The velocity is obtained by differentiation of the potential, e.g. for deep water:

$$v_x = \frac{\partial \phi}{\partial x} = \phi_x = \text{Re}(-\omega\widehat{h}e^{-kz}e^{i(\omega t - kx)}) \tag{4.8}$$

$$v_z = \frac{\partial \phi}{\partial z} = \phi_z = \text{Re}(i\omega\widehat{h}e^{-kz}e^{i(\omega t - kx)}) \tag{4.9}$$

The complex amplitudes of the velocities have the same absolute value and a phase shift of 90°. A water particle thus follows a circular track or orbital motion (from Latin *orbis* = circle). In water of finite depth, the motion of a water particle follows an ellipse. The vertical axis of each ellipse decreases with depth until at the water bottom $z = H$ the motion is only in the horizontal direction.

If we excite a group of waves (not elementary waves, but, say, ten wave crests and troughs) in initially calm water we will notice that the front of the wave crests decay while at the end of the wave packet new wave crests are formed (Fig. 4.2). The wave packet thus moves slower than the wave crests, i.e. with a speed slower than celerity c, namely with group velocity c_{gr}:

$$c_{gr} = \frac{1}{2}c \quad \text{for deep water} \tag{4.10}$$

$$c_{gr} = c\left(\frac{1}{2} + \frac{kH}{\sinh(2kH)}\right) \quad \text{for finite depth} \tag{4.11}$$

The linearized Bernoulli equation

$$p + \rho\frac{\partial \phi}{\partial t} - \rho g z = p_0 \tag{4.12}$$

and the wave potential give the difference pressure to atmospheric pressure at a point below the water surface (for deep water):

$$p - p_0 = \rho g z - \rho g\,\text{Re}\left(\widehat{h}e^{-kz}e^{i(\omega t - kx)}\right) \tag{4.13}$$

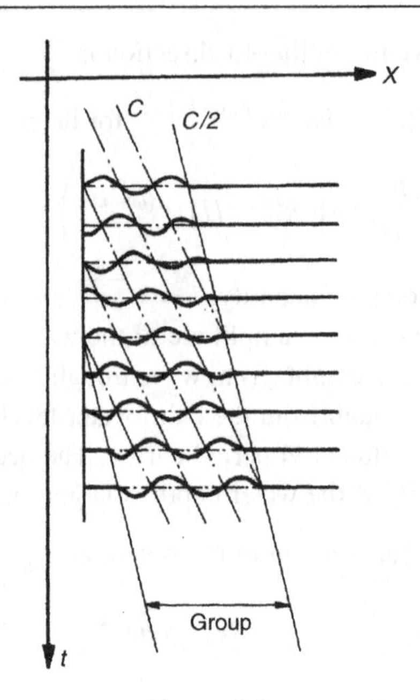

Figure 4.2:
Celerity and group velocity

p_0 is the atmospheric pressure, ρ the water density, and z the depth of the point below the calm-water surface. The first term represents the hydrostatic pressure in calm water. The second term represents the pressure change due to the wave. As with all wave effects, it decays exponentially with depth. The pressure gradient $\partial p/\partial z$ for the hydrostatic case is equal to the specific weight of the fluid and causes a buoyant lifting force on the immersed body that equals the weight of the displaced water. This lifting force changes in a wave! The lifting force is lower in a wave crest, higher in a wave trough. This is called the Smith effect.

The mechanical energy E per area of the water surface is composed of potential and kinetic energy. Let ζ be the momentary elevation of the free surface. Then the potential energy (per area) is:

$$E_{\text{pot}} = -\frac{\zeta}{2}\rho g(-\zeta) = \frac{1}{2}\rho g \zeta^2 \tag{4.14}$$

The potential energy is positive both in wave troughs and wave crests and oscillates in time and space between 0 and $\frac{1}{2}\rho g|\hat{h}|^2$. The time average is

$$\overline{E}_{\text{pot}} = \frac{1}{4}\rho g|\hat{h}|^2 \tag{4.15}$$

The kinetic energy per area is:

$$E_{\text{kin}} = \int_{\zeta}^{\infty} \frac{1}{2}\rho\left(v_x^2 + v_z^2\right)\,\mathrm{d}z = \int_{\zeta}^{\infty} \frac{1}{2}\rho\omega^2|\hat{h}|^2 \mathrm{e}^{-2kz}\,\mathrm{d}z \approx \int_{0}^{\infty} \ldots\,\mathrm{d}z = \frac{1}{4}\rho g|\hat{h}|^2 \tag{4.16}$$

Here Eqs. (4.8) and (4.9) have been used and in a linearization the wave elevation ζ was substituted by 0. The kinetic energy is constant in time and space. The time-averaged total energy per area for a deep-water wave is then:

$$\overline{E} = \frac{1}{2}\rho g|\hat{h}|^2 \tag{4.17}$$

The average energy travels with c_{gr} in the same direction as the wave. For finite-depth water the average energy remains the same but the kinetic energy also oscillates in time and space.

The elementary wave was so far described in an earth-fixed coordinate system. In a reference system moving with ship speed V in the direction of the ship axis x_s under an angle of encounter μ (Fig. 4.3), the wave seems to change its frequency. The (circular) frequency experienced by the ship is denoted encounter frequency:

$$\omega_e = |\omega - kV\cos\mu| = \left|\omega - \frac{\omega^2 V}{g}\cos\mu\right| \tag{4.18}$$

Figure 4.4 illustrates this phenomenon. For course against the sea ($\mu > 90°$) the encounter frequency is higher than the incident wave frequency ω. For course with the sea ($\mu < 90°$) the encounter frequency is usually lower than the incident wave frequency ω. An exception is short following seas which are passed by the ship. The condition for the ship passing the waves is:

$$F_n > \frac{0.4}{\cos\mu}\sqrt{\frac{\lambda}{L}} \tag{4.19}$$

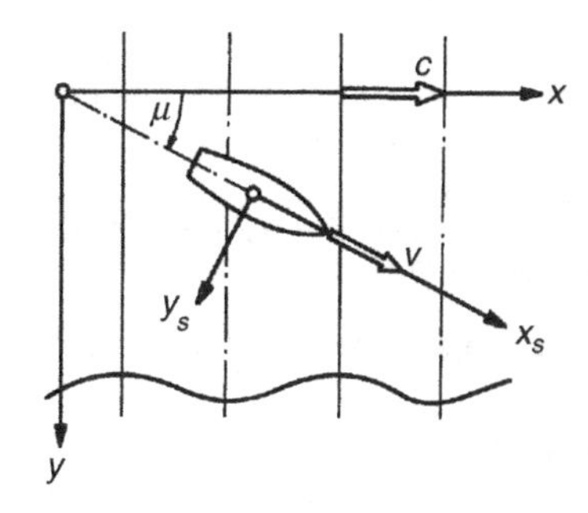

Figure 4.3:
Definition of angle of encounter

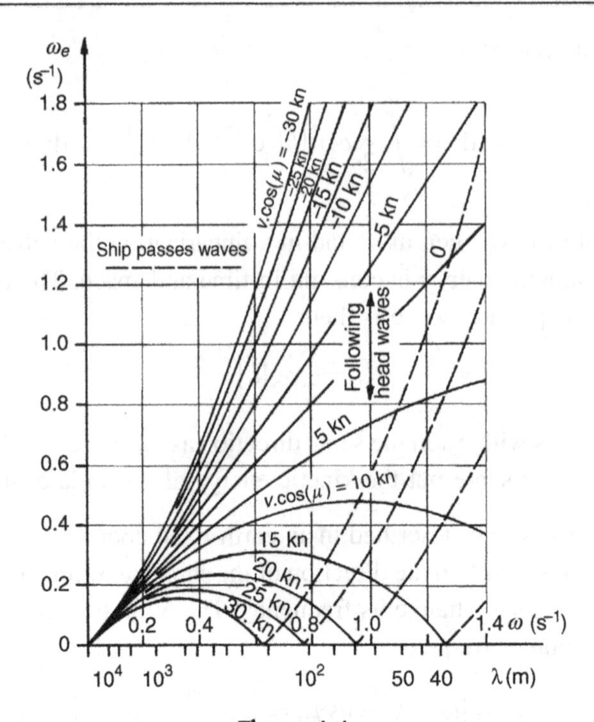

Figure 4.4:
Relation between wave frequency, wave length, and encounter frequency

An important parameter in this context is:

$$\tau = \frac{\omega_e V}{g} = \frac{\omega V}{g} - \left(\frac{\omega V}{g}\right)^2 \cos\mu \tag{4.20}$$

For following sea for cases with $\tau \cos\mu < 0.25$, for given speed V, encounter angle μ, and encounter frequency ω_e three possible ω values exist:

$$\omega_1 = \frac{g}{2V\cos\mu}(1 + \sqrt{1 + 4\tau\cos\mu}) \tag{4.21}$$

$$\omega_2 = \frac{g}{2V\cos\mu}(1 + \sqrt{1 - 4\tau\cos\mu}) \tag{4.22}$$

$$\omega_3 = \frac{g}{2V\cos\mu}(1 - \sqrt{1 - 4\tau\cos\mu}) \tag{4.23}$$

The potential of a deep-water wave in a coordinate system moving with ship speed is:

$$\phi = \mathrm{Re}(-ic\hat{h}e^{-kz}e^{-ik(x_s\cos\mu - y_s\sin\mu)}e^{i\omega_e t}) \tag{4.24}$$

The above formulae for velocities and pressures can correspondingly be derived in the coordinate system moving with ship speed.

4.3.2. Natural Seaway

Wind-excited seaway can be approximated with good accuracy as the superposition of many elementary waves of different wave lengths and propagation directions. The phase shifts between these elementary waves change with time and location and are taken as random quantities for the origin and time $t = 0$. The randomness of the phases – which corresponds to the randomness (irregularity) of the natural seaway – means that only statistical statements can be made, e.g. what the probability is that the wave height exceeds a given limit.

The initial assumptions are:

1. The seaway is stationary, i.e. its statistical properties (e.g. average wave height, average wave period, etc.) do not change within the considered time frame.
2. The seaway is not too steep so that linearized equations are still accurate enough. Then any linear superposition of two or more waves with the same or differing frequency or propagation direction will again be a possible form of the water surface.

Only those seaway properties which do not change for small variations of the registration location or the registration time are of interest for ship seakeeping. The procedure to obtain these properties is as follows. Assume we have a record of the wave elevation $\zeta(t)$ at a given point for the time interval $t = 0$ to T. Then ζ is decomposed in a Fourier analysis, i.e. the complex constants $\hat{A}_j$ are determined in a finite series:

$$\zeta(t) = A_0 + \sum_{j=1}^{J} \mathrm{Re}(\hat{A}_j e^{i\omega_j t}) \quad \text{with } \omega_j = j\Delta\omega,\ \Delta\omega = 2\pi/T \tag{4.25}$$

The average wave elevation A_0 is of no interest here. The phase angle ε_j of the complex amplitudes $\hat{A}_j = |\hat{A}_j|e^{i\varepsilon_j}$ would be different at a different (nearby) location and is therefore also of no interest here. The absolute value of $\hat{A}_j$ depends on the registration time T. Only the sea spectrum remains as constant and of interest in the above sense:

$$S_\zeta(\omega_j) = \frac{\text{Average value of } |\hat{A}_j|^2}{2\Delta\omega_j} \tag{4.26}$$

The averaging can be done:

- over many records of statistically equivalent seaways (e.g. at various locations spaced by a few kilometers at the same time), or
- over many records of time intervals of the total registration time T, or
- over several (10 to 30) 'neighboring' $|\hat{A}_j|^2$ (preferred choice in practice); e.g. for $j = 1$ to 10, 11 to 20, 21 to 30 etc., an average $|\hat{A}_j|^2$ can be found as the arithmetic average of ten $|\hat{A}_j|^2$ in each case.

The ω_j in the argument of the sea spectrum S_ζ is the (circular) frequency (in the last case the average frequency) on which the average is based.

The wave energy per horizontal area in an elementary wave is:

$$\overline{E} = \frac{1}{2}\rho g|\widehat{A}|^2 \tag{4.27}$$

$\rho g S_\zeta$ is thus the average seaway energy per frequency interval and area. Therefore S_ζ is also called the energy spectrum of a seaway. It describes the distribution of wave energy over the frequency ω. Its dimension is, e.g., $\mathrm{m}^2 \cdot \mathrm{s}$.

The spectrum can be used to reconstruct the time function $\zeta(t)$ given in Eq. (4.25):

$$\zeta(t) = \sum_{j=1}^{J} \sqrt{2S_\zeta(\omega_j)\Delta\omega_j} \cdot \cos(\omega_j t + \varepsilon_j) \tag{4.28}$$

(Instead of Re $e^{i\alpha}$ we simply write here cos α.) We substituted here $|\widehat{A}_j|^2$ by its average value; this usually has no significant effect. As the phase angle information is no longer contained in the spectrum (and we usually only have the spectrum information to reconstruct a seaway) the phase angles ε_j are chosen as random quantities equally distributed in the interval $[0, 2\pi]$. This creates various functions $\zeta(t)$ depending on the actual choice of ε_j, but all these functions have the same spectrum, i.e. the same characteristic (non-random) properties as the original seaway.

If all phase angles are chosen as zero the extremely unlikely (but not impossible) case results that all elementary waves have a wave trough at the considered location at time $t = 0$. The number of terms in the sum for $\zeta(t)$ in the above equation is taken as infinite in theoretical derivations. In practical simulations, usually 30 to 100 terms are taken.

Each elementary wave in a Fourier decomposition of natural seaway depends on time and space. The superposition of many elementary waves all propagating in the x direction, but

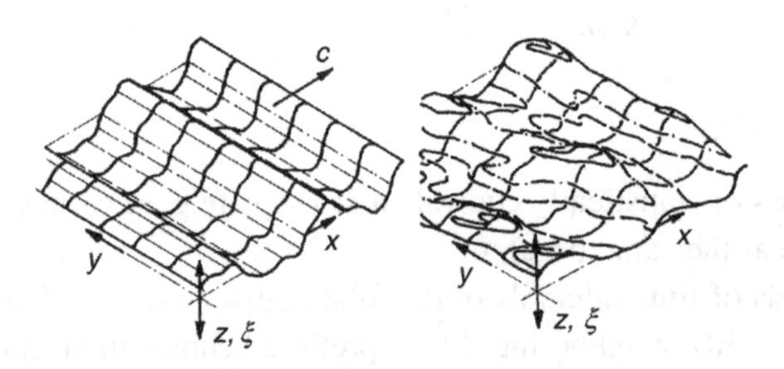

Figure 4.5:
Long-crested (left) and short-crested (right) seaways

having different frequencies, yields long-crested seaways as depicted in Fig. 4.5 (left). Long-crested seaway is described by:

$$\zeta(t) = \sum_{j=1}^{J} \sqrt{2S_\zeta(\omega_j)\Delta\omega_j} \cdot \cos(\omega_j t + k_j x + \varepsilon_j) \tag{4.29}$$

$k_j = \omega_j/g$ is the wave number corresponding to frequency ω_j.
Short-crested seaway (Fig. 4.5 (right)) is a better approximation to wind-excited seaway. Short-crested seaway is described if the wave energy is distributed not only over frequency, but also over wave propagation direction μ. The corresponding description is:

$$\zeta(t) = \sum_{j=1}^{J} \sum_{l=1}^{L} \sqrt{2S_\zeta(\omega_j, \mu_l)\Delta\omega_j\Delta\mu_l} \cdot \cos[\omega_j t - k_j(x \cos \mu_l - \sin \mu_l) + \varepsilon_{jl}] \tag{4.30}$$

$S_\zeta(\omega_j,\mu_l)$ is the directional or two-dimensional spectrum as opposed to the one-dimensional spectrum $S_\zeta(\omega_j)$.

At a ship, the wave elevation oscillates in a regular wave with encounter frequency ω_e. The encounter spectrum $S_e(\omega_e)$ describes the distribution of the wave energy in a seaway over ω_e instead of ω. The energy must be independent of the reference system:

$$S_\zeta(\omega) \cdot |\Delta\omega| = S_{\zeta e}(\omega_e) \cdot |\Delta\omega_e| \tag{4.31}$$

This yields:

$$S_{\zeta e}(\omega_e) = \frac{S_\zeta(\omega)}{\mathrm{d}\omega_e/\mathrm{d}\omega} = \frac{S_\zeta(\omega)}{\left|1 - \frac{2\omega}{g} V \cos \mu\right|} \tag{4.32}$$

If several ω result in the same ω_e the contributions of all three frequencies are added on the r.h.s. of this equation (Fig. 4.6). Correspondingly an encounter directional spectrum can also be determined. Because of the several possible contributions on the r.h.s. and the singularity at S_e – where the denominator on the r.h.s. in the above equation becomes zero – the encounter spectrum is not used in seakeeping computations. However, it is needed for the analysis of data if these were measured from a ship with forward speed.

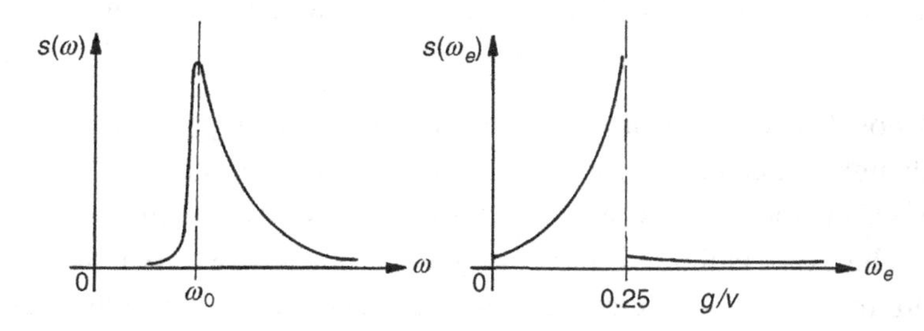

Figure 4.6:
Sea spectrum and corresponding encounter spectrum

4.3.3. Wind and Seaway

We distinguish between swell and wind sea. Swell waves have a celerity higher than the present wind speed (e.g. measured in 10 m height above mean sea level; only the component in the wave propagation direction is considered). Swell has been excited originally by some stronger winds at some other location and propagates without significant damping or excitation until it is damped in shallow-water regions or excited again to wind sea in stronger winds. By definition, wind sea has celerity less or equal to the wind speed. Due to the gustiness of wind and other factors, the distinction between swell and wind sea is not sharp.

Swell, sometimes also wind sea (for winds changing rapidly in time or space), can change the form of the spectrum considerably. On the other hand, a rather uniform form of a wind sea spectrum is achieved within ½ to 1 hour if the wind is constant in time and space. The relevant area in this context extends over a distance of (½ to 1 hour)/group velocity of waves in a downwind direction. In the following, we will consider only spectra developed in constant wind. The spectrum parameters, especially wave height and period, converge only after many hours or several days to constant values. The form of the spectrum is determined by the physical processes of:

- wave generation (e.g. the wind resistance of wave crests);
- dissipation (wave-breaking; in shallow water also friction at the ocean bottom);
- convection (transport of wave energy with group velocity);
- non-linear interaction between waves of different frequencies and direction.

The directional spectrum is described as the product of a one-dimensional spectrum $S_\zeta(\omega)$ with a function f. f describes the distribution of the wave energy over the propagation direction μ assumed to be symmetrical to a main propagation direction μ_0:

$$S_\zeta(\omega,\mu) = S_\zeta(\omega)\cdot f(\mu - \mu_0) \tag{4.33}$$

Söding and Bertram (1998) give a more modern form than the often cited Pierson–Moskowitz and JONSWAP spectra. The older spectra assume a stronger decay of the wave energy at higher frequencies (proportional to ω^{-5}, while more recent measurements indicate decay proportional to ω^{-4}).

The one-dimensional spectrum $S_\zeta(\omega)$ must be zero for small frequencies (where the wave celerity is much higher than the wind speed) and converge to zero for high frequencies, because high frequency means short waves, which in turn can only have small height as the wave steepness before breaking is limited. In between, there must be a maximum. The (circular) frequency where the spectrum assumes its maximum is called modal frequency or peak frequency ω_p. The function $S_\zeta(\omega)$ contains as an important parameter U_c/c_p. U_c is the component of the wind velocity in the main direction of wave propagation, measured in 10 m

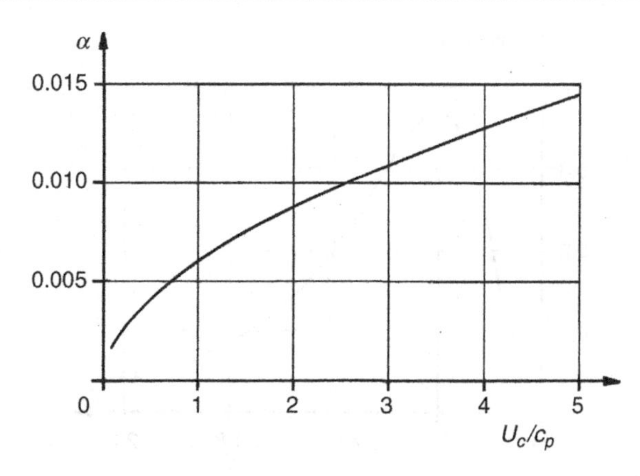

Figure 4.7:
Spectrum factor

height. c_p is the celerity of elementary waves of frequency ω_p. c_p is computed using the formula $c_p = g/\omega_p$ which is valid for elementary waves. In reality, waves of frequency ω_p travel some 5–10% faster due to their larger steepness. The ratio U_c/c_p usually lies between 1 (fully developed seaway) and 5 (strongly increasing seaway).

$S_\zeta(\omega)$ is written as the product of three factors:

- an initial factor $\alpha g^2/\omega_p^5$
- a 'base form' containing the ω dependency (corresponding to the Pierson–Moskowitz spectrum widely used previously)
- a peak enhancement factor γ^Γ independent of U_c/c_p:

$$S_\zeta(\omega) = \frac{\alpha g^2}{\omega_p^5} \cdot \left(\frac{\omega_p}{\omega}\right)^4 \exp\left[-\left(\frac{\omega_p}{\omega}\right)^4\right] \cdot \gamma^\Gamma \tag{4.34}$$

with $\alpha = 0.006(U_c/c_p)^{0.55}$.

Figure 4.7 illustrates α, Fig. 4.8 the base form, and Fig. 4.9 the peak enhancement for three representative values of U_c/c_p. The peak enhancement makes the maximum of the spectrum very pointed for a not fully developed seaway ($U_c/c_p > 1$), while fully developed seaways feature broader and less-pronounced maxima. γ describes the maximum of the peak enhancement over ω. It occurs at ω_p and increases the 'base form' by a factor of:

$$\gamma = 1.7 + \max[0.6 \log_{10}(U_c/c_p)] \tag{4.35}$$

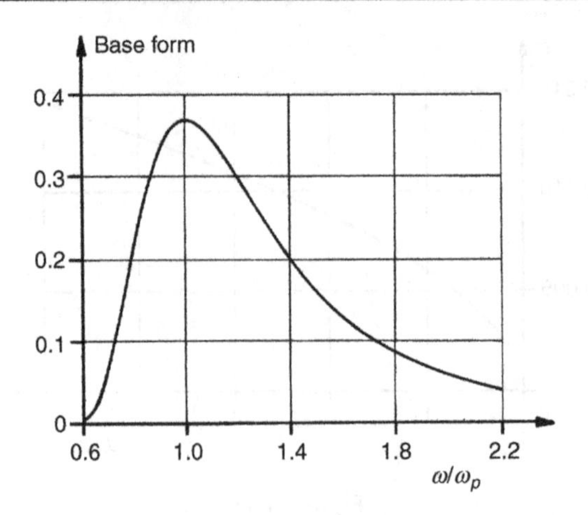

Figure 4.8:
Spectrum base form

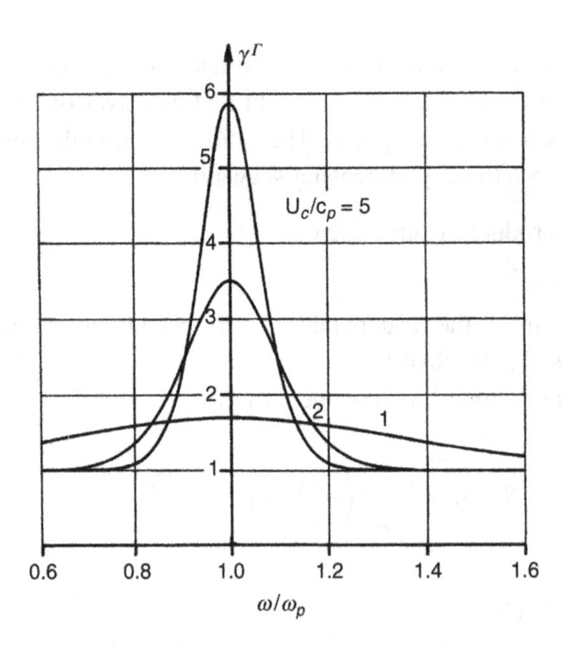

Figure 4.9:
Peak enhancement factor

Γ describes how the enhancement factor decays left and right of the model frequency ω_p; for this purpose a formula corresponding to a normal (Gaussian) distribution is chosen (but without a forefactor; thus the maximum of Γ is 1):

$$\Gamma = \exp\left(-\frac{(\omega/\omega_p - 1)^2}{2\sigma^2}\right) \quad \text{with} \quad \sigma = 0.08\left[1 + \frac{4}{(U_c/c_p)^3}\right] \tag{4.36}$$

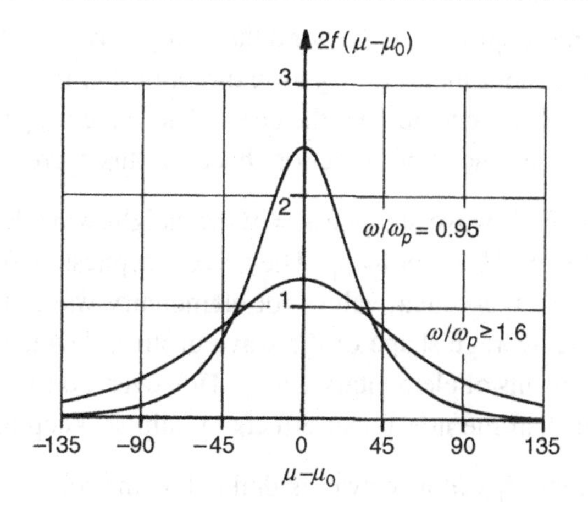

Figure 4.10:
Angular distribution of seaway energy

The distribution of the wave energy over the propagation direction $f(\mu - \mu_0)$ is independent of U_c/c_p. Instead, it depends on the non-dimensional frequency ω/ω_p:

$$f(\mu - \mu_0) = \frac{0.5\beta}{\cosh^2[\beta(\mu - \mu_0]} \text{ with} \tag{4.37}$$

$$\beta = \max\left(1.24, 2.61(\omega/\omega_p)^{1.3}\right) \quad \text{for } \omega/\omega_p < 0.95 \tag{4.38}$$

$$\beta = \max\left(1.24, 2.28(\omega/\omega_p)^{-1.3}\right) \quad \text{for } \omega/\omega_p \geq 0.95 \tag{4.39}$$

Figure 4.10 illustrates $f(\mu - \mu_0)$. Figure 4.11 illustrates $\beta(\omega/\omega_p)$.

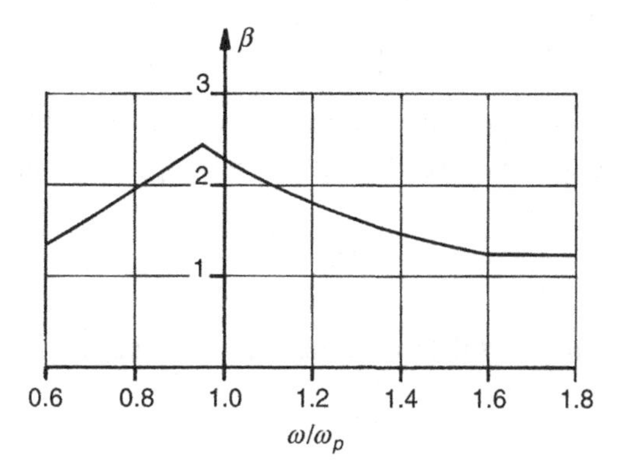

Figure 4.11:
Angular spreading β

Since short waves adapt more quickly to the wind than long waves, a changing wind direction results in a frequency-dependent main propagation direction μ_0. Frequency-dependent μ_0 are also observed for oblique offshore wind near the coast. The wave propagation direction here is more parallel to the coast than the wind direction, because this corresponds to a longer fetch.

The (only statistically defined) wave steepness = wave height/wave length does not depend strongly on the wind velocity, U_c/c_p, or ω/ω_p. The wave steepness is so large that the celerity deviates noticeably from the theoretical values for elementary waves (of small amplitude) as described above. Also, the average shape of the wave profiles deviates noticeably from the assumed sinusoidal wave forms of elementary waves. However, non-linear effects in the waves are usually much weaker than the non-linear effects of ship seakeeping in the seaway.

The significant wave height $H_{1/3}$ of a seaway is defined as the mean of the top third of all waves, measured from wave crest to wave trough. $H_{1/3}$ is related to the area m_0 under the sea spectrum:

$$H_{1/3} = 4\sqrt{m_0} \quad \text{with } m_0 = \int_0^\infty \int_0^{2\pi} S_\zeta(\omega, \mu)\, d\mu\, d\omega \tag{4.40}$$

For the above given wind sea spectrum, $H_{1/3}$ can be approximated by:

$$H_{1/3} = 0.21 \frac{U_c^2}{g} \left(\frac{U_c}{c_p} \right)^{-1.65} \tag{4.41}$$

The modal period is:

$$T_p = 2\pi/\omega_p \tag{4.42}$$

The periods T_1 and T_2, which were traditionally popular to describe the seaway, are much shorter than the modal period. T_1 corresponds to the frequency ω where the area under the spectrum has its center. T_2 is the average period of upward zero crossings.

If we assume that water is initially calm and then a constant wind blows for a duration t and over a distance x, the seaway parameter U_c/c_p becomes approximately:

$$\frac{U_c}{c_p} = \max(1, 18\xi^{-3/10}, 110\theta^{-3/7}) \tag{4.43}$$

ξ is the non-dimensional fetch x, θ the non-dimensional wind duration t:

$$\xi = gx/U_c^2; \quad \theta = gt/U_c \tag{4.44}$$

The fetch is to be taken upwind from the point where the seaway is considered, but of course at most to the shore. In reality, there is no sudden and then constant wind. But the seakeeping

Table 4.1: Sea spectra for various wind duration times for $U_c = 20$ m/s

Quantity	Case 1	Case 2	Case 3
Assumed wind duration time t	5 h	20 h	50 h
Non-dimensional duration time θ	8830	35000	88000
Maturity parameter U_c/c_p	2.24	1.24	1
Significant wave height $H_{1/3}$	2.26 m	6.00 m	8.56 m
$\omega_p = g/c_p = g/U_c \cdot U_c/c_p$	1.10 Hz	0.61 Hz	0.49 Hz
Modal period $2\pi/\omega_p$	5.7 s	10.3 s	12.8 s

parameters are not very sensitive towards x and t. Therefore it is possible to estimate the seaway with practical accuracy in most cases when the wind field is given.

Table 4.1 shows how the above formulae estimate the seaway parameters $H_{1/3}$ and T_p for various assumed wind durations t for an exemplary wind velocity $U_c = 20$ m/s. The fetch x was assumed to be so large that the center term in the 'max'-bracket in the above formula for U_c/c_p is always smaller than one of the other two terms. That is, the seaway is not fetch-limited, but either time-limited (for $110\theta^{-3/7} > 1$) or fully developed.

Figure 4.12 shows wind sea spectra for $U_c = 20$ m/s for various fetch values. Figure 4.13 shows the relation between wave period T_p and significant wave height $H_{1/3}$ for various values of U_c/c_p. c_p (lower scale) and U_c/c_p together yield the wind velocity U_c that has excited the wind

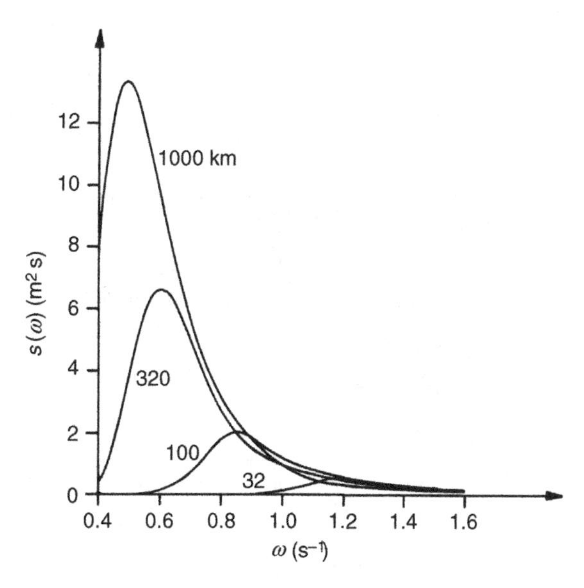

Figure 4.12:
Wind sea spectra for $U_c = 20$ m/s for various fetch values

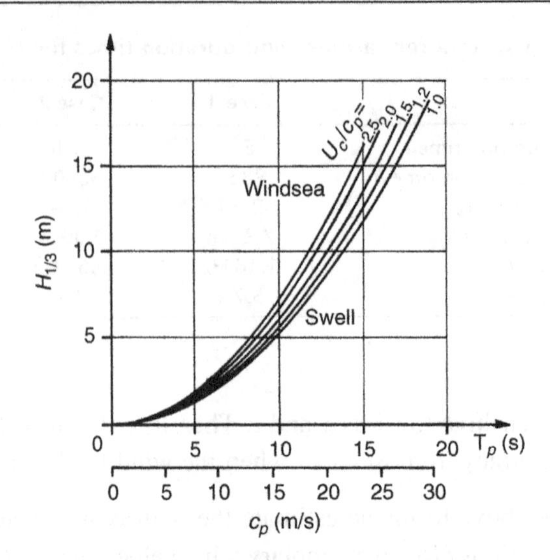

Figure 4.13:
Correlation between significant wave height $H_{1/3}$, modal period T_p, wind speed U_c, and wave celerity at modal frequency c_p

sea characterized by $H_{1/3}$ and T_p. For swell, we can assume $U_c \approx c_p$. Figure 4.14 shows the relation between various seaway parameters, the 'wind force' and the wind velocity U_c.

Programs to compute the given wind sea spectrum from either U_c, t, and x or $H_{1/3}$ and T_p are given by Söding (1997).

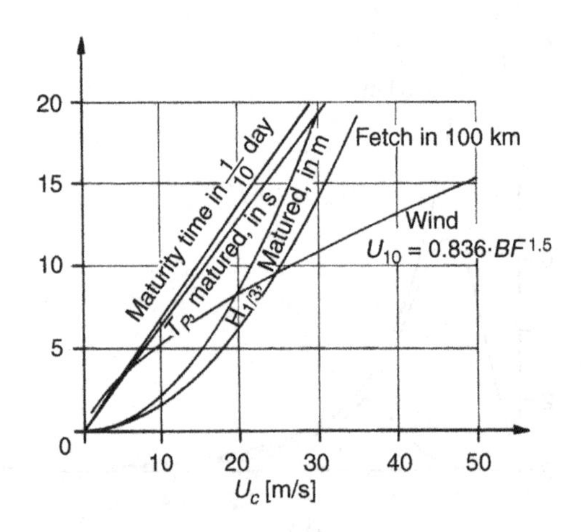

Figure 4.14:
Key wind sea parameters depending on the wind speed U_c (component in wave propagation direction in 10 m height)

Table 4.2: Relative occurrence · 10^6 of combinations of $H_{1/3}$ and T_1 in the North Atlantic

T_1 (s)			$H_{1/3}$ (m)																
from	to	FCUM	0.5	1.5	2.5	3.5	4.5	5.5	6.5	7.5	8.5	9.5	11.0	13.0	15.0	17.0	19.0	21.0	24.0
1.9	3.1	0.2	2040	0	0	0	0	0	0	0	0	0	0	0	0	0	0	0	0
3.1	4.3	0.6	2343	1324	0	0	0	0	0	0	0	0	0	0	0	0	0	0	0
4.3	5.3	5.3	21165	25562	306	0	0	0	0	0	0	0	0	0	0	0	0	0	0
5.3	6.2	14.3	17770	51668	20543	308	0	0	0	0	0	0	0	0	0	0	0	0	0
6.2	7.1	26.4	14666	38973	58152	8922	0	0	0	0	0	0	0	0	0	0	0	0	0
7.1	7.9	41.6	15234	29453	52102	49055	6093	304	0	0	0	0	0	0	0	0	0	0	0
7.9	9.0	57.0	9918	21472	33742	43660	36809	7464	715	0	0	0	0	0	0	0	0	0	0
9.0	10.1	75.9	7894	21221	26655	37214	39675	36189	17120	2768	307	0	0	0	0	0	0	0	0
10.1	11.1	85.4	3062	8167	11945	14497	15621	15314	13579	9188	3369	714	0	0	0	0	0	0	0
11.1	12.1	91.3	1672	4094	6034	7374	8208	8467	8121	6955	4845	2120	822	0	0	0	0	0	0
12.1	13.2	95.2	981	2185	3140	3986	4659	4948	4947	4726	4117	3062	2318	215	0	0	0	0	0
13.2	14.6	97.7	547	1038	1527	2122	2418	2633	2788	2754	2632	2385	3043	784	78	0	0	0	0
14.6	16.4	99.1	269	412	719	942	1069	1259	1312	1374	1358	1325	2246	1303	378	44	0	0	0
16.4	18.6	99.8	110	124	290	314	424	451	516	534	559	557	1072	908	544	197	43	3	0
18.6	21.0	100.0	32	32	71	86	106	126	132	151	154	162	327	314	268	187	86	27	5
		FCUM	9.8	30.3	51.9	68.7	80.2	87.9	92.9	95.7	97.4	98.5	99.5	99.8	99.9	100.0	100.0	100.0	100.0

4.3.4. Wave Climate

Predictions of maximum loads, load spectra for fatigue strength analyses, etc. require distributions of the significant seaway properties in individual ocean areas. The best sources for such statistics are computations of the seaway based on measured wind fields. ANEP-II (1983) gives such statistical data extensively for North Atlantic, North Sea, Baltic Sea, Mediterranean Sea, and Black Sea. Based on these data, Germanischer Lloyd derived distributions for $H_{1/3}$ and T_1 for all of the Atlantic between 50 and 60 longitudinal and the western Atlantic between 40 and 50 longitudinal (Table 4.2). The table is based on data for a period of 10 years. T_1 is the period corresponding to the center of gravity of the area under the sea spectrum. The modal period for this table is:

$$T_p = T_1/0.77$$

The values in the table give 10^6 the time share when T_1 was in the given time interval and $H_{1/3}$ in the interval denoted by its mean value, at an arbitrary point in the sea area. FCUM denotes the cumulated share in per cent. Similar tables can be derived from ANEP-II and other publications for special seaway directions, seasons, and other ocean areas.

Table 4.2 can also be used to approximate other ocean areas by comparing the wind field in the North Atlantic with the wind field in another ocean area, using data of Blendermann (1998), and employing the relation between wind and sea as given in the previous chapter.

4.4. Numerical Prediction of Ship Seakeeping

4.4.1. Overview of Computational Methods

If the effect of the wave amplitude on the ship seakeeping is significantly non-linear, there is little sense in investigating the ship in elementary waves, since these waves do not appear in nature and the non-linear reaction of the ship in natural seaways cannot be deduced from the reaction in elementary waves. In these non-linear cases, simulation in the time domain is the appropriate tool.

If the non-linearity is weak or moderate the seakeeping properties of a ship in natural seaways can be approximated by superposition of the reactions in elementary waves of different frequency and direction. In these cases, the accuracy can be enhanced by introducing some relatively simple corrections of the purely linear computations to account for force contributions depending quadratically on the water velocity or considering the time-dependent change of position and wetted surface of the ship, for example. Even if iterative corrections are applied the basic computations of the ship seakeeping is still based on its reaction in elementary waves, expressed by complex amplitudes of the ship reactions. The time dependency is then always assumed to be harmonic, i.e. sinusoidal.

For flows involving strong non-linearities, particularly breaking waves or green water on deck, free-surface RANSE simulations are the most appropriate tool. Such simulations have entered industry practice to an increasing extent since the year 2000.

In practice, potential flow solvers still dominate in seakeeping predictions. The most frequent application is the computation of the linear seakeeping properties of a ship in elementary waves. In addition to the assumption for Euler solvers potential flow assumes that the flow is irrotational. This is no major loss in the physical model, because rotation is created by the water adhering to the hull and this information is already lost in the Euler flow model. Relevant for practical applications is that potential flow solvers are much faster than Euler and RANSE solvers, because potential flows have to solve only one linear differential equation instead of four non-linear coupled differential equations. Also, potential flow solvers are usually based on boundary element methods and need only to discretize the boundaries of the domain, not the whole fluid space. This reduces the effort in grid generation considerably. On the other hand, potential flow methods require a simple, continuous free surface. Flows involving breaking waves and splashes cannot be analyzed properly by potential flow methods.

In reality, viscosity is significant in seakeeping, especially if the boundary layer separates periodically from the hull. This is definitely the case for roll and yaw motions. In practice, empirical corrections are introduced. Also, for flow separation at sharp edges in the aftbody (e.g. vertical sterns, rudder, or transoms) a Kutta condition is usually employed to enforce a smooth detachment of the flow from the relevant edge.

The theoretical basics and boundary conditions of linear potential methods for ship seakeeping are treated extensively in the literature, e.g. by Newman (1978). Therefore, we can limit ourselves here to a description of the fundamental results important to the naval architect.

The ship flow in elementary waves is described in a coordinate system moving with ship speed in the x direction, but not following its periodic motions. The derivatives of the potential give the velocity of water relative to such a coordinate system. The total velocity potential is decomposed:

$$\phi^t = (-Vx + \phi^s) + (\phi^w + \phi^I) \tag{4.45}$$

with
ϕ^t potential of total flow
$-Vx$ potential of (downstream) uniform flow with ship speed V
ϕ^s potential of the steady flow disturbance
ϕ^w potential of the undisturbed wave as given at the end of Section 4.3.1
ϕ^I remaining unsteady potential.

The first parenthesis describes the steady (time-independent) flow, the second parenthesis the periodic flow due to sea waves. The potentials can be superimposed, since the fundamental field equation (Laplace equation, describing continuity of mass) is linear with respect to ϕ^t:

$$\Delta\phi^t = \left(\frac{\partial^2}{\partial x^2} + \frac{\partial^2}{\partial y^2} + \frac{\partial^2}{\partial z^2}\right)\phi^t \tag{4.46}$$

Various approximations can be used for ϕ^s and ϕ^I which affect computational effort and accuracy of results. The most important linear methods can be classified as follows:

- *Strip method.* Strip methods are the standard tool for ship seakeeping computations. They omit ϕ^s completely and approximate ϕ^I in each strip x = constant, independently of the other strips. Thus in essence the three-dimensional problem is reduced to a set (e.g. typically 10–30) of two-dimensional boundary value problems. This also requires a simplification of the actual free surface condition. The method originated in the late 1950s with the work of Korvin-Kroukovsky and Jacobs. Most of today's strip methods are variations of the strip method proposed by Salvesen, Tuck, and Faltinsen (1970). These are sometimes also called STF strip methods where the first letter of each author is taken to form the abbreviation. The two-dimensional problem for each strip can be solved analytically or by panel methods, which are the two-dimensional equivalents of the three-dimensional methods described below. The analytical approaches use conform mapping to transform semicircles to cross-sections resembling ship sections (Lewis sections). Although this transformation is limited and, for example, submerged bulbous bow sections cannot be represented in satisfactory approximation, this approach still yields for many ships results of similar quality as strip methods based on panel methods (close-fit approach). A close-fit approach (panel method) to solve the two-dimensional problem is described on the website. Strip methods are – despite inherent theoretical shortcomings – fast, cheap, and for most problems sufficiently accurate. However, this depends on many details. Insufficient accuracy of strip methods often cited in the literature is often due to the particular implementation of a code and not due to the strip method in principle. But, at least in their conventional form, strip methods fail (as do most other computational methods) for waves shorter than perhaps one-third of the ship length. Therefore, the added resistance in short waves (being considerable for ships with a blunt waterline) can also only be estimated by strip methods if empirical corrections are introduced. Section 4.4.2 describes a linear strip method in more detail.

- *Unified theory.* Newman (1978) and Sclavounos developed at the MIT the 'unified theory' for slender bodies. Kashiwagi (1997) describes more recent developments of this theory. In essence, the theory uses the slenderness of the ship hull to justify a two-dimensional approach in the near field which is coupled to a three-dimensional flow in the far field. The far-field flow is generated by distributing singularities along the centerline of the ship. This approach is theoretically applicable to all frequencies, hence 'unified'. Despite its better theoretical foundation, unified theories failed to give significantly and consistently better results than strip theories for real ship geometries. The method therefore failed to be accepted in practice.
- *'High-speed strip theory' (HSST).* Several authors have contributed to the high-speed strip theory after the initial work of Chapman (1975). A review of work since then can be found in Kashiwagi (1997). HSST usually computes the ship motions in an elementary wave using linear potential theory. The method is often called the 2.5*d* or 2*d*+*t* method, since it considers the effect of upstream sections on the flow at a point x, but not the effect of downstream sections. Starting at the bow, the flow problem is solved for individual strips (sections) $x =$ constant. The boundary conditions at the free surface and the hull (strip contour) are used to determine the wave elevation and the velocity potential at the free surface and the hull. Derivatives in the longitudinal direction are computed as numerical differences to the upstream strip which has been computed in the previous step. The computation marches downstream from strip to strip and ends at the stern resp. just before the transom. HSST is the appropriate tool for fast ships with Froude numbers $F_n > 0.4$. For lower Froude numbers, it is inappropriate.
- *Green function method (GFM).* ISSC (1994) gives a literature review of these methods. GFM distributes panels on the average wetted surface (usually for calm-water floating position neglecting dynamical trim and sinkage and the steady wave profile) or on a slightly submerged surface inside the hull. The velocity potential of each panel (Green function) fulfills automatically the Laplace equation, the radiation condition (waves propagate in the right direction) and a simplified free-surface condition (omitting the ϕ^s completely). The unknown (either source strength or potential) is determined for each element by solving a linear system of equations such that for each panel at one point the no-penetration condition on the hull (zero normal velocity) is fulfilled. The various methods differ primarily in the way the Green function is computed. This involves the numerical evaluation of complicated integrals from 0 to ∞ with highly oscillating integrands. Some GFM approaches formulate the boundary conditions on the ship under consideration of the forward speed, but evaluate the Green function only at zero speed. This saves a lot of computational effort, but cannot be justified physically and it is not

recommended. As an alternative to the solution in the frequency domain (for excitation by elementary waves), GFM may also be formulated in the time domain (for impulsive excitation). This avoids the evaluation of highly oscillating integrands, but introduces other difficulties related to the proper treatment of time history of the flow in so-called convolution integrals. Both frequency and time domain solutions can be superimposed to give the response to arbitrary excitation, e.g. by natural seaway, assuming that the problem is linear. All GFMs are fundamentally restricted to simplifications in the treatment of ϕ^s. Usually ϕ^s is completely omitted, which is questionable for usual ship hulls. It will introduce, especially in the bow region, larger errors in predicting local pressures.

- *Rankine singularity method (RSM).* Bertram and Yasukawa (1996) give an extensive overview of these methods covering both frequency and time domains. RSMs, in principle, capture ϕ^s completely and also more complicated boundary conditions on the free surface and the hull. In summary, they offer the option for the best approximation of the seakeeping problem within potential theory. This comes at a price. Both ship hull and the free surface in the near field around the ship have to be discretized by panels. Capturing all waves while avoiding unphysical reflections of the waves at the outer (artificial) boundary of the computational domain poses the main problem for RSMs. Since the early 1990s, various RSMs for ship seakeeping have been developed. By the end of the 1990s, the time-domain SWAN code (SWAN = Ship Wave ANalysis) of MIT was the first such code to be used commercially.
- *Combined RSM–GFM approach.* GFMs are fundamentally limited in capturing the physics when the steady flow differs considerably from uniform flow, i.e. in the near field. RSMs have fundamental problems in capturing the radiation condition for low τ values. Both methods can be combined to overcome the individual shortcomings and to combine their strengths. This is the idea behind combined approaches. These are described as 'Combined Boundary Integral Equation Methods' by the Japanese, and as 'hybrid methods' by Americans. Initially only hybrid methods were used which matched near-field RSM solutions directly to far-field GFM solutions by introducing vertical control surfaces at the outer boundary of the near field. The solutions are matched by requiring that the potential and its normal derivative are continuous at the control surface between near field and far field. In principle methods with overlapping regions also appear possible.

4.4.2. Strip Method

This section presents the most important formulae for a linear frequency-domain strip method for slender ships in elementary waves. The formulae will be given without derivation. For a more extensive coverage of the theoretical background, the reader is referred to Newman (1978).

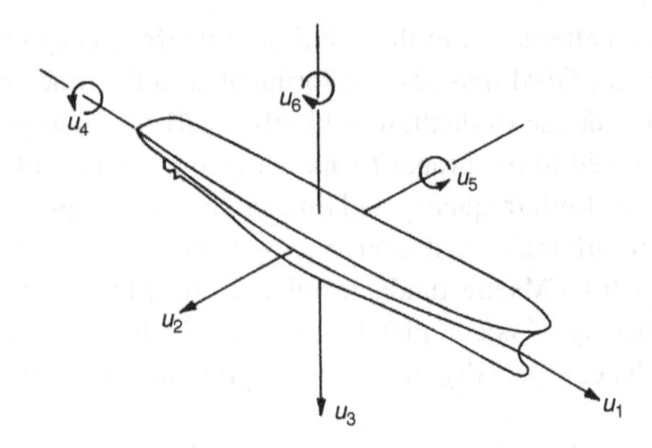

Figure 4.15:
Six degrees of freedom for motions

Two coordinate systems are used:

- The ship-fixed system x, y, z, with axes pointing from amidships forward, to starboard and downwards. In this system, the ship's center of gravity is time-independent x_g, y_g, z_g.
- The inertial system ξ, η, ζ. This system follows the steady forward motion of the ship with speed V and coincides in the time average with the ship-fixed system.

The main purpose of the strip method is to compute the ship's rigid-body motions, i.e. the three translations of the origin of ship-fixed system in the ξ, η, ζ direction and the three rotations around these axes. We denote (Fig. 4.15):

u_1	surge	u_4	roll
u_2	sway	u_5	pitch
u_3	heave	u_6	yaw

The motions are combined in a six-component vector $\vec{u}$. The forces and moments acting on the ship are similarly combined in a six-component vector $\vec{F}$. $\vec{u}$ and $\vec{F}$ are harmonic functions of time t oscillating with encounter frequency ω_e:

$$\vec{F} = \mathrm{Re}(\widehat{\vec{F}} e^{i\omega_e t}) \text{ and } \vec{u} = \mathrm{Re}(\widehat{\vec{u}} e^{i\omega_e t}) \tag{4.47}$$

The fundamental equation of motion is derived from $\vec{F} = M \cdot \ddot{\vec{u}}$:

$$[-\omega_e^2(M + A) + i\omega_e N + S]\ \widehat{\vec{u}} = \widehat{\vec{F}}_e \tag{4.48}$$

Here M, A, N, and S are real-valued 6×6 matrices. For mass distribution symmetrical to $y = 0$ the mass matrix M is:

$$M = \begin{bmatrix} m & 0 & 0 & 0 & mz_g & 0 \\ 0 & m & 0 & -mz_g & 0 & mx_g \\ 0 & 0 & m & 0 & -mx_g & 0 \\ 0 & -mz_g & 0 & \theta_{xx} & 0 & -\theta_{xz} \\ mz_g & 0 & -mx_g & 0 & \theta_{yy} & 0 \\ 0 & mx_g & 0 & -\theta_{xz} & 0 & \theta_{zz} \end{bmatrix} \tag{4.49}$$

The mass moments of inertia θ are related to the origin of the ship-fixed coordinate system:

$$\theta_{xx} = \int (y^2 + z^2)\,\mathrm{d}m\,; \quad \theta_{xz} = \int xz\,\mathrm{d}m\,;\ \text{etc.} \tag{4.50}$$

If we neglect contributions from a dry transom stern and other hydrodynamic forces due to the forward speed of the ship, the restoring forces matrix S is:

$$S = \begin{bmatrix} 0 & 0 & 0 & 0 & 0 & 0 \\ 0 & 0 & 0 & 0 & 0 & 0 \\ 0 & 0 & \rho g A_w & 0 & -\rho g A_w x_w & 0 \\ 0 & 0 & 0 & gm\overline{GM} & 0 & 0 \\ 0 & 0 & -\rho g A_w x_w & 0 & gm\overline{GM}_L & 0 \\ 0 & 0 & 0 & 0 & 0 & \theta_{zz}\omega_g^2 \end{bmatrix} \tag{4.51}$$

Here A_w is the waterline area, x_w the x coordinate of the center of the waterline, GM the metacentric height, GM_L the longitudinal metacentric height, and ω_g the circular natural frequency of yaw motions. ω_g is determined by the control characteristics of the autopilot and usually has little influence on the yaw motions in seaways. In computing GM_L, the moment of inertia is taken with respect to the origin of the coordinate system (usually amidships) and not, as usual, with respect to the center of the waterline. For corrections for dry transoms and unsymmetrical bodies reference is made to Söding (1987).

N is the damping matrix; it contains mainly the effect of the radiated waves. A is the added mass matrix. The decomposition of the force into hydrostatic (S) and hydrodynamic (A) components is somewhat arbitrary, especially for the ship with forward speed. Therefore, comparisons between computations and experiments are often based on the term $-\omega_e^2 A + S$.

$\vec{F}_e$ is the vector of exciting forces which a wave would exert on a ship fixed on its average position (diffraction problem). The exciting forces can be decomposed into a contribution due to the pressure distribution in the undisturbed incident wave (Froude–Krilov force) and the contribution due to the disturbance by the ship (diffraction force). Both contributions are of similar order of magnitude. To determine A and N, the flow due to the harmonic ship motions $\vec{u}$ must be computed (radiation problem). For small frequency of the motion (i.e. large wave length of the radiated waves), the hydrostatic forces dominate and the hydrodynamic forces are almost negligible. Therefore large relative errors in computing A and N are acceptable. For high frequencies, the crests of the waves radiated by the ship motions are near the ship almost parallel to the ship hull, i.e. predominantly in the longitudinal direction. Therefore the longitudinal velocity component of the radiated waves can be neglected. Then only the two-dimensional flow around the ship sections (strips) must be determined. This simplifies the computations a great deal.

For the diffraction problem (disturbance of the wave due to the ship hull), which determines the exciting forces, a similar reasoning does not hold: unlike radiation waves (due to ship motions), diffraction waves (due to partial reflection at the hull and distortion beyond the hull) form a similar angle (except for sign) with the hull as the incident wave. Therefore, for most incident waves, the diffraction flow will also feature considerable velocities in the longitudinal direction. These cannot be considered in a regular strip method, i.e. if we want to consider all strips as hydrodynamically independent. This error is partially compensated by computing the diffraction flow for wave frequency ω instead of encounter frequency ω_e, but a residual error remains. To avoid these residual errors, sometimes $\vec{F}_e$ is determined indirectly from the radiation potential following formulae of Newman (1965). However, these formulae are only valid if the waterline is also streamline. This is especially not true for ships with submerged transom sterns.

For the determination of the radiation and (usually also) diffraction (= exciting) forces, the two-dimensional flow around an infinite cylinder of the same cross-section as the ship at the considered position is solved (Fig. 4.16). The flow is generated by harmonic motions of the cylinder (radiation) or an incident wave (diffraction). Classical methods used analytical solutions based on multipole methods. Today, usually two-dimensional panel methods are preferred due to their (slightly) higher accuracy for realistic ship geometries. These two-dimensional panel methods can be based on GFM or RSM (see Chapter 3).

The flow and thus the pressure distribution depend on:

- for the radiation problem: hull shape, frequency ω_e, and direction of the motion (vertical, horizontal, rotational)
- for the diffraction problem: hull shape, wave frequency ω, and encounter angle μ.

For the radiation problem, we compute the pressure distributions for unit amplitude motions in one degree of freedom and set all other motions to zero and omit the incident wave. For the diffraction problem, we set all motions to zero and consider only the incident wave and its diffraction. We denote the resulting pressures by:

$\hat{p}_2$ for horizontal unit motion of the cylinder;
$\hat{p}_3$ for vertical unit motion of the cylinder;
$\hat{p}_4$ for rotational unit motion of the cylinder around the x axis;
$\hat{p}_0$ for the fixed cylinder in waves (only the pressure in the undisturbed wave);
$\hat{p}_7$ for the fixed cylinder in waves (only the disturbance of the pressure due to the body).

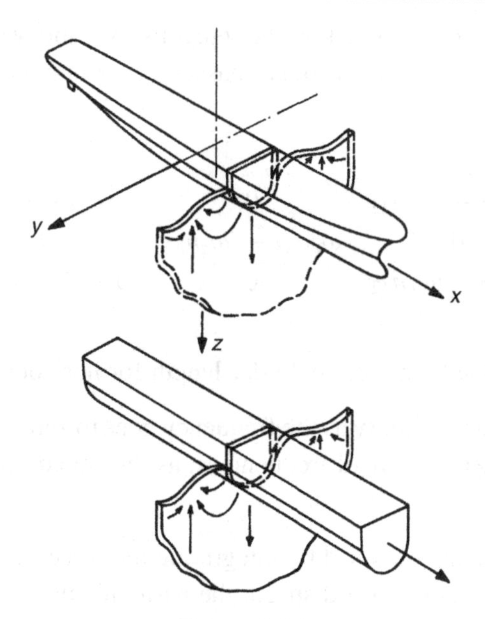

Figure 4.16:
Principle of strip method

Let the actual motions of the cylinder in a wave of amplitude $\widehat{h}_x$ be described by the complex amplitudes $\widehat{u}_{2,0x}$, $\widehat{u}_{3,0x}$, $\widehat{u}_{4,0x}$. Then the complex amplitude of the harmonic pressure is:

$$\widehat{p}_i = \widehat{p}_2\widehat{u}_{2,0x} + \widehat{p}_3\widehat{u}_{3,0x} + \widehat{p}_4\widehat{u}_{4,0x} + (\widehat{p}_0 + \widehat{p}_7)\widehat{h}_x \tag{4.52}$$

The amplitudes of the forces per length on the cylinder are obtained by integrating the pressure over the wetted surface of a cross-section (wetted circumference):

$$\begin{Bmatrix} \widehat{f}_2 \\ \widehat{f}_3 \\ \widehat{f}_4 \end{Bmatrix} = \int_0^l \begin{Bmatrix} n_2 \\ n_3 \\ yn_3 - zn_2 \end{Bmatrix} \cdot \widehat{p}_i \, d\ell = \int_0^l \begin{Bmatrix} n_2 \\ n_3 \\ yn_3 - zn_2 \end{Bmatrix} \cdot \begin{Bmatrix} \widehat{p}_2 \\ \widehat{p}_3 \\ \widehat{p}_4 \\ \widehat{p}_0 + \widehat{p}_7 \end{Bmatrix}^T d\ell \cdot \begin{Bmatrix} \widehat{u}_{2,0x} \\ \widehat{u}_{3,0x} \\ \widehat{u}_{4,0x} \\ \widehat{h}_x \end{Bmatrix} \tag{4.53}$$

$\{0, n_2, n_3\}$ is here in the inward unit normal on the cylinder surface. The index x in the last vector indicates that all quantities are taken at the longitudinal coordinate x at the ship, i.e. the position of the strip under consideration. ℓ is the circumferential length coordinate of the wetted contour. We can write the above equation in the form:

$$\widehat{\vec{f}} = \widehat{H} \cdot \left\{\widehat{u}_{2,0x}, \widehat{u}_{3,0x}, \widehat{u}_{4,0x}, \widehat{h}_x\right\}^T \tag{4.54}$$

The elements of the matrix $\widehat{H}$, obtained by the integrals over the wetted surface in the above original equation, can be interpreted as added masses a_{ij}, damping n_{ij} and exciting forces per wave amplitude $\widehat{f}_{ei}$:

$$\widehat{H} = \begin{bmatrix} \omega_e^2 a_{22} - i\omega_e n_{22} & 0 & \omega_e^2 a_{24} - i\omega_e n_{24} & \widehat{f}_{e2} \\ 0 & \omega_e^2 a_{33} - i\omega_e n_{33} & 0 & \widehat{f}_{e3} \\ \omega_e^2 a_{42} - i\omega_e n_{42} & 0 & \omega_e^2 a_{44} - i\omega_e n_{44} & \widehat{f}_{e4} \end{bmatrix} \tag{4.55}$$

For example, a_{22} is the added mass per cylinder length for horizontal motion.

The added mass tends towards infinity as the frequency goes to zero. However, the effect of the added mass also goes to zero for small frequencies, as the added mass is multiplied by the square of the frequency.

The forces on the total ship are obtained by integrating the forces per length (obtained for the strips) over the ship length. For forward speed, the harmonic pressure according to the linearized Bernoulli equation also contains a product of the constant ship speed $-V$ and the harmonic velocity component in the x direction. Also, the strip motions denoted by index x have to be converted to global ship motions in six degrees of freedom. This results in the global equation of motion:

$$\left[S - \omega_e^2(M + \widehat{B})\right] \widehat{u} = \widehat{E}h \tag{4.56}$$

$\widehat{B}$ is a complex matrix. Its real part is the added mass matrix A. Its imaginary part is the damping matrix N:

$$\omega_e^2 \widehat{B} = \omega_e^2 A - i\omega_e N = \int_L V(x)\cdot\left(1 + \frac{iV}{\omega_e}\frac{\partial}{\partial x}\right)\left(\widehat{H}_B \cdot W(x)\right) \mathrm{d}x \tag{4.57}$$

This equation can be used directly to compute $\widehat{B}$, e.g. using the trapezoidal rule for the integrals and numerical difference schemes for the differentiation in x. Alternatively, partial integration can remove the x derivatives. The new quantities in the above equations are defined as:

$$\widehat{\vec{E}} = \frac{\widehat{\vec{F}}_E}{h} = \int_L V(x)\cdot\left(\widehat{H}_E + \frac{iV}{\omega}\frac{\partial \widehat{H}_{E7}}{\partial x}\right) \mathrm{e}^{ikx\cos\mu}\, \mathrm{d}x \tag{4.58}$$

$$W(x) = \begin{bmatrix} 0 & 1 & 0 & t_x & 0 & x - V/(i\omega_e) \\ 0 & 0 & 1 & 0 & -x + V/(i\omega_e) & 0 \\ 0 & 0 & 0 & 1 & 0 & 0 \end{bmatrix} \tag{4.59}$$

t_x is the z coordinate (in the global ship system) of the origin of the reference system for a strip. (Often a strip reference system is chosen with origin in the waterline, while the global ship coordinate system may have its origin on the keel.)

$$V(x) = \begin{bmatrix} 1 & 0 & 0 & 0 & 0 \\ 0 & 1 & 0 & 0 & 0 \\ 0 & 0 & 1 & 0 & 0 \\ 0 & t_x & 0 & 1 & 0 \\ -t_x & 0 & -x & 0 & 1 \\ 0 & x & 0 & 0 & 0 \end{bmatrix} \tag{4.60}$$

$$\widehat{H}_B = \begin{bmatrix} 0 & 0 & 0 \\ \omega_e^2 a_{22} - i\omega_e n_{22} & 0 & \omega_e^2 a_{24} - i\omega_e n_{24} \\ 0 & \omega_e^2 a_{33} - i\omega_e n_{33} & 0 \\ \omega_e^2 a_{42} - i\omega_e n_{42} & 0 & \omega_e^2 a_{44} - i\omega_e n_{44} \\ 0 & 0 & 0 \end{bmatrix} \tag{4.61}$$

$$\widehat{H}_E = \begin{Bmatrix} -i\rho g k A_x \cos\mu \\ \widehat{f}_{e2} \\ \widehat{f}_{e3} \\ \widehat{f}_{e4} \\ -i\rho g k A_x s_x \cos\mu \end{Bmatrix} \tag{4.62}$$

A_x is the submerged section area at x; s_x is the vertical coordinate of the center of the submerged section area in the global system. $\widehat{H}_E$ contains both the Froude–Krilov part from the undisturbed wave (Index 0) and the diffraction part (Index 7), while $\widehat{H}_{E7}$ contains only the diffraction part.

The formulae for $\widehat{B}$ and $\widehat{\vec{E}}$ contain x derivatives. At locations x, where the ship cross-section changes suddenly (propeller aperture, vertical stem, submerged transom stern), this would result in extremely high forces per length. To a large extent, this is actually true at the bow, but not at the stern. If the cross-sections decrease rapidly there, the streamlines separate from the ship hull. The momentum (which equals added mass of the cross-section times velocity of the cross-section) then remains in the ship's wake while the above formulae would yield in strict application zero momentum behind the ship as the added mass is zero there. Therefore, the integration of the x derivatives over the ship length in the above formulae has to end at such locations of flow separation in the aftbody.

The global equation of motion above yields the vector of the response amplitude operators (RAOs) (= complex amplitude of reaction/wave amplitude) for the ship motions:

$$\frac{\hat{u}}{h} = \left(S - \omega_e^2[M + \hat{B}]\right)^{-1} \cdot \hat{\vec{E}} \tag{4.63}$$

The effect of rudder actions due to course deviations (yaw oscillations) was already considered in the matrix S. In addition, there are forces on the rudder (and thus the ship) due to ship motions (for centrally located rudders only due to sway, yaw, and roll) and due to the incident wave. Here it is customary to incorporate the rudder in the model of the rigid ship filling the gaps between rudder and ship. (While this is sufficient for the computation of the ship motions, it is far too crude if the forces on the rudder in a seaway are to be computed.)

Accurate computation of the motions, pressures, internal forces, etc. requires further additions and corrections, e.g. to capture the influence of non-linear effects especially for roll motion, treatment of low encounter frequencies, influence of bilge keels, stabilizing fins, etc. The special and often empirical treatment of these effects differs in various strip methods. Details can be found in the relevant specialized literature.

4.4.3. Rankine Singularity Methods

Bertram and Yasukawa (1996) give an extensive survey of these methods. A linear frequency-domain method is described briefly here to exemplify the general approach.

In principle, RSM can consider the steady potential completely. If ϕ^s is completely captured the methods are called 'fully three-dimensional' to indicate that they capture both the steady and the harmonic flow three-dimensionally. In this case, first the 'fully non-linear' wave resistance problem is solved to determine ϕ^s and its derivatives, including second derivatives of ϕ^s on the hull. The solution also yields all other steady flow effects, namely dynamic trim and sinkage, steady wave profile on the hull, and the steady wave pattern on the free surface. Then the actual seakeeping computations can be performed considering the interaction between steady and harmonic flow components. The boundary conditions for ϕ^I are linearized with regard to wave amplitude h and quantities proportional to h, e.g. ship motions. The Laplace equation (mass conservation) is solved subject to the boundary conditions:

1. Water does not penetrate the hull.
2. Water does not penetrate the free surface.
3. At the free surface there is atmospheric pressure.
4. Far away from the ship, the flow is undisturbed.
5. Waves generated by the ship radiate away from the ship.

6. Waves generated by the ship are not reflected at the artificial boundary of the computational domain.
7. For antisymmetric motions (sway, roll, yaw), a Kutta condition is enforced on the stern.
8. Forces (and moments) not in equilibrium result in ship motions.

For $\tau = \omega_e V/g > 0.25$ waves generated by the ship travel only downstream, similar to the steady wave pattern. Thus also the same numerical techniques as for the steady wave resistance problem can be used to enforce proper radiation, e.g. shifting source elements relative to collocation points downstream. Values $\tau < 0.25$ appear especially in following waves. Various techniques have been proposed for this case, as discussed in Bertram and Yasukawa (1996). However, there is no easy and accurate way in the frequency domain. In the time domain, proper radiation follows automatically and numerical beaches have to be introduced to avoid reflection at the outer boundary of the computational domain.

We split here the six-component motion vector of the section on the strip method approach into two three-component vectors. $\vec{u} = \{u_1, u_2, u_3\}^T$ describes the translations, $\vec{\alpha} = \{\alpha_1, \alpha_2, \alpha_3\}^T$ the rotations. The velocity potential is again decomposed as in Section 4.4.1:

$$\phi^t = (-Vx + \phi^s) + (\phi^w + \phi^I) \tag{4.64}$$

The steady potential ϕ^s is determined first. Typically, a 'fully non-linear' wave resistance code employing higher-order panels is also used to determine second derivatives of the potential on the hull. Such higher-order panels are described in the section on boundary elements. ϕ^w is the incident wave as in Section 4.3:

$$\phi^w = \mathrm{Re}(-ic\hat{h}e^{-kz}e^{-ik(x\cos\mu - y\sin\mu)}e^{i\omega_e t}) \tag{4.65}$$

The wave amplitude is chosen to $\hat{h} = 1$. The remaining unknown potential ϕ^I is decomposed into diffraction and radiation components:

$$\phi^I = \phi^d + \sum_{i=1}^{6} \phi^i u_i \tag{4.66}$$

The boundary conditions 1–3 and 7 are numerically enforced in a collocation scheme, i.e. at selected individual points. The remaining boundary conditions are automatically fulfilled in a Rankine singularity method. Combining 2 and 3 yields the boundary condition on the free surface, to be fulfilled by the unsteady potential $\phi^w + \phi^I$:

$$\left(-\omega_e^2 + Bi\omega_e\right)\hat{\phi}^{(1)} + \left(\left[2i\omega_e + B\right]\nabla\phi^{(0)} + \vec{a}^{(0)} + \vec{a}^g\right)\nabla\hat{\phi}^{(1)} + \nabla\phi^{(0)}(\nabla\phi^{(0)}\nabla)\nabla\hat{\phi}^{(1)} = 0 \tag{4.67}$$

with:

$$\phi^{(0)} = -Vx + \phi^s \quad \text{steady potential} \tag{4.68}$$

$$\vec{a}^{(0)} = (\nabla\phi^{(0)}\nabla)\nabla\phi^{(0)} \quad \text{steady particle acceleration} \tag{4.69}$$

$$\vec{a}^g = \vec{a} - \{0, 0, g\}^T \tag{4.70}$$

$$B = -\frac{1}{a_3^g}\frac{\partial(\nabla\phi^{(0)}\vec{a}^g)}{\partial z} \tag{4.71}$$

$$\nabla = \{\partial/\partial x, \partial/\partial y, \partial/\partial z\}^T \tag{4.72}$$

The boundary condition 1 yields on the ship hull:

$$\vec{n}\nabla\hat{\phi}^{(1)} + \hat{\vec{u}}(\vec{m} - i\omega_e\vec{n}) + \hat{\vec{\alpha}}[\vec{x} \times (\vec{m} - i\omega_e\vec{n}) + \vec{n} \times \nabla\phi^{(0)}] = 0 \tag{4.73}$$

Here the m-terms have been introduced:

$$\vec{m} = (\vec{n}\nabla)\nabla\phi^{(0)} \tag{4.74}$$

Vectors $\vec{n}$ and $\vec{x}$ are to be taken in the ship-fixed system.

The diffraction potential ϕ^d and the six radiation potentials ϕ^i are determined in a panel method that can employ regular first-order panels. The panels are distributed on the hull and on (or above) the free surface around the ship. The Kutta condition requires the introduction of additional dipole (or alternatively vortex) elements.

Test computations for a container ship (standard ITTC test case S-175) have shown a significant influence of the Kutta condition for sway, yaw, and roll motions for small encounter frequencies.

To determine ϕ^d, all motions (u_i, $i = 1$ to 6) are set to zero. To determine the ϕ^i, the corresponding u_i is set to 1, all other motion amplitudes, ϕ^d and ϕ^w to zero. Then the boundary conditions form a system of linear equations for the unknown element strengths which is solved, for example, by Gauss elimination. Once the element strengths are known, all potentials and derivatives can be computed.

For the computation of the total potential ϕ^t, the motion amplitudes u_i remain to be determined. The necessary equations are supplied by the momentum equations:

$$m(\ddot{\vec{u}} + \ddot{\vec{\alpha}} \times \vec{x}_g) = -\vec{\alpha} \times \vec{G} + \int (p^{(1)} - \rho[\vec{u}\vec{a}^g + \vec{\alpha}(\vec{x} \times \vec{a}^g)])\vec{n}\,\mathrm{d}S \tag{4.75}$$

$$m(\vec{x}_g \times \ddot{\vec{u}}) + I\ddot{\vec{\alpha}} = -\vec{x}_g \times (\vec{\alpha} \times \vec{G}) + \int (p^{(1)} - \rho[\vec{u}\vec{a}^g + \vec{\alpha}(\vec{x} \times \vec{a}^g)]) \times (\vec{x} \times \vec{n})\,\mathrm{d}S \tag{4.76}$$

$G = gm$ is the ship's weight, $\vec{x}_g$ its center of gravity and I the matrix of the moments of inertia of the ship (without added masses) with respect to the coordinate system. I is the lower-right 3×3 sub-matrix of the 6×6 matrix M given in the section for the strip method.

The integrals extend over the average wetted surface of the ship. The harmonic pressure $p^{(1)}$ can be decomposed into parts due to the incident wave, due to diffraction, and due to radiation:

$$p^{(1)} = p^w + p^d + \sum_{i=1}^{6} p^i u_i \tag{4.77}$$

The pressures p^w, p^d and p^i, collectively denoted by p^j, are determined from the linearized Bernoulli equation as:

$$p^j = -\rho(\phi_t^j + \nabla\phi^{(0)}\nabla\phi^j) \tag{4.78}$$

The two momentum vector equations above form a linear system of equations for the six motions, u_i, which is easily solved.

The explicit consideration of the steady potential s changes the results for computed heave and pitch motions for wavelengths of similar magnitude as the ship length – these are the wavelengths of predominant interest – by as much as 20–30% compared to total neglect. The results for standard test cases such as the Series-60 and the S-175 agree much better with experimental data for the 'fully three-dimensional' method. For the standard ITTC test case of the S-175 container ship, in most cases good agreement with experiments could be obtained (Fig. 4.17). Only for low encounter frequencies are the antisymmetric motions over-predicted, probably because viscous effects and autopilot were not modeled at all in the computations.

If the steady flow is approximated by double-body flow, similar results are obtained as long as the dynamic trim and sinkage are small. However, the computational effort is nearly the same.

Japanese experiments on a tanker model indicate that for full hulls the diffraction pressures in the forebody for short head waves ($\lambda/L = 0.3$ and 0.5) are predicted with errors of up to 50% if ϕ^s is neglected (as typically in GFM or strip methods). Computations with and without consideration of ϕ^s yield large differences in the pressures in the bow region for radiation in short waves and for diffraction in long waves.

4.4.4. Problems for Fast and Unconventional Ships

Seakeeping computations are problematic for fast and unconventional ships. Seakeeping plays a special role here, as fast ships are often passenger ferries, which need good seakeeping characteristics to attract passengers. This is the reason why, for instance, planing boats with their bad seakeeping are hardly ever used for commercial passenger transport. For fast cargo ships, the reduced speed in seaways can considerably influence transport

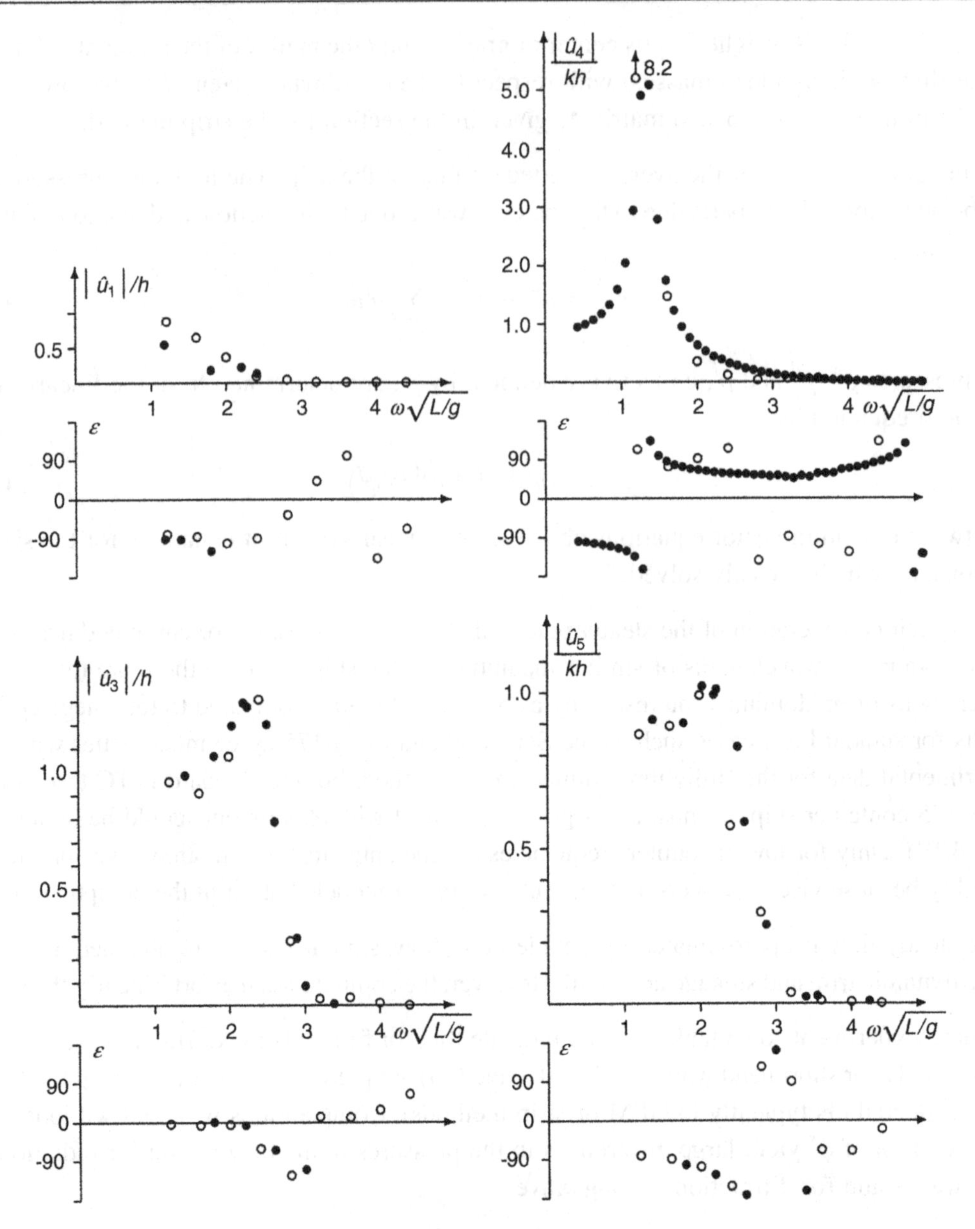

Figure 4.17:
Selected response amplitude operators of motions for the container ship S-175 at $F_n = 0.275$: experiment, computation surge (top left) for $\mu = 180°$; roll (top right) for $\mu = 120°$; heave (bottom left) for $\mu = 150°$; pitch (bottom right) for $\mu = 150°$

efficiency. A hull form, which is superior in calm water, may well become inferior in moderate seaways. Warships also often require good seakeeping to supply stable platforms for weapon systems, helicopters, or planes.

Unfortunately, computational methods for conventional ships are usually not at all or only with special modifications suitable for fast and unconventional ships. The special 'high-speed strip theory' (see Section 4.4.1) has been successfully applied in various forms to both fast monohulls and multihulls. Japanese validation studies showed that for a fast monohull with transom stern the HSST fared much better than both conventional strip methods and three-dimensional GFM and RSM. However, the conventional strip methods and the three-dimensional methods did not use any special treatment of the large transom stern of the test case. This impairs the validity of the conclusions. Researchers at the MIT have shown that at least for time-domain RSM the treatment of transom sterns is possible and also yields good results for fast ships, albeit at a much higher computational effort than the HSST. In most cases, HSST should yield the best cost–benefit ratio for fast ships.

It is often claimed in the literature that conventional strip methods are only suitable for low ship speeds. However, benchmark tests show that strip methods can yield good predictions of motion RAOs up to Froude numbers $F_n \approx 0.6$, provided that proper care is taken and the dynamic trim and sinkage and the steady wave profile at the hull are included to define the average submergence of the strips. The prediction of dynamic trim and sinkage is relatively easy for fast displacement ships, but difficult for planing boats. Neglecting these effects, i.e. computing for the calm-water wetted surface, may be a significant reason why often a lower Froude number limit of $F_n \approx 0.4$ is cited in the literature.

For catamarans, the interaction between the hulls plays an important role especially for low speeds. For design speed, the interaction is usually negligible in head seas. Three-dimensional methods (RSM, GFM) capture automatically the interaction as both hulls are simultaneously modeled. The very slender form of the demihulls introduces smaller errors for GFM catamaran computations than for monohulls. Both RSM and GFM applications to catamarans can be found in the literature, usually for simplified research geometries. Strip methods require special modifications to capture, at least in good approximation, the hull interaction, namely multiple reflection of radiation and diffraction waves. Simply using the hydrodynamic coefficients for the two-dimensional flow between the two cross-sections leads to strong overestimation of the interaction for $V > 0$.

Seakeeping computations for air-cushioned vehicles and surface effect ships are particularly difficult due to additional problems:

- The flexible skirts deform under the changing air cushion pressure and the contact with the free surface. Thus the effective cushion area and its center of gravity change.
- The flow and the pressure in the cushion contain unsteady parts which depend strongly on the average gap between free surface and skirts.
- The dynamics of fans (and their motors) influences the ship motions.

In particular the narrow gaps between skirts and free surface result in a strongly non-linear behavior that so far excludes accurate predictions.

4.4.5. Further Quantities in Regular Waves

Within a linear theory, the velocity and acceleration RAOs can be directly derived, once the motion RAOs are determined. The relative motion between a point on the ship and the water surface is important to evaluate the danger of slamming or water on deck. The RAOs for relative motion should incorporate the effect of diffraction and radiation, which is again quite simple once the RAOs for the ship motions are determined. However, effects of flared hull shape with outward forming spray for heave motion cannot be modeled properly within a linear theory, because these depend non-linearly on the relative motion. In practice, the section flare is important for estimating the amount of water on deck.

Internal forces on the ship hull (longitudinal, transverse, and vertical forces, torsional, transverse, and longitudinal bending moments) can also be determined relatively easily for known motions. The pressures are then only integrated up to a given cross-section instead of over the whole ship length. (Within a strip method approach, this also includes the matrix of restoring forces S, which contains implicitly many hydrostatic pressure terms.) Also, the mass forces (in matrix M) should only be considered up to the given location x of the cross-section. Stresses in the hull can then be derived from the internal forces. However, care must be taken that the moments are transformed to the neutral axis of the 'beam' ship hull. Also, stresses in the hull are often of interest for extreme loads where linear theory should no longer be applied.

The longitudinal force on the ship in a seaway is to first order within a linear theory also a harmonically oscillating quantity. The time average of this quantity is zero. However, in practice the ship experiences a significantly non-zero added resistance in seaways. This added resistance (and similarly the transverse drift force) can be estimated using linear theory. Two main contributions appear:

- Second-order pressure contributions are integrated over the average wetted surface.
- First-order pressure contributions are integrated over the difference between average and instantaneous wetted surface; this yields an integral over the contour of the water-plane.

If the steady flow contribution is completely retained (as in some three-dimensional BEM), the resulting expression for the added resistance is rather complicated and also involves second derivatives of the potential on the hull. Usually this formula is simplified assuming:

- uniform flow as the steady base flow;
- dropping a term involving x-derivatives of the flow;
- considering only heave and pitch as main contributions to added resistance.

4.4.6. Ship Responses in Stationary Seaway

Here the issue is how to get statistically significant properties in natural seaways from a response amplitude operator $Y_r(\omega,\mu)$ in elementary waves for an arbitrary response r depending linearly on wave amplitude. The seaway is assumed to be stationary with known spectrum $S_\zeta(\omega,\mu)$.

Since the spectrum is a representation of the distribution of the amplitude squared over ω and μ, and the RAO $\widehat{Y}_r$ is the complex ratio of r_A/ζ_A, the spectrum of r is given by:

$$S_r(\omega,\mu) = |Y_r(\omega,\mu)|^2 S_\zeta(\omega,\mu) \tag{4.79}$$

Values of r, chosen at a random point in time, follow a Gaussian distribution. The average of r is zero if we assume $r \sim \zeta_A$, i.e. in calm water $r = 0$. The probability density of randomly chosen r values is:

$$f(r) = \frac{1}{\sqrt{2\pi}\sigma_r} \exp\left(-\frac{r^2}{2\sigma_r^2}\right) \tag{4.80}$$

The variance σ_r^2 is obtained by adding the variances due to the elementary waves in which the natural seaway is decomposed:

$$\sigma_r^2 = \int_0^\infty \int_0^{2\pi} S_r(\omega,\mu)\, \mathrm{d}\mu\, \mathrm{d}\omega \tag{4.81}$$

The sum distribution corresponding to the frequency density $f(r)$ above is:

$$F(r) = \int_{-\infty}^{r} f(\rho)\, \mathrm{d}\rho = \frac{1}{2}[1 + \phi(r/\sigma_r)] \tag{4.82}$$

The probability integral ϕ is defined as:

$$\phi = \frac{2}{\sqrt{2\pi}} \int_{-\infty}^{x} \mathrm{e}^{-t^2/2}\, \mathrm{d}t \tag{4.83}$$

$F(r)$ gives the percentage of time when a response (in the long-term average) is less or equal to a given limit r. $1 - F(r)$ is then the corresponding percentage of time when the limit r is exceeded.

More often the distribution of the amplitudes of r is of interest. We define here the amplitude of r (differing from some authors) as the maximum of r between two following upward zero crossings (where $r = 0$ and $\dot{r} > 0$). The amplitudes of r are denoted by r_A. They have approximately (except for extremely 'broad' spectra) the following probability density:

$$f(r_A) = \frac{r_A}{\sigma_r^2} \exp\left(-\frac{r_A^2}{2\sigma_r^2}\right) \tag{4.84}$$

The corresponding sum distribution is:

$$F(r_A) = 1 - \exp\left(-\frac{r_A^2}{2\sigma_r^2}\right) \tag{4.85}$$

σ_r follows again from Eq. (4.81). The formula for $F(r_A)$ describes a so-called Rayleigh distribution. The probability that a randomly chosen amplitude of the response r exceeds r_A is:

$$1 - F(r_A) = \exp\left(-\frac{r_A^2}{2\sigma_r^2}\right) \tag{4.86}$$

The average frequency (occurrences/time) of upward zero crossings is derived from the r spectrum to:

$$f_0 = \frac{1}{2\pi\sigma_r} \sqrt{\int_0^{\infty} \int_0^{2\pi} \omega_e^2 S_r(\omega, \mu) \, d\mu \, d\omega} \tag{4.87}$$

Together with Eq. (4.86) this yields the average occurrence of r amplitudes which exceed a limit r_A during a period T:

$$z(r_A) = T f_0 \exp\left(-\frac{r_A^2}{2\sigma_r^2}\right) \tag{4.88}$$

Often we are interested in questions such as, 'What is the probability that during a period T a certain stress is exceeded in a structure or an opening is flooded?' Generally, the issue is then the probability $P_0(r_A)$ that during a period T the limit r_A is never exceeded. In other words, $P_0(r_A)$ is the probability that the maximum amplitude during the period T is less than r_A. This is given by the sum function of the distribution of the maximum of r during T. We make two assumptions:

- $z(r_A) << T f_0$; this is sufficiently well fulfilled for $r_A \geq 2\sigma_r$.
- An amplitude r_A is statistically nearly independent of its predecessors. This is true for most seakeeping responses, but not for the weakly damped amplitudes of elastic ship vibration excited by seaway, for example.

Under these assumptions we have:

$$P_0(r_A) = e^{-z(r_A)} \tag{4.89}$$

If we insert here the above expression for $z(r_A)$ we obtain the 'double' exponential distribution typical for the distribution of extreme values:

$$P_0(r_A) = e^{-Tf_0 \exp(-r_A^2/(2\sigma_r^2))} \tag{4.90}$$

The probability of exceedence is then $1 - P_0(r_A)$. Under the (far more limiting) assumption that $z(r_A) << 1$ we obtain the approximation:

$$1 - P_0(r_A) \approx z(r_A) \tag{4.91}$$

The equations for $P_0(r_A)$ assume neither a linear correlation of the response r from the wave amplitude nor a stationary seaway. They can therefore also be applied to results of non-linear simulations or long-term distributions.

4.4.7. Time-Domain Simulation Methods

The appropriate tools to investigate strongly non-linear ship reactions are simulations in the time domain. The seaway itself is usually linearized, i.e. computed as superposition of elementary waves. The frequencies of the individual elementary waves ω_j may not be integer multiples of a minimum frequency $\omega_{\min}$. In this case, the seaway would repeat itself after $2\pi/\omega_{\min}$ unlike a real natural seaway. Appropriate methods to choose the ω_j are:

- The ω_j are chosen such that the area under the sea spectrum between ω_j and ω_{j+1} is the same for all j. This results in constant amplitudes for all elementary waves regardless of frequency.
- The frequency interval of interest for the simulation is divided into intervals. These intervals are larger where S_ζ or the important RAOs are small and vice versa. In each interval a frequency ω_j is chosen randomly (based on constant probability distribution). One should not choose the same ω_j for all the L encounter angles under consideration. Rather each combination of frequency ω_j and encounter angle μ_l should be chosen anew and randomly.

The frequencies, encounter angles, and phase angles chosen before the simulation must be kept during the whole simulation.

Starting from a realistically chosen start position and velocity of the ship, the simulation computes in each time step the forces and moments acting from the moving water on the ship. The momentum equations for translations and rotations give the translational and rotational accelerations. Both are three-component vectors and are suitably expressed in a ship-fixed coordinate system. The momentum equations form a system of six scalar, coupled ordinary second-order differential equations. These can be transformed into a system of 12 first-order differential equations which can be solved by standard methods, e.g. fourth-order Runge–Kutta integration. This means that the ship position and velocity at the end of a small time interval, e.g. 1 second, are determined from the corresponding data at the beginning of this interval using the computed accelerations.

The forces and moments can be obtained by integrating the pressure distribution over the momentary wetted ship surface. Three-dimensional methods are usually too expensive for this purpose. Therefore modified strip methods are most frequently used. A problem is that the pressure distribution depends not only on the momentary position, velocity, and acceleration, but also on the history of the motion which is reflected in the wave pattern. This effect is especially strong for heave and pitch motions. In computations for the frequency domain, the historical effect is expressed in the frequency dependency of the added mass and damping. In time-domain simulations, we cannot consider a frequency dependency because there are many frequencies at the same time and the superposition principle does not hold. Therefore, the historical effect on the hydrodynamic forces and moments $\vec{F}$ is either expressed in convolution integrals ($\vec{u}$ contains here not only the ship motions, but also the incident waves):

$$\vec{F}(t) = \int_{-\infty}^{t} K(\tau)\vec{u}(\tau)\,d\tau \tag{4.92}$$

Or one considers 0 to n time derivatives of the forces $\vec{F}$ and 1 to $(n+1)$ time derivatives of the motions $\vec{u}$:

$$B_0\vec{F}(t) + B_1\dot{\vec{F}}(t) + B_2\ddot{\vec{F}}(t) + \ldots = A_0\dot{\vec{u}}(t) + A_1\ddot{\vec{u}}(t) + A_0\dddot{\vec{u}}(t) + \ldots \tag{4.93}$$

The matrix $K(\tau)$ in the first alternative and the scalars A_i, B_i in the second alternative are determined in potential flow computations for various sinkage and heel of the individual strips.

The second alternative is called the state model and appears to be far superior to the first alternative. Typical values for n are 2–4; for larger n, problems occur in the determination of the constants A_i and B_i resulting, for example, in numerically triggered oscillations. Pereira (1988) gives details of such a simulation method, namely SIMBEL. The simulation method has been extended considerably in the meantime and can also consider simultaneously the flow of water through a damaged hull, sloshing of water in the hull, or water on deck.

A far simpler and far faster approach is described, e.g., in Söding (1987). Here only the strongly non-linear surge and roll motions are determined by a direct solution of the equations of motion in the time-domain simulation (code ROLLS). The other four degrees of freedom are linearized and then treated similarly as the incident waves, i.e. they are computed from RAOs in the time domain. This is necessary to couple the four linear motions to the two non-linear motions. (Roll motions are often simulated as independent from the other motions, but this yields totally unrealistic results.) The restriction to surge and roll much simplifies the computation, because the history effect for these degrees of freedom is negligible. Extensive validation studies for this approach with model tests gave excellent agreement for capsizing of damaged ro-ro vessels drifting without forward speed in transverse waves (Chang and Blume 1998).

Simulations often aim to predict the average occurrence $z(r_A)$ of incidents where in a given period T a seakeeping response $r(t)$ exceeds a limit r_A. A new incident is then counted when after a previous incident another zero crossing of r occurred. The average occurrence is computed by multiple simulations with the characteristic data, but other random phases ε_{jl} for the superposition of the seaway. Alternatively, the simulation time can be chosen as nT and the number of occurrences can be divided by n. Both alternatives yield the same results except for random fluctuations.

Often seldom (extremely unlikely) incidents are of interest which would require simulation times of weeks to years to determine $z(r_A)$ directly if the occurrences are determined as described above. However, these incidents are expected predominantly in the presence of one or several particularly high waves. One can then reduce the required simulation time drastically by substituting the real seaway of significant wave height H_{real} by a seaway with larger significant wave height H_{sim}. The periods of both seaways shall be the same. The following relation between the incidents in the real seaway and in the simulated seaway exists (Söding 1987):

$$\frac{H_{sim}^2}{H_{real}^2} = \frac{\ln[z_{real}(r_A)/z(0)] + 1.25}{\ln[z_{sim}(r_A)/z(0)] + 1.25} \tag{4.94}$$

This equation is sufficiently accurate for $z_{sim}/z(0) < 0.03$. In practice, one determines in simulated seaway, e.g. with 1.5–2 times larger significant wave height, the occurrences $z_{sim}(r_A)$ and $z(0)$ by direct counting; then Eq. (4.94) is solved for the unknown $z_{real}(r_A)$:

$$z_{real}(r_A) = z(0)\exp\left(\frac{H_{sim}^2}{H_{real}^2}\{\ln[z_{sim}(r_A)/z(0)] + 1.25\} - 1.25\right) \tag{4.95}$$

4.4.8. Long-Term Distributions

Section 4.4.6 treated ship reactions in stationary seaway. This section will cover probability distributions of ship reactions r during periods T with changing sea spectra. A typical example for T is the total operational time of a ship. A quantity of interest is the average occurrence $z_L(r_A)$ of cases when the reaction $r(t)$ exceeds the limit r_A. The average can be thought of as the average over many hypothetical realizations, e.g. many equivalently operated sister ships.

First, one determines the occurrence $z(r_A;H_{1/3},T_p,\mu_0)$ of exceeding the limit in a stationary seaway with characteristics $H_{1/3}$, T_p, and μ_0 during total time T. (See Section 4.4.6 for linear ship reactions and Section 4.4.7 for non-linear ship reactions.) The weighted average of the occurrences in various seaways is formed. The weighing factor is the probability $p(H_{1/3}, T_p, \mu_0)$ that the ship encounters the specific seaway:

$$z_L(r_A) = \sum_{\text{all } H_{1/3}} \sum_{\text{all } T_p} \sum_{\text{all } \mu_0} z(r_A; H_{1/3}, T_p, \mu_0) \cdot p(H_{1/3}, T_p, \mu_0) \tag{4.96}$$

Usually, for simplification, it is assumed that the ship encounters seaways with the same probability under n_μ encounter angles μ_0:

$$z_L(r_A) = \frac{1}{n_\mu} \sum_{\text{all } H_{1/3}} \sum_{\text{all } T_p} \sum_{i=1}^{n_\mu} z(r_A; H_{1/3}, T_p, \mu_{0i}) \cdot p(H_{1/3}, T_p) \tag{4.97}$$

The probability $p(H_{1/3}, T_p)$ for encountering a specific seaway can be estimated using data as given in Table 4.2. If the ship were to operate exclusively in the ocean area for Table 4.2, the table values (divided by 10^6) could be taken directly. This is not the case in practice and requires corrections. A customary correction then is to base the calculation only on 1/50 or 1/100 of the actual operating time of the ship. This correction considers, e.g.:

- The ship usually operates in areas with not quite so strong seaways as given in Table 4.2.
- The ship tries to avoid particularly strong seaways.
- The ship reduces speed or changes course relative to the dominant wave direction, if it cannot avoid a particularly strong seaway.
- Some exceedence of r_A is not important, e.g. for bending moments if they occur in load conditions when the ship has only a small calm-water bending moment.

The sum distribution of the amplitudes r_A, i.e. the probability that an amplitude r is less than a limit r_A, follows from z_L:

$$P_L(r_A) = 1 - \frac{z_L(r_A)}{z_L(0)} \tag{4.98}$$

$z_L(0)$ is the number of amplitudes during the considered period T. This distribution is used for seakeeping loads in fatigue strength analyses of the ship structure. It is often only slightly different from an exponential distribution, i.e. it has approximately the sum distribution:

$$P_L(r_A) = 1 - e^{-r_A/r_0} \tag{4.99}$$

r_0 is a constant describing the load intensity. (In fatigue strength analyses, often the logarithm of the exceedence probability $\log(1 - P_L)$ is plotted over r_A; since for an exponential distribution the logarithm results in a straight line, this is called a log-linear distribution.)

The probability distribution of the largest loads during the period T can be determined from (see Section 4.4.6 for the underlying assumptions):

$$P_0(r_A) = e^{-z(r_A)} \tag{4.100}$$

The long-term occurrence $z_L(r_A)$ of exceeding the limit r_A is inserted here for $z(r_A)$.

4.5. Slamming

In rough seas with large relative ship motion, slamming may occur with large water impact loads. Usually, slamming loads are much larger than other wave loads. Sometimes ships suffer local damage from the impact load or large-scale buckling on the deck. For high-speed ships, even if each impact load is small, frequent impact loads accelerate fatigue failures of hulls. Thus, slamming loads may threaten the safety of ships. The expansion of ship size and new concepts in fast ships have decreased relative rigidity, causing in some cases serious wrecks.

A rational and practical estimation method of wave impact loads is one of the most important prerequisites for safety design of ships and ocean structures. Wave impact has challenged many researchers since von Karman's work in 1929. Today, mechanisms of wave impacts are correctly understood for the two-dimensional case, and accurate impact load estimation is possible for the deterministic case. The long-term prediction of wave impact loads can also be given in the framework of linear stochastic theories. However, our knowledge on wave impact is still insufficient. A fully satisfactory theoretical treatment has been prevented so far by the complexity of the problem:

- Slamming is a strongly non-linear phenomenon, which is very sensitive to relative motion and contact angle between body and free surface.
- Predictions in natural seaways are inherently stochastic; slamming is a random process in reality.
- Since the duration of wave impact loads is very short, hydro-elastic effects are large.
- Air trapping may lead to compressible, partially supersonic flows where the flow in the water interacts with the flow in the air.

Most theories and numerical applications are for two-dimensional rigid bodies (infinite cylinders or bodies of rotational symmetry), but slamming in reality is a strongly three-dimensional phenomenon. We will here briefly review the most relevant theories. Further recommended literature includes:

- Tanizawa and Bertram (1998) for practical recommendations translated from the Kansai Society of Naval Architects, Japan.
- Mizoguchi and Tanizawa (1996) for stochastic slamming theories.

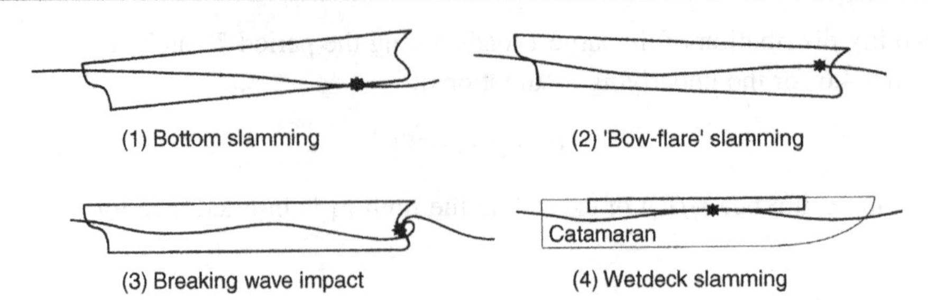

Figure 4.18:
Types of slamming impact of a ship

- Korobkin (1996) for theories with strong mathematical focus.
- SSC (1995) for a comprehensive compilation (more than 1000 references) of older slamming literature.

The wave impact caused by slamming can be roughly classified into four types (Fig. 4.18):

1. Bottom slamming occurs when emerged bottoms re-enter the water surface.
2. Bow-flare slamming occurs for high relative speed of bow-flare to the water surface.
3. Breaking wave impacts are generated by the superposition of incident wave and bow wave hitting the bow of a blunt ship even for small ship motion.
4. Wet-deck slamming occurs when the relative heaving amplitude is larger than the height of a catamaran's wet-deck.

Both bottom and bow-flare slamming occur typically in head seas with large pitching and heaving motions. All four water impacts are three-dimensional phenomena, but have been treated as two-dimensional for simplicity. For example, types 1 and 2 were idealized as two-dimensional wedge entry to the calm-water surface. Type 3 was also studied as a two-dimensional phenomenon similar to wave impact on breakwaters. We will therefore review two-dimensional theories first.

- *Linear slamming theories based on expanding thin-plate approximation*
  Classical theories approximate the fluid as inviscid, irrotational, incompressible, and free of surface tension. In addition, it is assumed that gravity effects are negligible. This allows a (predominantly) analytical treatment of the problem in the framework of potential theory.

  For bodies with small deadrise angle, the problem can be linearized. Von Karman (1929) was the first to study theoretically water impact (slamming). He idealized the impact as a two-dimensional wedge entry problem on the calm-water surface to estimate the water impact load on a seaplane during landing (Fig. 4.19). Mass, deadrise angle, and initial penetrating velocity of the wedge are denoted as m, β and V_0. Since the impact is so rapid, von Karman assumed very small water surface elevation during impact and negligible gravity effects. Then the added mass is approximately $m_v = 1/2\pi\rho c^2$. ρ is the water

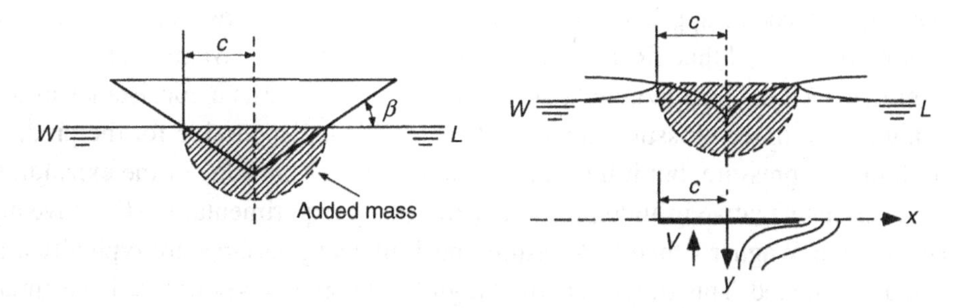

Figure 4.19:
Water impact models of von Karman (left) and Wagner (right)

density and c the half width of the wet area implicitly computed from $dc/dt = \cot\beta$. The momentum before the impact mV_0 must be equal to the sum of the wedge momentum mV and added mass momentum $m_v V$, yielding the impact load as:

$$P = \frac{V_0^2/\tan\beta}{\left(1 + \frac{\rho\pi c^2}{2m}\right)^3} \cdot \rho\pi c \tag{4.101}$$

Since von Karman's impact model is based on momentum conservation, it is usually referred to as momentum impact, and because it neglects the water surface elevation, the added mass and impact load are underestimated, particularly for small deadrise angle. Wagner derived a more realistic water impact theory in 1932. Although he assumed still small deadrise angles in his derivation, the theory was found to be unsuitable for $\beta < 3°$, since then air trapping and compressibility of water play an increasingly important role. If β is assumed small and gravity neglected, the flow under the wedge can be approximated by the flow around an expanding flat plate in uniform flow with velocity V (Fig. 4.19). Using this model, the velocity potential and its derivative with respect to y on the plate $y = 0^+$ is analytically given as:

$$\phi = \begin{cases} V\sqrt{c^2 - x^2} & \text{for } x < c \\ 0 & \text{for } x > c \end{cases} \tag{4.102}$$

$$\frac{\partial\phi}{\partial y} = \begin{cases} 0 & \text{for } x < c \\ V/\sqrt{1 - c^2/x^2} & \text{for } x > c \end{cases} \tag{4.103}$$

The time integral of the last equation gives the water surface elevation and the half width of the wetted area c. The impact pressure on the wedge is determined from Bernoulli's equation as:

$$\frac{p(x)}{\rho} = \frac{\partial\phi}{\partial t} - \frac{1}{2}(\nabla\phi)^2 = \sqrt{c^2 - x^2}\frac{dV}{dt} + V\frac{c}{\sqrt{c^2 - x^2}}\frac{dc}{dt} - \frac{1}{2}\frac{V^2 x^2}{c^2 - x^2} \tag{4.104}$$

Wagner's theory can be applied to arbitrarily shaped bodies as long as the deadrise angle is small, but not so small that air trapping plays a significant role. Wagner's theory is simple and useful, even if the linearization is sometimes criticized for its inconsistency as it retains a quadratic term in the pressure equation. This term is indispensable for the prediction of the peak impact pressure, but it introduces a singularity at the edge of the expanding plate ($x = \pm c$) giving negative infinite pressure there. Many experimental studies have checked the accuracy of Wagner's theory. Measured peak impact pressures are typically a little lower than estimated. This suggested that Wagner's theory gives conservative estimates for practical use. However, a correction is needed on the peak pressure measured by pressure gauges with finite gauge area. Special numerical FEM analyses of the local pressure in a pressure gauge can be used to correct measured data. The corrected peak pressures agree well with estimated values by Wagner's theory. Today, Wagner's theory is believed to give accurate peak impact pressure for practical use, albeit only for suitable hull forms with small deadrise angles.

The singularity of Wagner's theory can be removed taking spray into account. An 'inner' solution for the plate is asymptotically matched to an 'outer' solution of the spray region, as, for example, proposed by Watanabe in Japan in the mid-1980s (Fig. 4.20). The resulting equation for constant falling velocity is consistent and free from singularities. Despite this theoretical improvement, Watanabe's and Wagner's theories predict basically the same peak impact pressure (Fig. 4.21).

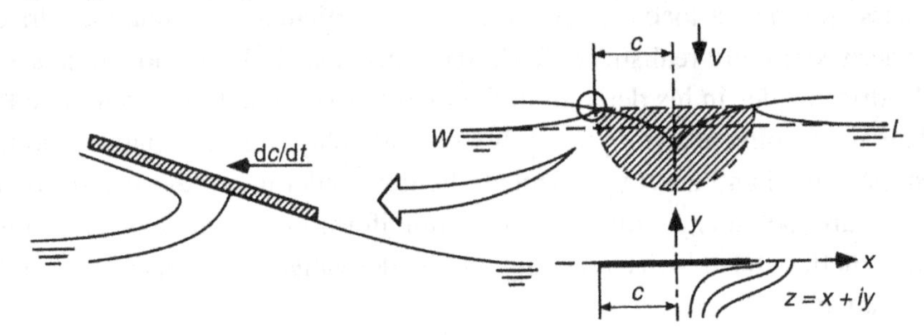

Figure 4.20:
Water impact model of Watanabe

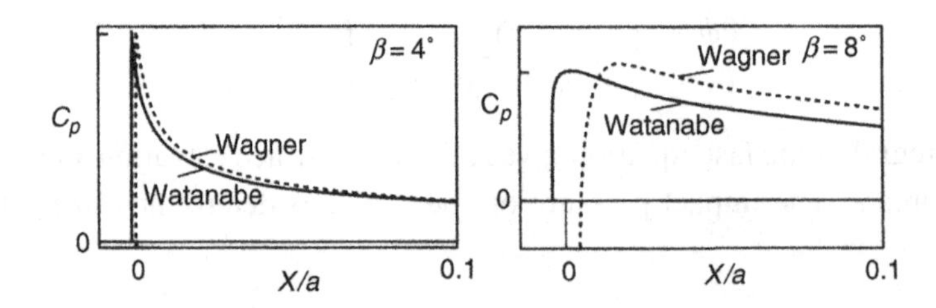

Figure 4.21:
Spatial impact pressure distribution

- *Simple non-linear slamming theories based on self-similar flow*
  We consider the flow near the vertex of a two-dimensional body immediately after water penetration. We can assume:
  - Near the vertex, the shape of the two-dimensional body can be approximated by a simple wedge.
  - Gravity accelerations are negligible compared to fluid accelerations due to the impact.
  - The velocity of the body V_0 is constant in the initial stage of the impact.

  Then the flow can be considered as self-similar depending only on x/V_0t and y/V_0t, where x, y are Cartesian coordinates and t is time. Russian scientists have converted the problem to a one-dimensional integral equation for $f(t)$. The resulting integral equation is so complicated that it cannot be solved analytically. However, numerically it has been solved by Faltinsen in Norway up to deadrise angles $\beta \geq 4°$. The peak impact pressure for $\beta = 4°$ was almost identical (0.31% difference) to the value given by Wagner's theory.
- *Slamming theories including air trapping*
  So far slamming theories have neglected the density of air, i.e. if a deformation of the free surface was considered at all it occurred only after the body penetrated the water surface. The reality is different. The body is preceded by an air cushion that displaces water already before the actual body entry. Air plays an even bigger role if air trapping occurs. This is especially the case for breaking wave impacts. In the 1930s, Bagnold performed pioneering work in the development of theories that consider this effect. Bagnold's impact model is simply constructed from added mass, a rigid wall, and a non-linear air cushion between them (Fig. 4.22). This model allows qualitative predictions of the relation between impact velocity V_0, air cushion thickness H, and peak impact pressure. For example, the peak impact pressure is proportional to V and $\sqrt{H}$ for slight impact and weak non-linearity of the air cushion; but for severe impact, the peak impact pressure is proportional to V^2 and H. These scaling laws were validated by subsequent experiments.

Trapped air bottom slamming is another typical impact with air cushion effect. For two-dimensional bodies, air trapping occurs for deadrise angles $\beta \leq 3°$. Chuang's (1967) experiment for two-dimensional wedges gave peak impact pressures as in Table 4.3. The

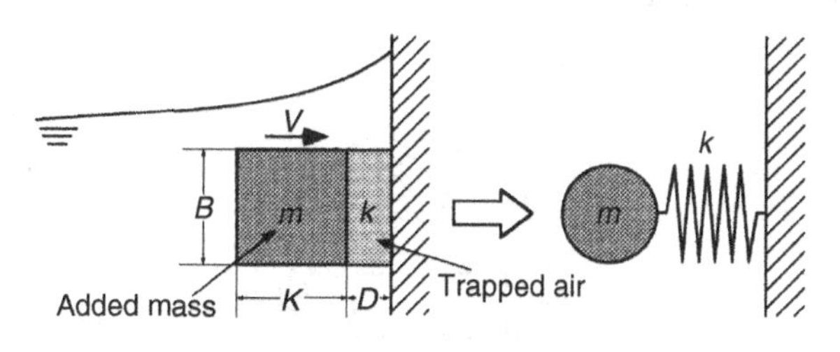

Figure 4.22:
Bagnold's model

Table 4.3: Chuang's (1967) relation for peak impact pressures

β	0°	1°	3°	6°	10°	15°	≥18°
P_{peak} (kPa)	$102V$	$115V^{1.4}$	$189V^{1.6}$	$64.5V^2$	$31V^2$	$17.8V^2$	Wagner's theory

impact velocity V is given in m/s. For $\beta = 0°$ air trapping is significant and the peak impact pressure is proportional to V. Increasing the deadrise angle reduces the amount of air trapping and thus the non-linearity. For practical use, the peak impact pressure is usually assumed to be proportional to V^2 for all β. This results in a conservative estimate.

Johnson and Verhagen developed two-dimensional theories for bottom impact with air trapping considering one-dimensional air flow between water surface and bottom to estimate the water surface distortion and the trapped air volume (Fig. 4.23).

The peak impact pressure thus estimated was much higher than measured. This disagreement results from the boundary condition at the edge of the flat bottom, where a jet emits to the open air. The theory assumes that the pressure at the edge is atmospheric pressure. This lets the air between water surface and bottom escape too easily, causing an underestimated trapped air volume. Experiments showed that the pressure is higher than atmospheric. Yamamoto has therefore proposed a modified model using a different boundary condition.

Experiments at the Japanese Ship Research Institute observed the trapped air impact with high-speed cameras and measured the initial thickness of air trapping. It was much thicker than the estimates of both Verhagen and Yamamoto. The reason is that a mixed area of air and water is formed by the high-speed air flow near the edge. Since the density of this mixed area is much higher than that of air, this area effectively chokes the air flow increasing air trapping.

The mechanism of wave impact with air trapping is in reality much more complicated. Viscosity of air, the effect of air leakage during compression, shock waves inside the air flow, and the complicated deformation of the free surface are all effects that may play an important role. Computational fluid dynamics may be the key to significant success here, but has not yet progressed sufficiently.

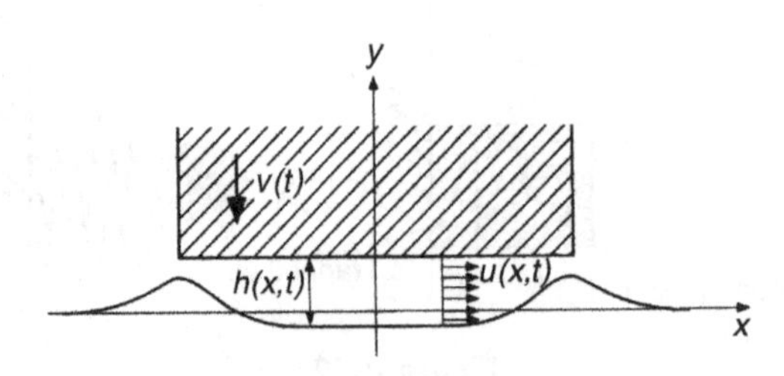

Figure 4.23:
One-dimensional air flow model of Verhagen

- *Effect of water compressibility*
  When a blunt body drops on calm water or a flat bottom drops on a smooth wave crest, usually no air trapping occurs. Nevertheless, one cannot simply use Wagner's theory, because at the top of such a blunt body or wave crest the relative angle between body and free surface becomes zero. Then both Wagner's and Watanabe's theories give infinite impact pressure. In reality, compressibility of liquid is important for a very short time at the initial stage of impact, when the expansion velocity of the wet surface dc/dt exceeds the speed of sound for water ($c_w \approx 1500$ m/s) producing a finite impact pressure. Korobkin (1996) developed two-dimensional theories which consider compressibility and free-surface deformation. For parabolic bodies dropping on the calm-water surface, he derived the impact pressure simply as $P = \rho c_w V$. Korobkin's theory is far more sophisticated, also yielding the time history of the pressure decay, but will not be treated here.
- *Three-dimensional slamming theories*
  All slamming theories treated so far were two-dimensional, i.e. they were limited to cross-sections (of infinite cylinders). Slamming for real ships is a strongly three-dimensional phenomenon due to, for example, pitch motion and cross-sections in the foreship changing rapidly in the longitudinal direction. Traditionally, approaches were used that obtain quasi three-dimensional solutions based on strip methods or high-speed strip methods. At the University of Michigan, Troesch developed a three-dimensional boundary element method for slamming. However, the method needs to simplify the physics of the process and the geometry of body and free surface and failed to show significant improvement over simpler strip-method approaches when compared to experiments.

Limiting oneself to axisymmetric bodies dropping vertically into the water makes the problem de facto two-dimensional. The study of three-dimensional water impact started from the simple extension of Wagner's theory to such cases. The water impact of a cone with small deadrise angle can then be treated in analogy to Wagner's theory as an expanding circular disk. A straightforward extension of Wagner's theory by Chuang over-predicts the peak impact pressure. Subsequent refinements of the theory resulted in a better estimate of the peak impact pressure:

$$p(r) = \frac{1}{2}\rho V^2 \left(\frac{2}{\pi}\right)^2 \left[\frac{4\cot\beta}{\sqrt{1 - r^2/c^2}} - \frac{r^2/c^2}{1 - r^2/c^2}\right] \tag{4.105}$$

r and c correspond to x and c in Fig. 4.19. This equation gives about 14% lower peak impact pressures than a straightforward extension of Wagner's theory. Experiments confirmed that the impact pressure on a cone is lower than that on a two-dimensional wedge of the same deadrise angle. So the three-dimensional effect reduces the impact pressure at least for convex bodies. This indicates that Wagner's theory gives conservative estimates for practical purposes. Since the impact on a ship hull is usually a very local phenomenon, Wagner's equation has also been used for three-dimensional surfaces using local relative velocity and angle between ship hull and water surface.

Watanabe (1986) extended his two-dimensional slamming theory to three-dimensional oblique impact of flat-bottomed ships. This theory was validated in experiments observing three-dimensional bottom slamming with a high-speed video camera and transparent models. Watanabe classified the slamming of flat-bottomed ships into three types:

1. Slamming due to inclined re-entry of the bottom. The impact pressure runs from stern to bow. No air trapping occurs.
2. Slamming due to vertical (orthogonal) re-entry of the bottom to a wave trough with large-scale air trapping.
3. Slamming due to vertical (orthogonal) re-entry of the bottom to a wave crest with only small-scale, local air trapping.

Type 1 (typical bottom impact observed for low ship speed) can be treated by Watanabe's three-dimensional theory. Type 3 (typical for short waves and high ship speed) corresponds to Chuang's theory for very small deadrise angle. Type 2 (also typical for short waves and high ship speed) corresponds to Bagnold's approach, but the air trapping and escaping mechanisms are different to simple two-dimensional models.

- *Hydro-elastic approaches in slamming*

  It is important to evaluate not only peak impact pressures but also structural responses to the impact, to consider the impact pressure in the design of marine structures. Whipping (large-scale, weakly dampened oscillations of the longitudinal bending moment) is a typical elastic response to impact. In the late 1960s and 1970s, slamming and whipping resulted in some spectacular shipwrecks, e.g. bulkers and container ships breaking amidships. The disasters triggered several research initiatives, especially in Japan, which eventually contributed considerably to the development of experimental and numerical techniques for the investigation of slamming and whipping.

Let us denote the slamming impact load as $Z(t)$ and the elastic response of a ship as $S(t)$. Assuming a linear relation between them, we can write:

$$S(t) = \int_0^\infty h(t-\tau)Z(\tau)\,\mathrm{d}\tau \tag{4.106}$$

$h(\tau)$ is the impulse response function of the structure. An appropriate modeling of the structure is indispensable to compute $h(\tau)$. For example, the large-scale (whipping) response can be modeled by a simple beam, whereas small-scale (local) effects can be modeled as panel responses. For complicated structures, FEM analyses determine $h(\tau)$.

When the duration of the impact load is of the same order as the natural period of the structure, the hydro-elastic interaction is strong. The impact load on the flexible bottom can be about twice that on the rigid bottom. Various theories have been developed, some including the effect of air trapping, but these theories are not powerful enough to explain experimental data quantitatively. Coupling free-surface RANSE solvers and

FEM to analyze both fluid and structure simultaneously should improve considerably our capability to analyze hydro-elastic slamming problems within the next decade.

- *CFD for slamming*
  For most practical impact problems, the body shape is complex, the effect of gravity is considerable, or the body is elastic. In such cases, analytical solutions are very difficult or even impossible. This leaves CFD as a tool. Due to the required computer resources, CFD applications to slamming appeared only since the 1980s. While the results of boundary element methods for water entry problems agree well with analytical results, it is doubtful whether they are really suited to this problem. Real progress is only likely with field methods. Various researchers have approached slamming problems, usually employing surface-capturing methods. The three-dimensional treatment of slamming has benefited greatly from the rapid increase in computing power. State-of-the-art analyses by 2010 used three-dimensional, free-surface RANSE simulations for rigid-body motions. These capture impact forces well enough for whipping analyses (hull girder vibration triggered by slamming impacts). Local pressure peaks are still not captured well, as local hydro-elasticity is not considered.

4.6. Roll Motion

4.6.1. Linear, Undamped Free Roll

A heeled ship in smooth water will return to its original upright position due to the restoring (or righting) moment $m \cdot g \cdot h(\varphi)$. However, due to its kinetic energy, the ship will roll beyond the upright position to a heel angle on the other side and from there back, etc. In the absence of damping, this oscillatory motion would continue forever. For small roll angles, the roll motion of such an undamped free roll motion in calm water is characterized by:

$$(m_{44} + a_{44}) \cdot \ddot{\varphi} + m \cdot g \cdot GM \cdot \varphi = 0 \qquad (4.107)$$

$m_{44} = \theta_{xx}$ is the mass moment of inertia for roll, a_{44} the added (hydrodynamic) mass moment of inertia, typically 10% of m_{44}. The natural roll frequency is thus:

$$\omega_n = \sqrt{\frac{m \cdot g \cdot GM}{m_{44} + a_{44}}} = \sqrt{\frac{g \cdot GM}{k_{xx}'^2}} \qquad (4.108)$$

The formula is valid up to roll angles $\varphi < 5°$. $k'_{xx} = \sqrt{(m_{44} + a_{44})/m}$ is the radius of inertia (with respect to the roll axis). The corresponding natural roll period (= period between two maximum positive roll angles) is $T_n = 2\pi/\omega_n$. Section 3.6 gives empirical formulae to estimate T_n. The relation for T_n is used to determine *GM* experimentally. The seaway changes the average metacentric height *GM*. In addition, larger roll angles introduce non-linear effects, changing the roll period considerably; e.g. the roll period tends towards infinity if the roll angle is close to angles where the righting lever is again zero.

For symmetric ships, within linear ship seakeeping theories, the roll motion is coupled only to yaw and sway motions. The roll axis (i.e. the axis where the sway and yaw motions disappear, leaving pure roll) is typically approximately halfway between waterline and center of gravity, with slightly higher values aft and lower values forward.

4.6.2. Capsizing in Waves

Few cases of capsizing are attributed directly to wave-excited roll motions, but capsizing has quite often been attributed to cargo shifts triggered by strong roll. While only numerical methods like non-linear strip methods can give detailed quantitative information, simplified considerations help in giving some quick estimates and general guidelines.

In regular waves from abeam, for wave length much longer than the ship width, the ship response is quasi-static. Within linear theory, the roll angle is given by:

$$|\varphi| = |\hat{u}_4| = \frac{1}{\sqrt{\left(1-\left[\frac{\omega_e}{\omega_n}\right]^2\right)^2+\left(2D\frac{\omega_e}{\omega_n}\right)^2}}\cdot k\cdot h \quad \text{with} \quad 2D = \frac{n_{44}}{(m_{44}+a_{44})\cdot\omega_n} \tag{4.109}$$

The response amplitude operator $|\varphi|/(kh)$ features a maximum for $D \le \sqrt{2}/2$ (at resonance) (Fig. 4.24). For $D > 1$ (very small *GM*), the damping prevents any oscillation. *GM* can then no longer be measured in a roll experiment as a roll period. Model tests show that the roll damping n_{44} is nearly constant up to roll angles of 10° and then increases.

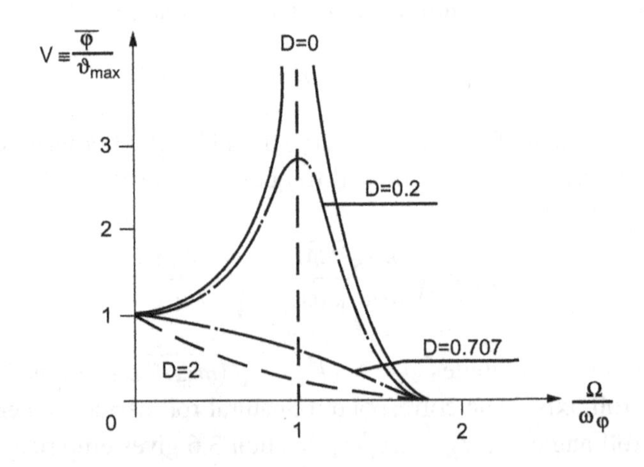

Figure 4.24:
Response amplitude operator for roll motion in waves from abeam

Non-linear effects often cannot be neglected in roll motion. For example, the restoring moment is only approximated by $m \cdot g \cdot GM$ for small angles, but for larger angles the real restoring moment curve has to be considered (Biran 2003). The solution of the resulting non-linear problem should be solved numerically. Different roll responses (roll angles) may then be obtained at a given exciting frequency, depending on whether the exciting frequency is approached from higher or from lower frequencies.

Following seas (and sometimes also head waves) may cause severe roll and even capsize for ships. In fact, following seas by themselves are more dangerous than beam seas. The resulting 'parametric excitation' can lead to severe rolling within a few roll periods, if the exciting frequency is near twice the natural roll frequency and metacentric heights vary greatly between hogging condition (ship in wave crest) and sagging condition (ship in wave trough). The righting lever in waves changes (for most ship hulls) with time, depending on the current waterline shape (Fig. 4.25). The slope of the curve at the origin is the metacentric height. Thus, for a ship in a seaway, there is no unique 'metacentric height' as for the ship in calm water. If people still use the word they implicitly mean the calm-water metacentric height.

Assuming a linear restoring moment, we write the fundamental differential equation for a free, undamped roll motion as:

$$(m_{44} + a_{44}) \cdot \ddot{\varphi} + m \cdot g \cdot GM \cdot \varphi = 0 \tag{4.110}$$

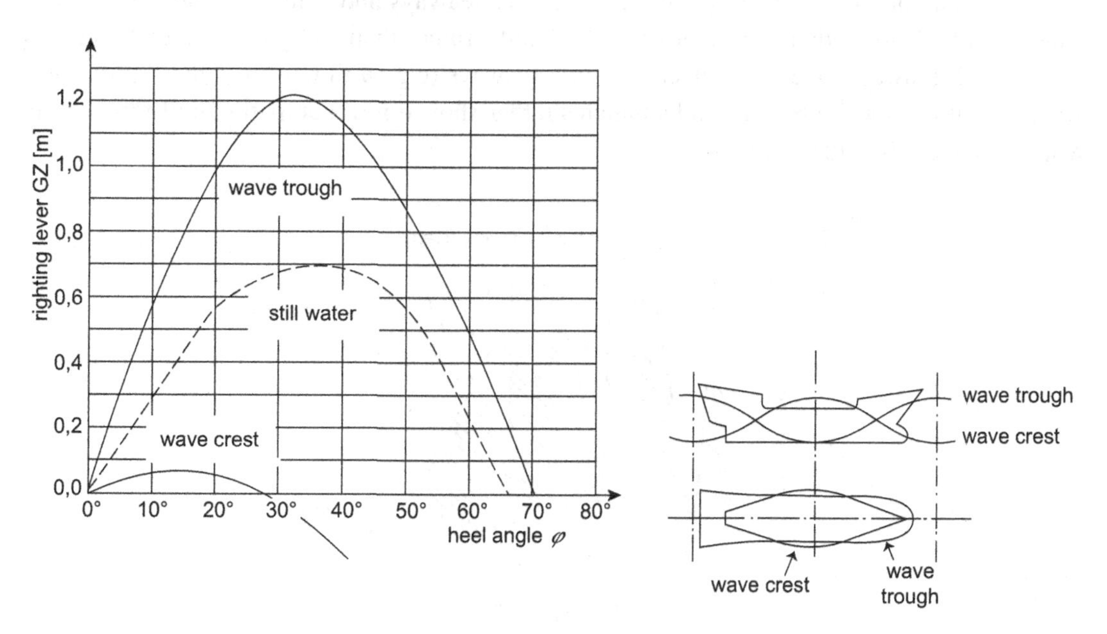

Figure 4.25:
Fluctuation of righting lever for ship in waves

The restoring moment now depends on a parameter, namely time t. The metacentric height is approximated to oscillate harmonically with exciting frequency ω_e:

$$GM(t) = GM_0 + \Delta GM \cdot \sin \omega_e t \tag{4.111}$$

This yields a so-called Mathieu equation:

$$\ddot{\varphi} + \omega_n^2 \cdot \left(1 + \frac{\Delta GM}{GM_0} \cdot \sin \omega_e t\right) \cdot \varphi = 0 \tag{4.112}$$

The solution of this differential equation features unstable areas where infinite amplitudes can be reached. For a ship, 'unstable response' means the ship capsizes. If ω_n/ω_e is close to a multiple of 1/2, the roll amplitudes can get infinitely large (resonance). The instability region increases as the fluctuation $\Delta GM/GM_0$ increases. Unstable areas can be plotted in a stability map (Strutt–Ince diagram, Fig. 4.26).

In reality, roll damping and non-linear restoring moments (righting moment curve) decrease the instability regions and roll amplitudes are no longer 'infinite'. With increasing frequency ratio, the amplitude decreases, making $\omega_n/\omega_e = 0.5$ (i.e. $\omega_e = 2\ \omega_n$) most critical.

The irregularity of real seaway makes parametric excitation less critical compared to regular waves in laboratory conditions, but still accidents due to parametric rolling have been reported at a rate showing that the phenomenon is not considered enough. A modern approach consists of selecting assorted time histories of representative seaways and using time-domain simulation tools to predict rolling of ships. Typically, rather than using long simulation times for 'normal' seaways, one then selects extreme seaways (e.g. with ten times the significant wave height for a given region) and simulates rather short times, comparing hull forms with respect to how often they capsize.

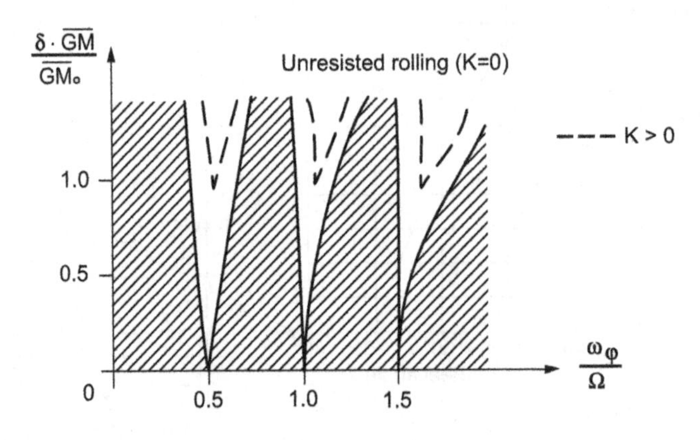

Figure 4.26:
Ince–Strutt diagram (stability map of Mathieu equation) plotting stable regions as shaded areas, linear restoring moment, without damping (solid lines) and with damping (dotted lines)

Parametric roll can be in practice a concrete danger if all the following conditions coincide:

1. The hull features large fluctuations of GM (between sagging and hogging conditions) (Fig. 4.4). Critical with respect to parametric roll excitation are hull forms with low block coefficient and large sectional flare at the ship ends like many modern hulls including container ships, ro-ro ferries, combatants, etc. The large change in waterline area between sagging and hogging then results in large changes of GM. The fluctuations are largest for wave lengths near ship length.
2. The ship speed U is such that the maximum of the encounter spectrum is near twice the natural roll frequency. For most ships, following and quartering seas are most critical. For large container ships, head waves can be critical. These ships are then excited to large pitch motions which increase the fluctuations of the metacentric height.

Large roll motions and accelerations, harmful to ship, cargo and humans (crew and passengers), may be avoided by:

(a) avoiding hull shapes with large difference in GM between ship in wave crest and ship in wave trough;
(b) shifting natural roll frequency to prevent resonance (changing GM);
(c) shifting exciting frequency (changing course or speed);
(d) increasing damping by active systems (foils, tanks).

Advance warning systems combining information on sea state and ship data with some simple rules are commercially available.

4.6.3. Roll Damping

Roll damping is usually weak. As a consequence, response amplitude operators for roll have a pronounced maximum near natural roll frequency. This is different for pitch and heave response amplitude operators which feature typically only weak and sometimes no local maxima. All computational methods, even simple strip methods, consider wave radiation and the associated damping. However, wave radiation is only for multi-hulls, an effective damping mechanism. For rotational bodies rolling around their axis of rotation, the wave radiation and associated damping is zero. For usual ship geometries, it is negligibly small.

The shear stress (tangential friction) on the hull is also negligible at zero speed. At forward speed, the damping moment can be estimated as:

$$M_{\text{roll},f} = R_f \frac{\omega_r u_4}{V} R^2 \tag{4.113}$$

R_f is the frictional resistance of the ship following ITTC'57, ω_r the actual roll frequency, V the ship speed, u_4 the roll amplitude, and R is the average distance of the hull surface to the roll axis.

The roll motion induces an oblique flow at the rudder (at center position). This in turn creates a rudder force which dampens the roll motion. The angle of attack is approximately

$$\alpha = \frac{\omega_r \cdot z}{V} \cdot u_4 \tag{4.114}$$

where z is the distance of a point on the rudder from the roll axis. We assume that the effects of wake (reducing the inflow speed) and propeller slipstream (increasing the inflow speed) cancel each other approximately. In addition, we neglect the oblique flow induced by the rolling ship and the propeller in oblique flow. We employ the usual estimate for the lift coefficient at the rudder (in rough approximation as this formula is valid for uniform flow with constant angle of attack over the height). Then we get for the roll damping moment due to the rudder:

$$M_{\text{roll,rudder}} = \omega_r \cdot u_4 \cdot \frac{\Lambda(\Lambda + 0.7)}{(\Lambda + 1.7)^2} \cdot \pi \cdot \rho \cdot V \cdot I_R \tag{4.115}$$

I_R is the areal moment of inertia of the rudder area with respect to the roll axis.

For a rectangular rudder, $I_R = c(z_2^3 - z_1^3)/3$; c is the chord length of the rudder. z_1 indicates the upper edge of the rudder, z_2 its lower edge. Λ is the rudder aspect ratio, c the chord length. Controlled rudder action can be used to actively dampen roll motions. Some course interference and added resistance must then be accepted. Similarly, Voith–Schneider propellers can be used to dampen roll motions. Unlike rudders, the VSP is also effective at zero forward speed.

Similar to the rudder, an immersed transom stern creates a roll damping moment for the ship at forward speed. We can use an equivalent formula as for the rudder, but employ the immersed beam instead of the rudder height. However, as the hull has water only on the underside, a factor 0.5 has to be applied:

$$M_{\text{roll,transom}} = \frac{1}{2} \omega_r \cdot u_4 \cdot \frac{\Lambda_t(\Lambda_t + 0.7)}{(\Lambda_t + 1.7)^2} \cdot \pi \cdot \rho \cdot V \cdot I_{R,t} \tag{4.116}$$

$\Lambda_t = B_t/(2L_{pp})$ (where B_t is the transom beam in the waterline) and $I_{R,t} = B_t^3 L_{pp}/12$.

Because the damping mechanisms discussed so far are rather weak (particularly at low speed), ships typically employ additional means to increase roll damping. These are discussed in the following.

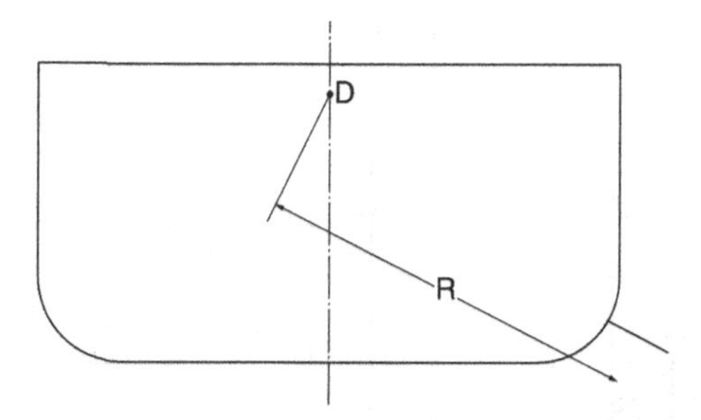

Figure 4.27:
Definition of R. D indicates the roll axis

Bilge keels are fitted on most ships. Bilge keels are narrow strips extending along the central part of the ship in the bilge region. They project no further than the width and depth of the ship to prevent contact damage. The effect of the bilge keels depends hardly on the ship speed. The damping moment can be estimated following:

$$M_{\text{roll,bilge}} = 2 \cdot \frac{\rho}{2} \cdot w^2 \cdot C_D \cdot l_{bk} \cdot h_{bk} \cdot R \qquad (4.117)$$

l_{bk} is the length of the bilge keel, h_{bk} its height, C_D a resistance coefficient. Figure 4.27 shows the definition of R. The factor 2 considers that we have bilge keels on port and on starboard. w is the transverse relative flow speed found approximately at half keel height if there were no bilge keel:

$$w \approx \omega_r \cdot u_4 \cdot R \cdot k \qquad (4.118)$$

The factor k considers the local flow changes in the bilge region (Fig. 4.28). Bilge keels are not very effective in comparison to even the passive roll damping of the rudder at design speed, but are still necessary for zero or low speed. Bilge keels also increase exciting forces (and resistance). For ships with effective alternative damping mechanisms (fins, tanks), one should then rather omit bilge keels. If the roll motion is largely suppressed, only the negative effect of increased exciting forces remains. The only argument left is then having a back-up in case of failure of the other more complex systems.

Blume (1979) gives the following values for C_D, depending on the amplitude of relative motion between bilge keel and water, $x_0 \approx u_4 \cdot R \cdot k$:

x_0/h_{bk}	0.4	0.8	1.2	1.6	2	3	4	6	8
C_D	11.7	9.6	7.8	7.0	6.5	5.0	4.3	3.6	3.2

Fin stabilizers are usually arranged symmetrically near the bilge, approximately amidships. The fins are tilted around an axis perpendicular to the ship to create a roll damping moment.

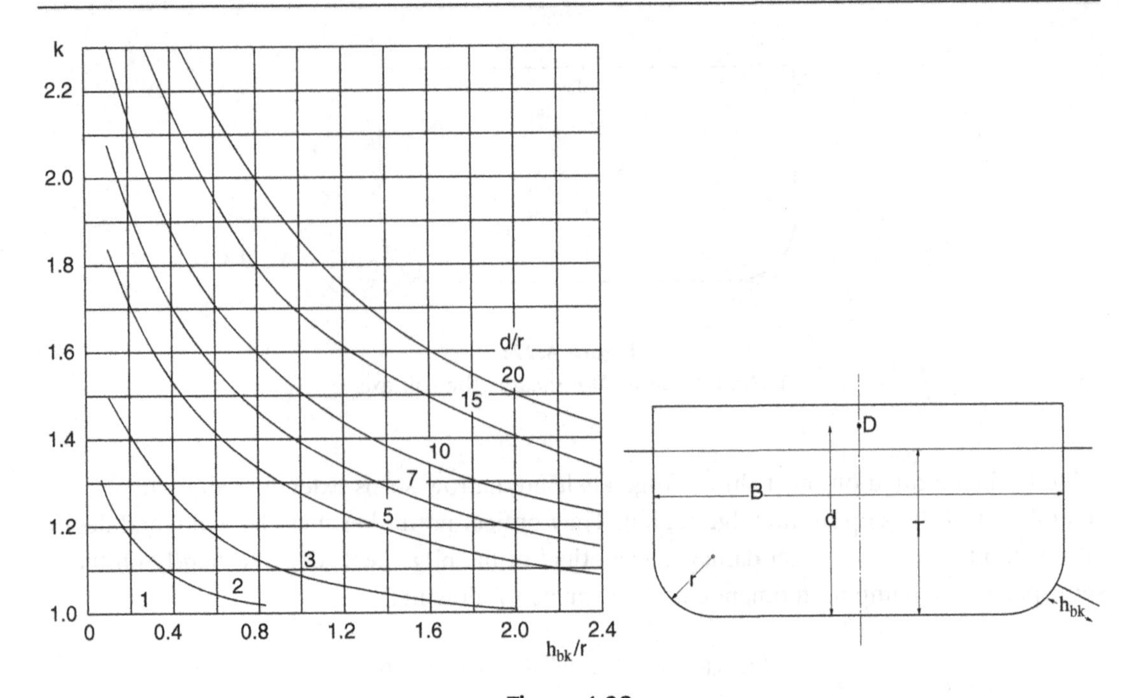

Figure 4.28:
Factor k for local flow speed in bilge region with dimensions as appearing in the diagram

The fins are usually retractable (to avoid damage in port). The damping moment furnished by a pair of fins can be estimated as

$$M_{\text{roll,fin}} = 2 \cdot \frac{\rho}{2} \cdot V^2 \cdot C_L \cdot A_{\text{fin}} \cdot R \tag{4.119}$$

R is the leverage as shown in Fig. 4.27, A_{fin} the fin area. The lift coefficient as function of angle of attack α can be estimated by:

$$C_L = 1.6 \left(2\pi \frac{\Lambda_{\text{eff}} \cdot (\Lambda_{\text{eff}} + 0.7)}{(\Lambda_{\text{eff}} + 1.7)^2} \cdot \sin \alpha + \sin \alpha \cdot |\sin \alpha| \cdot \cos \alpha \right) \tag{4.120}$$

Λ_{eff} is the effective side ratio of the fin. If there is (almost) no flow around the edge of the fin at the hull (due to small gap), we have $\Lambda_{\text{eff}} \approx 2\Lambda$. The factor 1.6 considers that stabilizing fins are usually flapped rudders where the aft flap turns by approximately 2δ if the main forward foil turns by δ. For single-foil fins the factor is 1. The maximum lift coefficient $C_{L,\max}$ lies typically between 3 and 3.5, provided that sufficiently high angles of attack are obtained.

The fin angle δ differs from the angle of attack α. Considering just the roll motion of the ship and assuming δ in phase with the roll velocity, we have:

$$\alpha = \delta + \arctan \frac{\omega_r \cdot u_4 \cdot R \cdot k}{V} \tag{4.121}$$

The fins are rather ineffective at low speed. At high speeds, a theoretical maximum $C_{L,\max}$ cannot be obtained due to structural overloading of the shaft and its supports. Therefore δ has to be limited to smaller values at higher speeds.

Roll stabilizing tanks are cheaper than fins and also effective at low speeds. This comes at the expense of larger weight (including the necessary water in the tanks), larger volume and a reduction of the metacentric height due to the free-surface effects. Unless special measures are taken, roll stabilizing tanks can also cause noise, which is particularly disturbing if the ship is transporting passengers. There are in principle two types of tanks: U-shaped tanks consist of two narrow tanks located at port and starboard, connected via the double bottom. Flume tanks are tanks with a free surface over the complete ship's width. In either case the tanks are partially filled, allowing the water to slosh from one side to the other. If the lowest natural frequency of the water sloshing coincides with the roll natural frequency of the ship and the ship is excited at this frequency by the waves the ship is excited to roll motion with a phase shift of 90° to the exciting waves and the sloshing water with another phase shift of 90° to the roll motion, yielding a total phase shift of 180° between exciting wave moment and damping tank moment. Ideally, the ship rests almost calm and the seaway excites only an oscillation of the water in the tank.

U-shaped tanks create a roll damping moment:

$$M_{\text{roll,tank}} = \rho \cdot g \cdot A_0 \cdot h_{\text{col}} \cdot B_1 \tag{4.122}$$

A_0 is the horizontal cross-section area of one side of the symmetric tank. The water level rises and falls by $\pm h_{\text{col}}$, without touching the top or the connecting pipe at the bottom. $B_1 < B$ is the horizontal distance of the tank centers on both sides. The natural frequency of such a tank can be estimated within the framework of a simple flow tube theory:

$$\omega_{n,\text{tank}} = \sqrt{\frac{2g}{A_0 \int_0^S \frac{1}{A} \, ds}} \tag{4.123}$$

$A(s)$ is the local cross-section area at the local one-dimensional flow coordinate s, S the total length of the flow tube. The formula shows that the dimensions of the connecting pipe influence the natural frequency. Once installed, different filling heights allow only small changes in natural frequency and come at the possible expense of reducing the maximum sloshing height h. A better strategy is therefore to design the tank such that the natural frequency is above the highest natural roll frequency and then retard the tank water motion in operation. The retard can be realized either by direct valves in the connecting pipe or (better) by controlling the air in the tanks above the water.

Flume tanks are typically rectangular tanks which extend over the whole ship width and are partially filled. They require more space and weight than U-shaped tanks, but can dampen wider ranges of frequencies without active control due to the effective damping in wave breaking. The natural frequency of a flume tank is approximately (for small water depth H compared to tank width b):

$$\omega_n = \frac{\pi\sqrt{gH}}{b} \tag{4.124}$$

CHAPTER 5

Vibrations

Chapter Outline

5.1. Introduction

Ship vibrations consider the ship hull and its structural members as *elastic* structures. Vibrations are important in the structural design due to the following design trends:

- Lightweight construction (with low stiffness and mass)
- Arrangement of living and working quarters near the propeller to optimize stowage space
- High propulsion power
- Small tip clearance of the propeller (to increase propeller efficiency)
- Fuel-efficient, slow-running main engines.

It has become standard practice to regulate vibration aspects for a newbuilding on a contractual basis. Therefore, vibration analyses are performed already during the preliminary or structural design stage for many ship types.

Practical Ship Hydrodynamics. DOI: 10.1016/B978-0-08-097150-6.10005-3

Vibrations cover the frequency range of 1 to 80 Hz according to ISO 6954. Table 5.1 shows typical natural frequencies of ship structures. Lower frequencies appear in 'ship motions' (classical ship seakeeping), higher frequencies in structure-borne noise (some overlap exists; noise may be perceived from 20 Hz upwards). Ship vibrations can become problematic if the exciting frequency is close to a natural frequency of the structure (resonance). Practical measures to avoid vibration problems include reduction of excitation amplitudes (e.g. elastic support for diesel engines) and avoidance of resonance at the lowest natural frequencies. As excitation frequencies of engines and propellers fluctuate with changing rpm, and natural frequencies of ship and some local structures (e.g. for tanks with various filling height) change with loading conditions, resonance at certain speeds often appears unavoidable.

Advances in computer methods have made many classical advanced beam models rather obsolete. Finite element analyses using rather large three-dimensional models are today standard tools, although simple beam theory allows understanding of certain typical relations and simple, fast (but often inaccurate) estimates.

5.2. Theory

Ship vibrations can generally be described by a linear equation of motion, allowing superposition of harmonic oscillations at different frequencies. The deflection (vector) $z(x)$ at point (vector) x of the vibrating structure follows then:

$$z(x,t) = \mathrm{Re}[\hat{z}(x,t)\mathrm{e}^{i\omega t}] \tag{5.1}$$

The circular frequency ω is connected to the frequency f by $\omega = 2\pi f$.

The motion equation for vibration problems contains the deflection and its first and second time derivatives:

$$K(z) + D(\dot{z}) + M(\ddot{z}) = F \tag{5.2}$$

Table 5.1: Natural frequency ranges in shipbuilding applications

	Min	Max
Global hull structures	0.5 Hz	10 Hz
Local structures	10.0 Hz	50 Hz
Deckhouse and aftbody structures	4.0 Hz	15 Hz
Structures above propeller	18.0 Hz	> 100 Hz
Large deck-panel structures	6.0 Hz	20 Hz
Engine foundations	20.0 Hz	> 100 Hz
Mast structures	7.0 Hz	21 Hz
Slow-running engines	4.5 Hz	12 Hz
Medium-speed engines, realistically supported	20.0 Hz	60 Hz
Medium-speed engines, mounted resiliently	1.5 Hz	7 Hz
Propeller shaft lines	4.0 Hz	19 Hz

$K(z)$ is the stiffness operator, $D(\dot{z})$ the damping operator, and $M(\ddot{z})$ the mass operator, which may include added mass terms. $F(x,t)$ is the excitation force. For usually assumed harmonic excitation, we have:

$$F(x,t) = \mathrm{Re}[\widehat{F}(x)e^{i\omega t}] \tag{5.3}$$

For linear operators K, D, and M, we then have:

$$K(\widehat{z}) + i\omega D(\widehat{z}) - \omega^2 M(\widehat{z}) = \widehat{F}(x) \tag{5.4}$$

The natural frequency of a beam with flexible support on both ends (Fig. 5.1) is:

$$f = \frac{\pi}{2\ell^2} \cdot \sqrt{\frac{EI}{\rho A}} \tag{5.5}$$

Equation (5.5) can also be used to estimate the lowest natural frequency of a longitudinal stiffener (with plate) supported by many equidistant transverse large stiffeners. The transverse stiffeners increase both the stiffness of the support and the vibrating mass and the effects cancel each other largely, making the above formula applicable for each segment of such a continuous beam.

For a single beam, the next highest natural frequency appears for the natural mode of a full sinusoidal wave between the supports. The beam then vibrates as in lowest natural mode for half the beam length. The natural frequencies are thus four times as high. Generally the natural frequencies of the single beam on two supports increase as 1:4:9:16, etc. This is not the case for a continuous beam on equidistant supports (of distance ℓ). Here, higher natural modes can appear with more nodes in only one or several segments. Therefore the next highest natural frequencies of very long continuous beams are only a little higher than the lowest natural frequency and above the lowest natural frequency there is practically no resonance-free region.

A beam under compression close to the buckling limit will deflect largely even under minimum transverse load. Thus, such a beam has vanishing bending stiffness. Correspondingly its natural frequency will approach zero. For longitudinal stiffeners in ships, the natural frequency is thus changed depending on the global bending moments due to static (weight/buoyancy) and dynamic (seaway) bending moments, which induce compressive stresses particularly at the top deck or bottom of the ship.

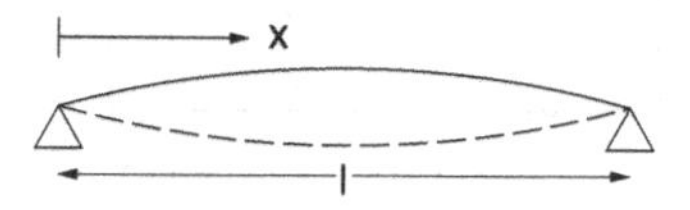

Figure 5.1:
Beam with flexible support on both ends

Consider a rectangular plane plate field (e.g. a plate between two longitudinal stiffeners and two transverse frames), considerably longer than wide (Fig. 5.2). Here the support from the longitudinal stiffeners (long side) is much stronger than the support from the transverse frames (short side). Each strip then vibrates almost like a beam with cross-section $b \cdot t$, where t is the plate thickness. However, unlike in a beam, the transverse contraction is suppressed by the adjacent strips. The effect is like an increased Young's modulus:

$$E^* = \frac{E}{1-\nu^2} \tag{5.6}$$

The material constants for steel are $E = 2.1 \cdot 10^{11}\ \mathrm{N/m^2}$, $\rho = 7800\ \mathrm{kg/m^3}$, $\nu = 0.3$. With $I = b \cdot t^3/12$ and $A = b \cdot t$, we then get the lowest natural frequency of a plate field:

$$f = \frac{\pi}{2 \cdot \ell^2} \cdot \sqrt{\frac{Et^2}{12\rho(1-\nu^2)}} \tag{5.7}$$

The remarks concerning continuous beams and influence of compressive stress apply likewise for the case of continuous plate fields, but the effect is less pronounced.

Plate curvature influences stiffness. The effect depends on the support (boundary conditions). Example: A rectangular plate of side ratio 1:5, flexibly supported at all edges, vibrates in lowest natural mode (half wave in each direction) and has an initial deflection of the same form as the vibration mode. The lowest natural frequency is then increased as follows:

Deflection at plate center/plate thickness	0.25	0.50	0.75	1.00
Natural frequency increased by factor	1.04	1.15	1.30	1.50

In stiffened plate fields (with longitudinal and transverse stiffeners), one should then consider the following options to avoid resonance:

- vibration of plate fields, as treated above;
- vibration of the longitudinal stiffeners (modeling the transverse frames as supports), as treated above;

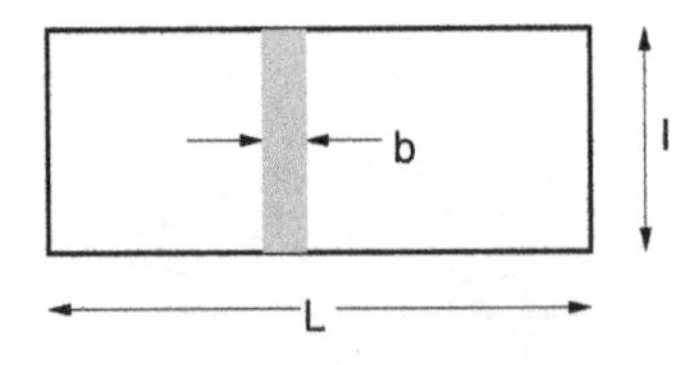

Figure 5.2:
Strip in a vibrating plate

- vibration of the transverse stiffeners. These are usually not slender enough to follow simple beam theory. Instead, bending and shear have to be considered together. Also, the support by the longitudinal stiffeners is usually too weak for the above simple models. More sophisticated analyses, typically employing finite element methods, are recommended.

The most important numerical techniques available to solve numerical vibration problems are:

- Finite element methods (FEM) approximate deflections by first-degree (for simple elements like trusses) and higher-order (for beams and plates) polynoms, typically Hermite polynoms.
- Spectral method, e.g. Doyle and Loh (1997): exact solutions are used for an idealized part of the structure, e.g. for a truss of constant section area.
- Rayleigh method, energy method: typically only one function is chosen to approximate the deflection, namely an estimated natural mode.

FEM are most popular, probably because the same software can be employed for static and vibration analysis, although spectral methods can be more efficient, particularly for higher frequencies.

5.3. Global Ship Hull Vibrations

The ship hull may perform global vibrations. The ship hull is usually (almost) symmetric to the centerplane in geometry and mass distribution. Then we can distinguish two types of natural modes:

- Vertical oscillations where points on the centerplane are displaced only within the centerplane. Vertical oscillations may induce significant longitudinal oscillations far away from the neutral layer (particularly on the bridge).
- Horizontal and torsional oscillations where points on the centerplane are displaced normal to the centerplane. For tankers, where the center of gravity and the shear center of cross-sections are close together, the horizontal vibrations are only weakly coupled to the torsional vibrations. Strong coupling is found in ships with large deck opening (container ships).

The lowest natural frequency for vertical vibrations appears for the two-node natural mode where two cross-sections remain (nearly) at rest. For a simple beam idealization, the next highest mode has three nodes (roughly twice the frequency), then four nodes (roughly three times the frequency), etc. A more detailed analysis shows further natural frequencies between these beam natural frequencies. These are due to local vibrations, e.g. of superstructures. The lowest natural frequency in vertical vibrations (two nodes) for ships of 150 m length lies typically around 1.5 Hz, for ships of 300 m length around 0.5 Hz. Because ships are typically wider than high, they are stiffer in the horizontal direction and the natural frequencies for horizontal vibrations are thus higher.

For a quick estimate, approximate formulae based on the simple beam models are useful. Lehmann (2000) gives for the lowest natural frequency (two nodes) for the steel ship hull girder in vertical bending:

$$f_0 = 1.62 \cdot 10^6 \cdot \sqrt{\frac{I}{\Delta_i \cdot L^3}} \tag{5.8}$$

The frequency is in Hz . I is the section moment of inertia amidships (m^4), Δ_i the displacement including a hydrodynamic mass (effect of surrounding water on vibrations) in kg:

$$\Delta_i = \left(1.2 + \frac{1}{3}\frac{B}{T}\right) \cdot \Delta \tag{5.9}$$

Δ is the mass of the ship (kg), L the ship length (m), B its width (m), T its draft (m).

For higher frequencies with n nodes we have:

$$f = (n-1)^{\mu} \cdot f_0 \tag{5.10}$$

The exponent μ accounts for the effect of shear stiffness:

$\mu = 1.02$ for tankers;
$\mu = 1.0$ for bulkers;
$\mu = 0.845$ for cargo ships.

These formulae cannot give more than a rough indication, particularly for the higher-order vibration modes.

Fast numerical computations of the ship hull vibrations employ beam models. The ship hull is then divided into beam segments. For each segment, mass per length, moments of inertia, etc. are taken as constant. The actual computation employs the method of transfer matrices. Vertical bending vibrations are relatively easy to analyze this way. They require 'only' the correct determination of bending stiffness and mass distribution. For horizontal vibrations, torsion and bending are strongly coupled, particularly for container ships. In practice, finite element programs are then employed, often using the services of classification societies.

The computations require longitudinal mass and stiffness distribution as input. The mass distribution considers the ship, the cargo and the hydrodynamic 'added' mass. The added mass reflects the effect of the surrounding water and depends on the frequency. Its determination is problematic. One can use estimates based on experience or employ sophisticated hydrodynamic simulations. Determination of the stiffness is also not trivial. Stress distributions in the stiffened bottom and deck plates depend on vibrational modes. Again either estimates based on experience or complex finite element analyses are employed. Estimates based on experience often work well. Of course the quality of the results depends on the input, which in turn depends on experience. The large classification societies usually have enough experience to give good estimates for estimating stiffness and added mass.

If the hull is modeled in (relatively) great detail in a finite element analysis, the effective width does not have to be specified explicitly. Then the added mass matrix is best determined for all degrees of freedom using special potential flow codes. The finite element models have typically 20 000 to 40 000 degrees of freedom (Fig. 5.3). The primary structural components including large web frames are typically modeled using plane stress elements. The grids are not fine enough to reflect explicitly smaller structural details, such as stiffeners. If considered at all, these stiffeners are approximated by increasing the plate thickness. This reflects only the effect on the membrane stresses (in the plane of the plate), but not the change in bending stiffness. These models yield 50–150 vibration natural modes in the range up to 20 Hz (examples are given in Fig. 5.4). The container ship features particularly low natural frequencies in torsional vibrations due to large deck-opening (= low torsional stiffness). After the analysis, one then has to check whether the chosen model for the predicted natural modes is appropriate. Consideration of bending stiffness of the deck grillages in the finite element model requires representation of transverse and longitudinal deck girders at least in the form of beam elements. Such models have typically 40 000 to 80 000 degrees of freedom (Fig. 5.5), yielding 300 to 500 natural modes in the range up to 20 Hz (examples given in Fig. 5.6).

The preparation of the finite element input (elements, associated values, added masses) involves considerable experience and man-time (Fricke 2002), typically outsourced to special consultants or classification societies. Natural frequencies change with loading conditions. Typical loading conditions (mass distributions) should be selected rather than extreme conditions.

Despite the considerable effort, the employed finite element models are still not really satisfactory. Differences between computed and on-board measured vibration amplitudes by a factor of 3 are not uncommon. The reasons for this disappointing performance are not completely clear. One factor is modeling errors for curved shells, found particularly at the ship ends. The stiffness of these shells with respect to longitudinal bending depends very much on the arrangement of internal bulkheads and stiffeners, as well as the longitudinals between these transverse structural elements. These finer details of the structure are lost in the 'coarse' finite element grids typically employed.

5.4. Vibrations of Local Structures

Resonance problems often appear for local ship structures. This can affect human comfort, but also induce fatigue problems of structures. The vibration analysis of these local structures is similar to that for the ship hull and nowadays is often based on finite element methods.

Figure 5.3:
FEM grids of some cargo ships. *Source: Germanischer Lloyd*

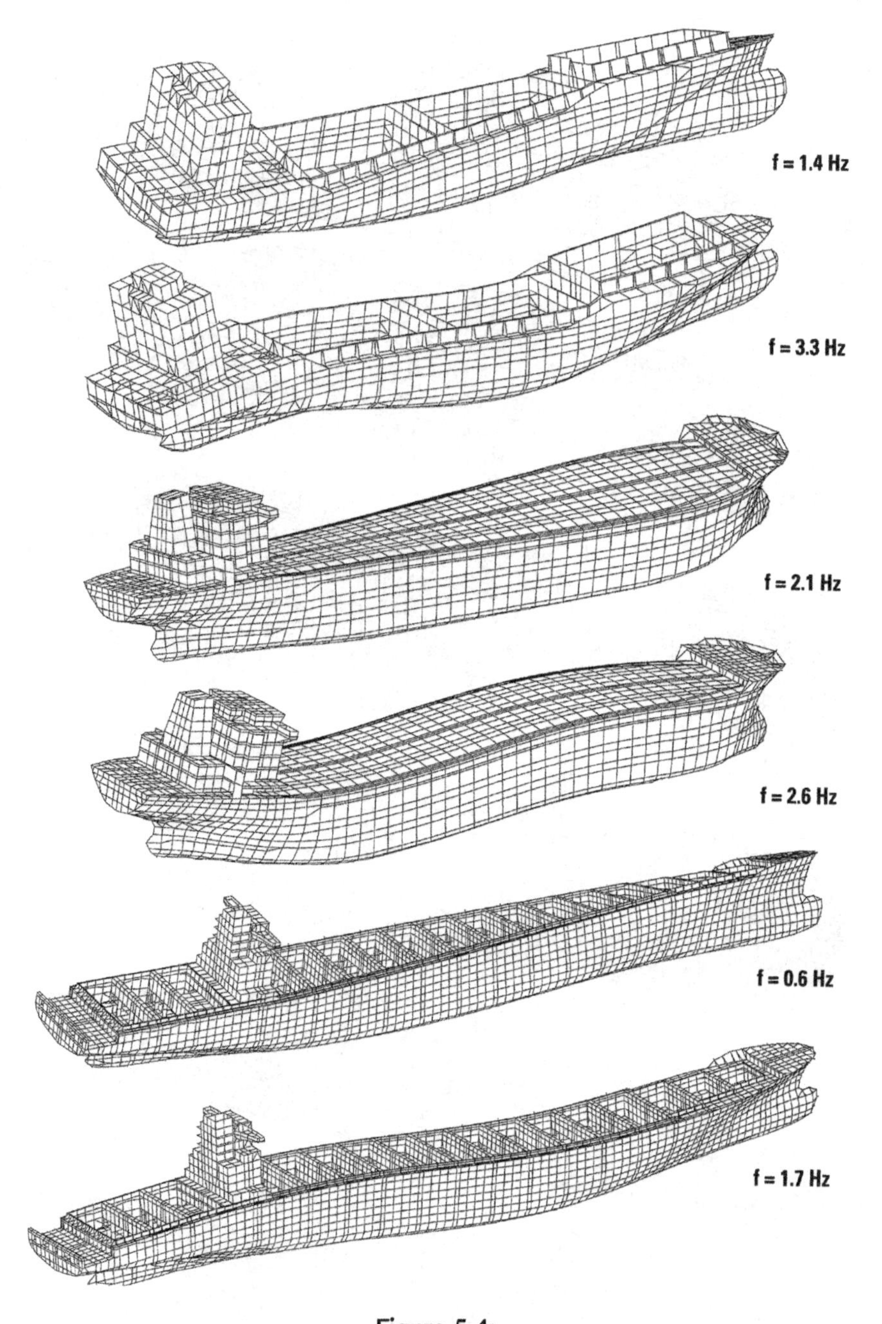

Figure 5.4:
Computed first torsional and second vertical vibration natural modes and corresponding natural frequencies for the ships in Fig. 5.3. *Source: Germanischer Lloyd*

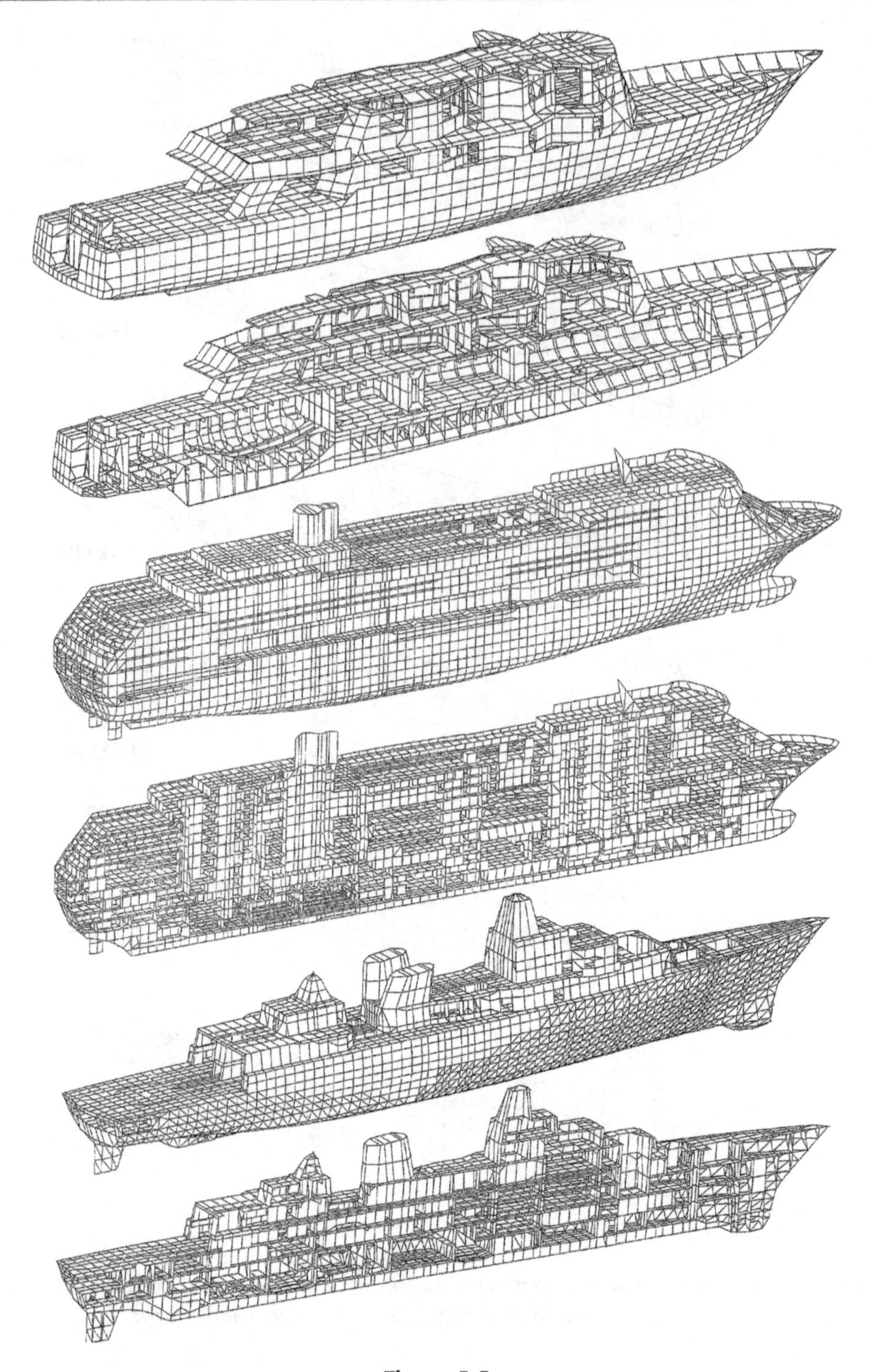

Figure 5.5:
FEM grids of a yacht, a passenger vessel, and a frigate. *Source: Germanischer Lloyd*

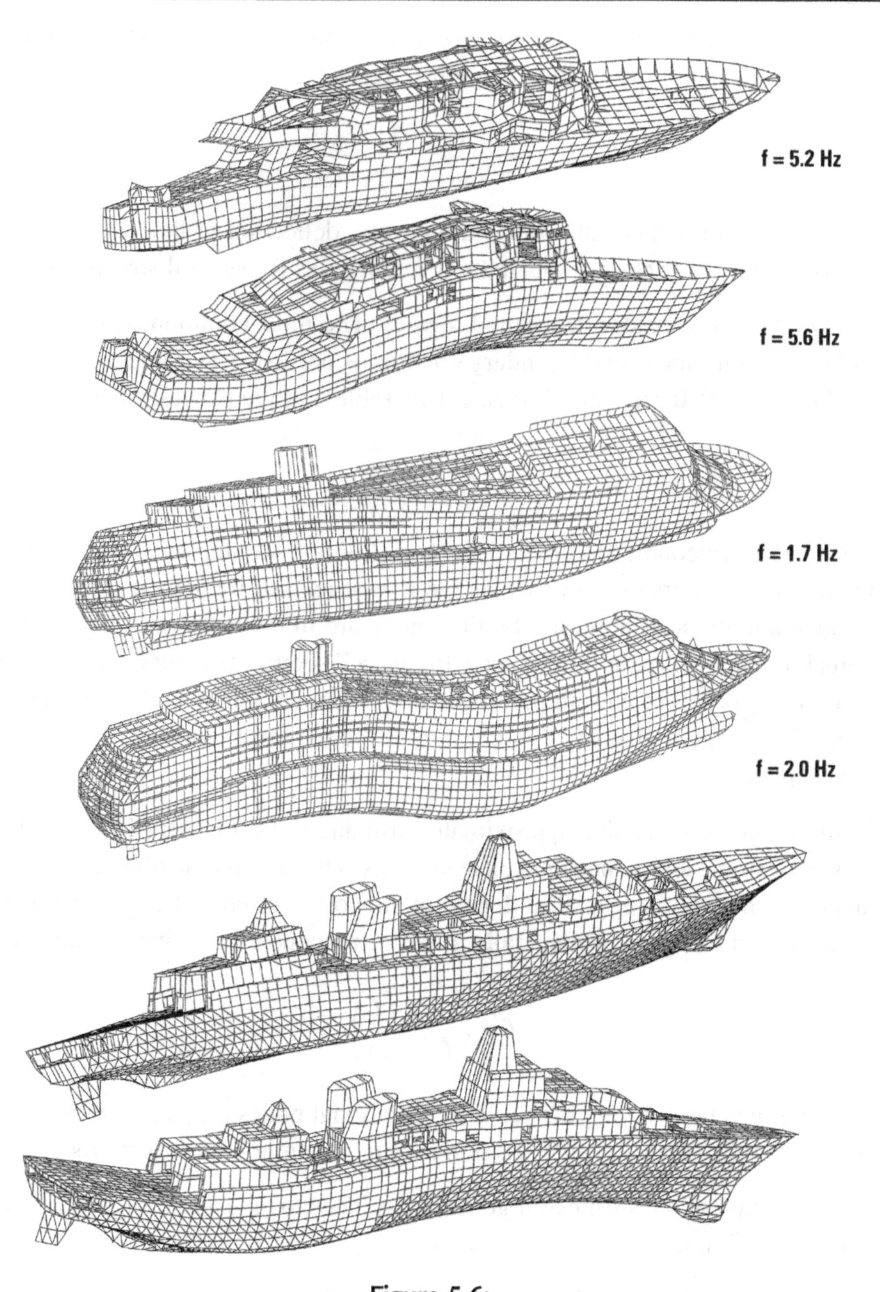

Figure 5.6:
Computed first torsional and second vertical vibration natural modes and corresponding natural frequencies for the ships in Fig. 5.5. *Source: Germanischer Lloyd*

For individual concentrated masses, the surrounding structure can be modeled in a single spring constant. The natural frequency of such a single mass-spring system is:

$$f = \frac{1}{2\pi}\sqrt{\frac{c}{m}} = \frac{1}{2\pi}\sqrt{\frac{g}{\delta_{\text{static}}}} \tag{5.11}$$

c is the spring constant, which follows from the static deflection of the system δ_{static} under a single mass load $m \cdot g$. Table 5.2 lists natural frequencies for several structures.

More often we have to consider distributed masses. The most common cases and corresponding natural modes and boundary conditions are listed in Table 5.3 with $\alpha = [\rho A/(EI)]^{-4}\sqrt{2\pi f}$ for a beam. For case 1 in Table 5.3, the (lowest) natural frequency is:

$$f_0 = \frac{\pi}{2\ell^2}\sqrt{\frac{EI}{\rho A}} \tag{5.12}$$

For all other boundary conditions we get $f = C \cdot f_0$. Table 5.4 compiles the constants C and natural modes. The free-free support yields the lowest natural frequencies. The lowest vibration mode and the next highest vibration mode are in this case rigid-body motions. The end supports influence the natural frequency. Figure 5.7 shows the natural frequency factor for a beam on two supports. The end support then varies from flexible ($\varepsilon = 0$) to fixed ($\varepsilon = 1$). The natural frequency is again given by $f = C \cdot f_0$. f_0 is the natural frequency of the beam with flexible end supports.

Some classification societies give approximate formulae to estimate the lowest natural frequencies of isotropic and orthotropic plate systems. These formulae often inherently assume partial support of the plate edges. The degree of support is often difficult to estimate, but influences the natural frequency. Generally, natural frequencies are given as functions of:

$$f \approx \left(\frac{1}{\ell^2}, \sqrt{\frac{I}{m}}\right) \tag{5.13}$$

Errors in estimating effective lengths between vibrational nodes propagate strongly (quadratic dependence on length). Errors in stiffness or mass are less important (square root dependence).

For the determination of the stiffness, it generally suffices to take an average plate width in the computations as follows:

$$B_m = 0.3 \cdot \ell \quad \text{for} \quad \ell/B \leq 3$$

$$B_m = B \quad \text{for} \quad \ell/B \geq 3$$

Here ℓ is the length between vibrational nodes. Stiffness and mass form a fraction. Thus errors usually cancel each other. Generally, accuracy decreases with higher vibrational modes. There is little sense in computing natural frequencies higher than the second or third harmonic.

Table 5.2: Natural frequencies *f* of typical structures (Lehmann 2000)

No.	System	Direction of vibration	Natural frequency
1		⇕	$\frac{1}{2\pi}\sqrt{\frac{3EI}{\ell^3 m}}$
2		⇔	$\frac{1}{2\pi}\sqrt{\frac{EA}{\ell \cdot m}}$
3		⇕	$\frac{1}{2\pi}\sqrt{\frac{3EI\ell^3}{\ell_1^3 \cdot \ell_2^3 \cdot m}}$
4		⇕	$\frac{1}{2\pi}\sqrt{\frac{4EI\ell^2}{\ell_1^2 \cdot \ell_2^3 \cdot \left[1+\frac{\ell_1}{3\ell}\right] m}}$
5		⇕	$\frac{1}{2\pi}\sqrt{\frac{3EI\ell}{\ell_1^2 \cdot \ell_2^2 \cdot m}}$
6	$k = \frac{l_1 \ell_2}{l_2 \ell_1}$	⇕	$\frac{1}{2\pi}\sqrt{\frac{12EI_1}{\ell_1^3 \cdot (4+3k) \cdot m}}$
7	$k = \frac{l_1 \ell_2}{l_2 \ell_1}$	⇕	$\frac{1}{2\pi}\sqrt{\frac{3EI_1}{\ell_1^3 \cdot (1+k) \cdot m}}$

(Continued)

Table 5.2: Continued

No.	System	Direction of vibration	Natural frequency
8	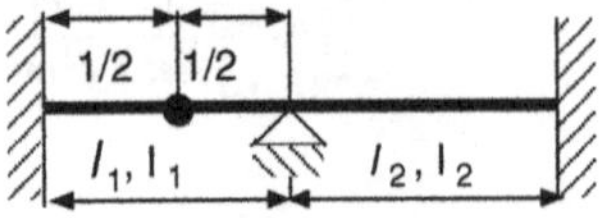 $k=\frac{I_1\ell_2}{I_2\ell_1}$	⇕	$\frac{1}{2\pi}\sqrt{\frac{768EI_1}{\ell_1^3\cdot m}\cdot\left(\frac{k+1}{7k+4}\right)}$
9	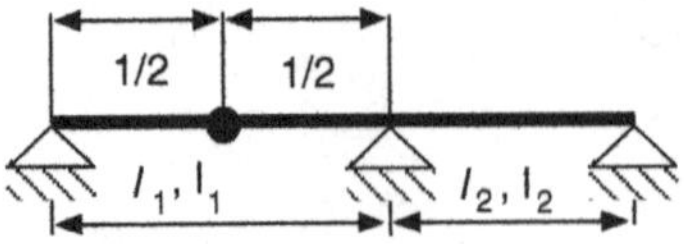 $k=\frac{I_1\ell_2}{I_2\ell_1}$	⇕	$\frac{1}{2\pi}\sqrt{\frac{768EI_1}{\ell_1^3\cdot m}\cdot\left(\frac{k+1}{16k+7}\right)}$
10	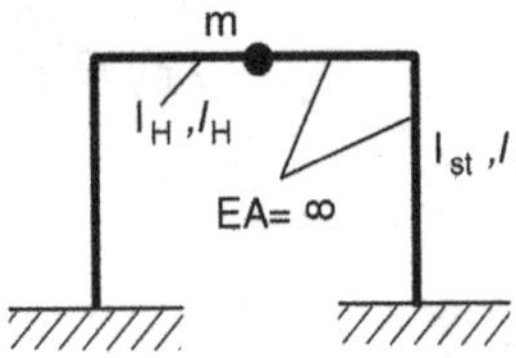 $k=\frac{I_H\ell_{St}}{I_{St}\ell_H}$	⇕	$\frac{1}{2\pi}\sqrt{\frac{192EI_H}{\ell_H^3\cdot m}\cdot\left(\frac{k+2}{4k+2}\right)}$
11	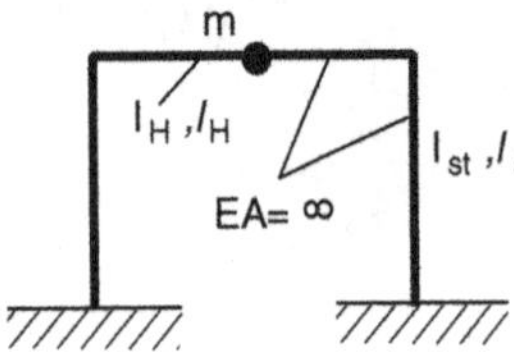 $k=\frac{I_H\ell_{St}}{I_{St}\ell_H}$	⇔	$\frac{1}{2\pi}\sqrt{\frac{24EI_{St}}{\ell_{St}^3\cdot m}\cdot\left(\frac{6k+1}{6k+4}\right)}$
12	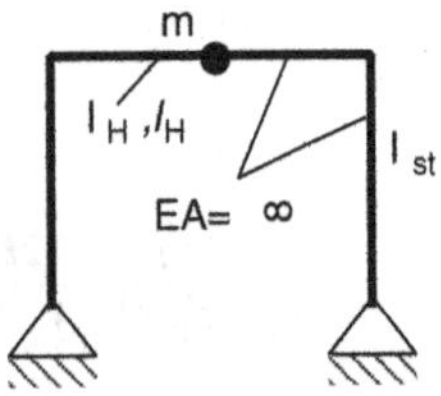 $k=\frac{I_H\ell_{St}}{I_{St}\ell_H}$	⇕	$\frac{1}{2\pi}\sqrt{\frac{192EI_H}{\ell_H^3\cdot m}\cdot\left(\frac{2k+3}{8k+3}\right)}$
13	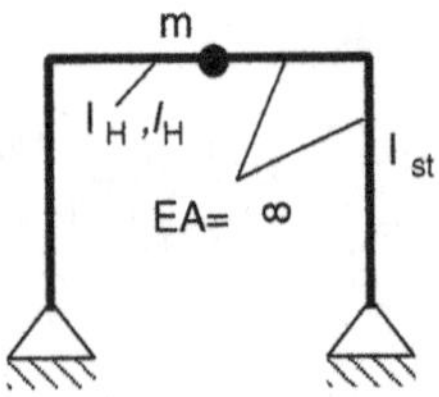 $k=\frac{I_H\ell_{St}}{I_{St}\ell_H}$	⇔	$\frac{1}{2\pi}\sqrt{\frac{6EI_{St}}{\ell_{St}^3\cdot m}\cdot\left(\frac{2k}{2k+1}\right)}$

Table 5.2: Continued

No.	System	Direction of vibration	Natural frequency
14	springs parallel:	⇳	$\frac{1}{2\pi}\sqrt{\frac{\delta_1+\delta_2}{\delta_1\delta_2}}$ $\delta_1 = \frac{\ell_1^3 m}{48EI_1};\ \delta_2 = \frac{\ell_2^3 m}{48EI_2}$
15		⇳	$\frac{1}{2\pi}\sqrt{\frac{\delta_1+\delta_2}{\delta_1\Delta_2}}$ $\delta_1 = \frac{\ell_1^3 m}{192EI_1};\ \delta_2 = \frac{\ell_2^3 m}{192EI_2}$
16	springs sequential:	⇳	$\frac{1}{2\pi\sqrt{\delta_1+\delta_2}}$ $\delta_1 = \frac{\ell_1^3 m}{3EI_1}\cdot\left(1+3\frac{I_1\ell_2}{I_2\ell_1}\right);$ $\delta_2 = \frac{\ell_2^3 m}{3EI_2}\cdot\left(1+3\frac{\ell_1}{\ell_2}\right)$

For vibrations of plates, we can use Fig. 5.8 to determine the lowest natural frequency for assorted side ratios and end supports with $f = C\cdot f_0$:

$$f_0 = \frac{\pi}{2b^2}\sqrt{\frac{Eh^2}{12(1-\nu^2)\rho}} \tag{5.14}$$

The simple formulae given above yield predictions which are good enough for practical purposes, provided that the following conditions are met (Asmussen et al. 1998):

- freely rotatable, fixed support at the edges;
- rectangular shape of grillage systems and plates;

Table 5.3: Natural modes for distributed mass systems (Lehmann 2000)

Case	System	Natural mode
1	x	$A \cdot \sin \alpha x$
2	x	$A \cdot \left[\sinh \alpha x - \sin \alpha x - \frac{\sinh \alpha \ell - \sin \alpha \ell}{\cosh \alpha \ell - \cos \alpha \ell} \cdot (\cosh \alpha x - \cos \alpha x)\right]$
3	x	$A \cdot [\sinh \alpha x - \sin \alpha x - \tan \alpha \ell \cdot (\cosh \alpha x - \cos \alpha x)]$
4	x	$A \cdot \left[\sinh \alpha x - \sin \alpha x - \frac{\sinh \alpha \ell - \sin \alpha \ell}{\cosh \alpha \ell - \cos \alpha \ell} \cdot (\cosh \alpha x - \cos \alpha x)\right]$

- regular arrangement of stiffeners;
- no pillars or stanchions within the considered system;
- uniform distribution of added mass.

If these conditions are violated, finite element analyses must be employed. Because of the relatively high natural frequencies of local structures, the finite element models must be quite detailed, including also the bending stiffness of structural elements. The amount of work required for the creation of such models is considerable (and often underestimated) despite modern pre-processors with parameterized input possibilities and graphic support. Beam grillage models usually suffice for the lowest natural modes; for higher natural modes, three-dimensional models of higher precision are needed.

The distribution of effective masses in such models is often impossible to specify accurately. Asmussen et al. (1998), based on the experience of many vibration analyses for ship structures at Germanischer Lloyd, recommend taking an effective additional mass of 40 kg/m^2 into account for decks in living and working spaces and 20 kg/m^2 for bulkheads.

Ideally, ship structures have natural frequencies (well) above the main exciting frequencies (subcritical design). Asmussen et al. (1998) give as guidelines:

- natural frequency greater than 1.2 times, twice the propeller blade frequency or main engine ignition frequency in the ship's aftbody, engine room, and deckhouse area;

Table 5.4: Natural modes and natural frequency for beams (Lehmann 2000)

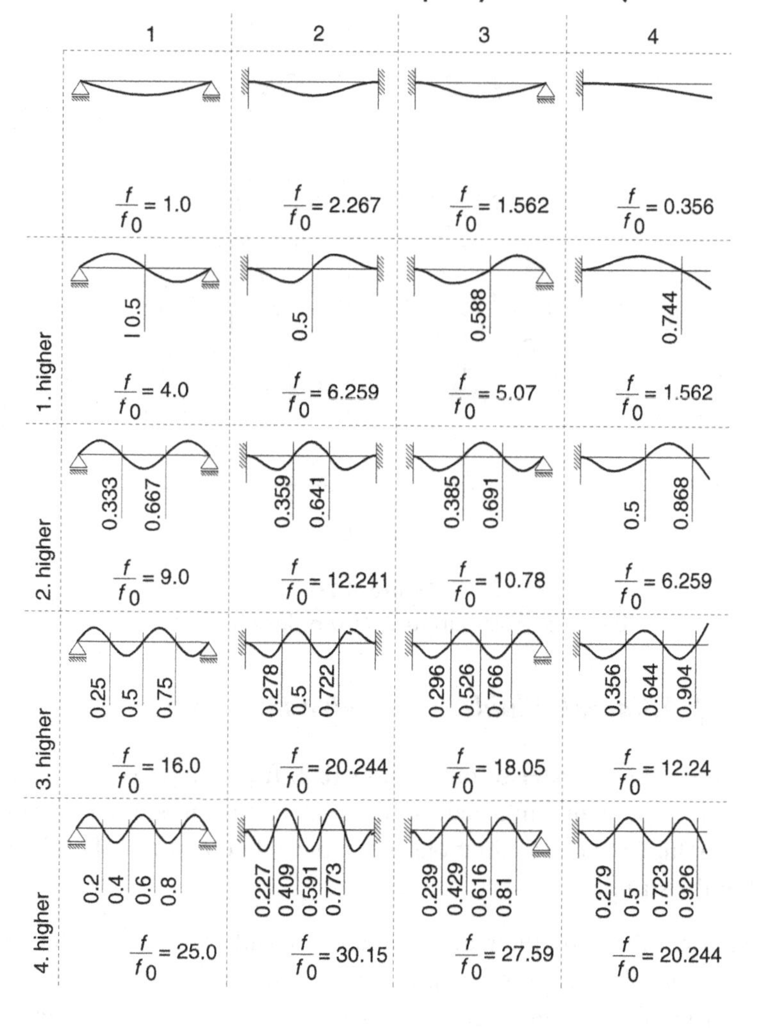

- natural frequency greater than 1.1 times the propeller blade frequency for the ship's shell structure directly above the propeller.

The assumption of simply supported edges is conservative, as any constraining effect increases the natural frequency further. To avoid in turn over-dimensioning, stiff bracket connections are sometimes considered by taking 50–70% of the bracket length as 'effective', reducing correspondingly the lengths of beam elements.

In most cases, it is sufficient to design natural frequencies of local structures subcritically up to 35 Hz. Further increases of natural frequencies usually come at exhibitive cost. A supercritical design (natural frequency (well) below exciting frequency) or a 'design in frequency windows' (between exciting frequencies) is then chosen for such high frequencies.

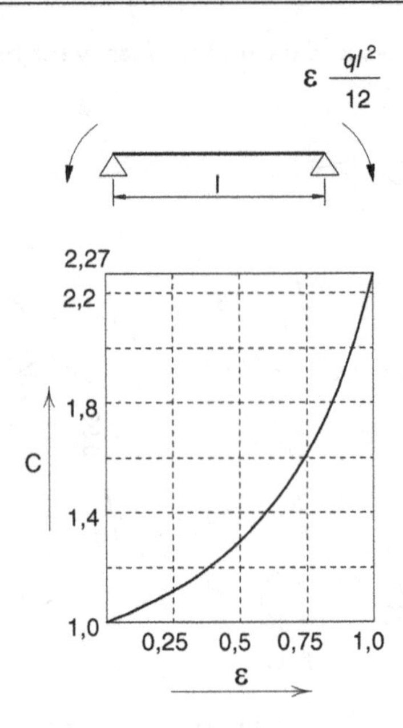

Figure 5.7:
Beam on two supports (from flexible to fixed)

5.5. Effects of Adjacent Fluids: Hydrodynamic Mass

Ship structures often border water or other fluids (e.g. fuel), either on one side (outer hull plates) or both sides (fuel tank bulkhead). Fluid immersion can be total or partial. If the structure borders a fluid, structure vibrations will induce fluid motions. This induces a pressure from the fluid to the structure.

The effect of adjacent fluids on the mass operator is usually large, its effect on damping and stiffness small. The small effect on damping can nevertheless be important, because other damping mechanisms are also small. Changes in displacement volume (increased or decreased immersion) affect the stiffness operator (restoring forces). The 'hydrostatic stiffness' increases the natural frequency of vertical bending vibrations of the hull girder. The effect is usually negligible, except perhaps for particularly 'soft' ships such as inland cargo vessels.

In the following we discuss the influence on the mass operator. The 'added mass' or 'hydrodynamic mass' is the equivalent mass one would have to fix to the structure to obtain the same effect on the structure as the adjacent accelerated fluid has.

The fundamental equation for the hydrodynamic mass A is:

$$A = \sum \int_S \frac{1}{\omega^2} \cdot p_a \cdot \vec{n} \cdot \vec{q}_j \ \mathrm{d}S \tag{5.15}$$

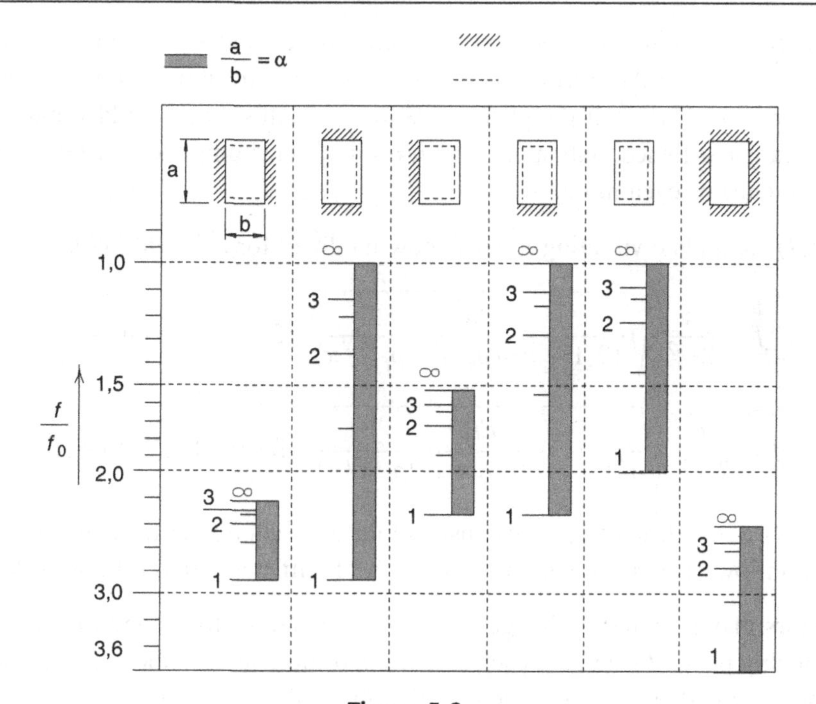

Figure 5.8:
Natural frequencies of rectangular plates; //// fixed support, - - - - flexible support (Lehmann 2000)

$\vec{n}$ is the unit normal pointing from the fluid to the body, $\vec{q}_j$ the jth deflection form, and p_a the amplitude of the oscillating part of the pressure (due to a certain natural mode).

For an infinite plate with the fluid of density ρ above the plate (Fig. 5.9) the added mass for a plate strip of length a and width b orthogonal to the paper plane is:

$$A = \frac{\rho a^2 b}{\pi} \tag{5.16}$$

The formula is valid for infinite plate extension with all plate segments vibrating sinusoidally and no obstacles in the flow. The formula is based on potential flow considerations. Viscosity generally plays a negligible role in plate vibrations in ships. For typical ship plates covered by water or fluids of similar density (like oil), the added mass exceeds the structural mass.

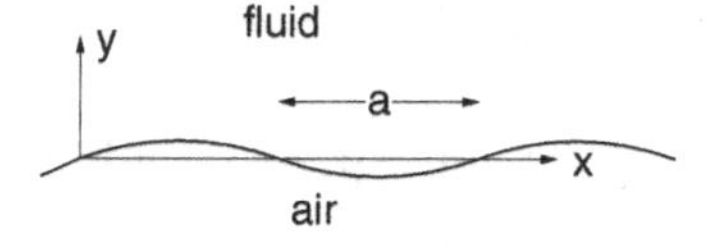

Figure 5.9:
Plate with adjacent fluid on one side

For example, for a 10-mm steel plate and distance between nodes 700 mm, the steel mass per area is $7.8\ \text{t/m}^3 \cdot 0.01\ \text{m} = 0.078\ \text{t/m}^2$; the added mass due to fresh water on one side is already $1\ \text{t/m}^3 \cdot 0.7\ \text{m}/\pi = 0.223\ \text{t/m}^2$. If the plate has water on both sides, the added mass value doubles. For larger stiffened plate areas which oscillate including their stiffeners, the added mass effect dominates even more.

Equation (5.7) for a plate vibrating in air is now modified for adjacent fluids:

$$f = \frac{\pi}{2 \cdot \ell^2} \cdot \sqrt{\frac{Et^3}{12(1-\nu^2)(\rho_{st} \cdot t + \rho_f \cdot \ell/\pi)}} \quad \text{for fluid on one side} \tag{5.17}$$

$$f = \frac{\pi}{2 \cdot \ell^2} \cdot \sqrt{\frac{Et^3}{12(1-\nu^2)(\rho_{st} \cdot t + 2\rho_f \cdot \ell/\pi)}} \quad \text{for fluid on both sides} \tag{5.18}$$

ρ_{st} is the density of the structure, ρ_f the density of the adjacent fluid, ℓ the length between nodes (plate length in lowest frequency mode = spacing of stiffeners), and t the plate thickness.

Various factors may in practice change the effect of adjacent fluids and thus the natural frequency of ship plates. Generally, each constraint of fluid motion increases the added mass, each relaxation (e.g. due to three-dimensional effects) decreases the added mass. In particular:

- Rectangular plates (limited side ratio)
  The previous formulae for plates assumed large side ratio ($a << b$). For the general case, the lowest natural frequency is:

$$f = \left(\frac{\pi}{2 \cdot a^2} + \frac{\pi}{2 \cdot b^2}\right) \sqrt{\frac{Et^3}{12(1-\nu^2)(\rho_{st} \cdot t + \rho_f \cdot d)}} \tag{5.19}$$

  d is the thickness of an equivalent 'water layer' (a mass layer attached to the structure giving the same mass effect as the added mass). In this case:

$$d = \frac{1}{\pi\sqrt{a^{-2} + b^{-2}}} \tag{5.20}$$

- A rigid wall at distance h parallel to the vibrating plate modifies d by a factor:

$$\frac{1}{\tanh(\pi h \sqrt{a^{-2} + b^{-2}})} \tag{5.21}$$

- A constant flow parallel to the plate with speed V (hull plating neglecting boundary layer) modifies d by the factor $1 + \left(\frac{V\pi}{\omega a'}\right)^2$. a' is the distance between vibration nodes in the flow direction.

- Vibrating plate between rigid walls orthogonal to the plate
  The added mass becomes very large, unless the plate moves in a vibration mode keeping the fluid volume constant (i.e. lowest natural mode is full sine wave between nodes). This case appears typically for tank walls. As a consequence of these natural modes, the added masses are much lower.
- Plate curvature (shells)
  Plate curvature decreases/increases added mass if the fluid is on the convex/concave side. The effect is usually small.
- Perforated plates
  If the structure is perforated, the added mass is reduced by multiplying the added mass of the full plate by a reduction factor, which can be simply approximated following Lehmann (2000):

  $C_{\text{red}} = 1 - 8.44\alpha + 27.6\alpha^2 - 30.2\alpha^3$, where α is the ratio of area of hole/area of the plate.

5.5.1. Hydrodynamic Mass and Damping at Rudders

We consider first the two-dimensional case (infinite rudder height) in uniform flow of speed U. The rudder makes harmonic transverse motions q_1 and rotational motions q_2 around the center of the chord (Fig. 5.10).

Let the non-dimensional frequency be $\pi \cdot f \cdot c/U > 2$. Then the transverse force F_1 and the transverse F_2 (around the point on half chord length) are almost the same as for infinite frequency. The hydrodynamic force vector is then expressed as usual, with components proportional to accelerations, velocities, and deflections:

$$\begin{Bmatrix} F_1 \\ F_2 \end{Bmatrix} = -A\begin{Bmatrix} \ddot{q}_1 \\ \ddot{q}_2 \end{Bmatrix} - D''\begin{Bmatrix} \dot{q}_1 \\ \dot{q}_2 \end{Bmatrix} - K''\begin{Bmatrix} q_1 \\ q_2 \end{Bmatrix} \tag{5.22}$$

The hydrodynamic stiffness matrix K″ is negligibly small in comparison to the structural stiffness. Added mass and damping matrices are:

$$A = \begin{bmatrix} \pi\rho_f c^2/4 & 0 \\ 0 & \pi\rho_f c^4/128 \end{bmatrix} \quad \text{and} \quad D'' = \begin{bmatrix} \pi\rho_f c|U|/2 & 3\pi\rho_f c^2|U|/8 \\ -\pi\rho_f c^2|U|/8 & \pi\rho_f c^3|U|/32 \end{bmatrix} \tag{5.23}$$

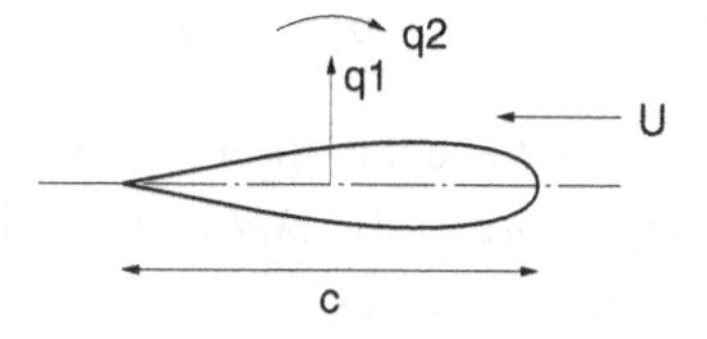

Figure 5.10:
Notation for rudder

For three-dimensional flow (finite rudder height h), the top left element of the added mass matrix is reduced by a factor as follows:

h/c	10	5	4	2.5	2	1.5	1	0.5
Reduction factor	0.94	0.90	0.87	0.80	0.76	0.70	0.48	0.38

Reduction factors for other matrix elements are not known. If required they should be determined by more elaborate numerical simulations. If the coordinate system is not at half chord length, but shifted upstream by a distance e (e.g. at the center of the profile area), we have (neglecting the small hydrodynamic stiffness):

$$\begin{Bmatrix} F_1 \\ F_2 \end{Bmatrix}_{\text{w.r.t. } e} = -\begin{bmatrix} 1 & 0 \\ e & 1 \end{bmatrix} M \begin{bmatrix} 1 & e \\ 0 & 1 \end{bmatrix} \begin{Bmatrix} \ddot{q}_1 \\ \ddot{q}_2 \end{Bmatrix}_{\text{w.r.t. } e} - \begin{bmatrix} 1 & 0 \\ e & 1 \end{bmatrix} D \begin{bmatrix} 1 & e \\ 0 & 1 \end{bmatrix} \begin{Bmatrix} \dot{q}_1 \\ \dot{q}_2 \end{Bmatrix}_{\text{w.r.t. } e} \quad (5.24)$$

5.5.2. Hydrodynamic Mass and Damping for Propellers

Schwanecke (1963) used an unsteady lifting-line method to compute mass and damping matrices for propellers vibrating in six degrees of freedom as rigid bodies. x points forward, y to port, z upward. Then forces and moments are expressed as:

$$\begin{Bmatrix} F_x \\ M_x \\ F_y \\ M_y \\ F_z \\ M_z \end{Bmatrix} = -\begin{bmatrix} a_{11} & a_{12} & 0 & 0 & 0 & 0 \\ a_{12} & a_{22} & 0 & 0 & 0 & 0 \\ 0 & 0 & a_{33} & a_{34} & 0 & 0 \\ 0 & 0 & a_{34} & a_{44} & 0 & 0 \\ 0 & 0 & 0 & 0 & a_{33} & a_{34} \\ 0 & 0 & 0 & 0 & a_{34} & a_{44} \end{bmatrix} \begin{Bmatrix} \ddot{u}_x \\ \ddot{\alpha}_x \\ \ddot{u}_y \\ \ddot{\alpha}_y \\ \ddot{u}_z \\ \ddot{\alpha}_z \end{Bmatrix} - \begin{bmatrix} b_{11} & b_{12} & 0 & 0 & 0 & 0 \\ b_{12} & b_{22} & 0 & 0 & 0 & 0 \\ 0 & 0 & b_{33} & b_{34} & b_{35} & b_{36} \\ 0 & 0 & b_{34} & b_{44} & b_{45} & b_{46} \\ 0 & 0 & -b_{35} & -b_{36} & b_{33} & b_{34} \\ 0 & 0 & -b_{45} & -b_{46} & b_{34} & b_{44} \end{bmatrix} \begin{Bmatrix} \dot{u}_x \\ \dot{\alpha}_x \\ \dot{u}_y \\ \dot{\alpha}_y \\ \dot{u}_z \\ \dot{\alpha}_z \end{Bmatrix} \quad (5.25)$$

u denotes deflections, α rotations. Schwanecke approximates the elements of the mass and damping matrices as functions of area ratio A_E/A_0, propeller pitch P, blade number Z, diameter D, and propeller circular frequency ω_w:

$$a_{11} = 0.209\,\pi\rho\, D^3 (A_E/A_0)^2/Z \quad a_{12} = -0.105\,\rho\, D^4 (P/D)(A_E/A_0)^2/Z$$

$a_{22} = 0.052\ \rho\ D^5(P/D)^2(A_E/A_O)^2/(\pi Z)$ $\quad a_{33} = 0.566\ \rho\ D^3(P/D)^2(A_E/A_O)^2/(\pi Z)$

$a_{34} = 0.052\ \rho\ D^4(P/D)(A_E/A_O)^2/Z$ $\quad a_{44} = 0.009\ \pi\rho\ D^5(A_E/A_O)^2/Z$

$b_{11} = 0.066\ \pi\rho\omega_w D^3\ (A_E/A_O)$ $\quad b_{12} = -0.033\ \rho\omega_w D^4(A_E/A_O)$

$b_{22} = 0.017\rho\omega_w D^5(P/D)^2(A_E/A_O)/\pi$ $\quad b_{33} = 0.124\ \rho\omega_w D^3(P/D)^2(A_E/A_O)/\pi$

$b_{44} = 0.004\ \pi\rho\omega_w D^5(A_E/A_O)$ $\quad b_{34} = 0.017\ \rho\omega_w D^4(P/D)(A_E/A_O)$

$b_{35} = 0.566\ \rho\omega_w D^3(P/D)^2(A_E/A_O)^2/(\pi Z)$ $\quad b_{36} = 0.105\ \rho\omega_w D^4(P/D)\ )(A_E/A_O)^2/Z$

$b_{45} = 0.052\ \rho\omega_w D^4(P/D)(A_E/A_O)^2/Z$ $\quad b_{46} = 0.017\ \pi\rho\omega_w D^5(A_E/A_O)^2/Z$

5.5.3. Computation of Hydrodynamic Mass for Ships

Because ships are longitudinal, slender structures, we assume (approximately) that the water flows around individual strips (rather than in the x direction). This assumption for the ship vibrations is the same as for the rigid-body motions of the ship in ship seakeeping.

Lewis (1929) already computed the motion of water around vertically oscillating ship cross-sections using conformal mapping for the limiting case of infinite frequency. This approximation is already sufficiently accurate for the lowest vibration natural mode with two nodes. In this case, the free-surface water motion is limited to a vertical motion. (However, for lower frequencies as in rigid-body ship seakeeping, we should also consider the horizontal water motion in a more sophisticated model.) The result of the Lewis approach can be summarized as follows. The added mass (per length) of a cross-section is:

$$m''_{33} = \rho \cdot \frac{\pi}{2} r^2 [(1-a)^2 + 3b^2] \tag{5.26}$$

ρ is the water density. The three parameters r, a, and b characterize size and shape of the cross-section. They follow from the following non-linear system of equations:

$$\frac{B}{2T} = \frac{1-a+b}{1+a+b} \tag{5.27}$$

$$C_M = \frac{\pi}{4} \frac{1-a^2-3b^2}{(1+b)^2-a^2} \tag{5.28}$$

$$r = \frac{B}{2(1-a+b)} \tag{5.29}$$

B is the cross-section width in the waterline, T its draft (at $y = 0$), and C_M its cross-section coefficient = cross-section area/$(B \cdot T)$. This system of equations has two solutions. The

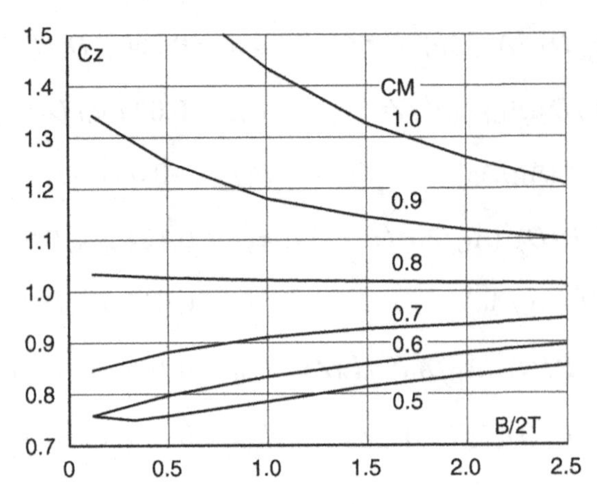

Figure 5.11:
Coefficient of hydrodynamic mass of ship cross-section $C_z = m''/(\rho\pi B^2/8)$ for high-frequency vertical motion at the water surface

resulting minimum added mass m'' is taken. Figure 5.11 shows results for a non-dimensional coefficient of the added mass.

For high-frequency horizontal vibrations of a cross-section at a free surface, the hydrodynamic mass (per length) is almost independent of the cross-section shape:

$$m''_{22} = 0.205\pi\rho T^2 \tag{5.30}$$

The hydrodynamic mass for roll motion can be approximated for a surface piercing cross-section:

$$m''_{44} = \left[1 - \left(\frac{2T}{B}\right)^2\right]^2 \cdot \pi \frac{\rho B^4}{256} \tag{5.31}$$

For submerged cross-sections near the free surface, this value needs to be doubled. For cross-sections with sharp bilge corners or bilge keels, the hydrodynamic added mass can be considerably larger.

All these values are for two-dimensional strips, assuming that the water moves only in the plane of the strip. This assumption is increasingly less valid as the distance between vibration nodes decreases with respect to strip width (for vertical and torsional vibrations) and draft (for horizontal vibrations). Figure 5.12 gives necessary reduction factors J to account for three-dimensional effects. These curves were determined numerically by a boundary element method.

5.5.4. Damping of Ship Hull Vertical Vibrations

Numerical (deformation) methods compute the vertical vibrations of the ship hull using mass, damping, and stiffness matrices. The damping matrix is generally expressed as follows:

$$D_{ij} = \int_L d(x) \cdot q_i(x) \cdot q_j(x)\ \mathrm{d}x \tag{5.32}$$

$d(x)$ is the damping force per length, divided by the vertical velocity of the cross-section. q_i and q_j are the shape functions for the vertical deflections of the ship cross-sections.

If we use the vibration natural modes for q_i, the mass matrix and the stiffness matrix are simple diagonal matrices. The damping of vertical ship hull vibrations is weak. Therefore, the off-diagonal elements in the damping matrix can be neglected and the individual vibration modes can be considered separately.

We can estimate the hydrodynamic damping as follows:

$$\begin{aligned} D_{ii} = & \int_L \left(q_i^2 + \frac{U^2}{\omega^2}(q_i')^2 \right) N\ \mathrm{d}x + \rho \int_L c_w B|w| q_i^2\ \mathrm{d}x \\ & + \frac{U^2}{\omega^2} q_{iT} q_{iT}'\ N_T + U q_{iT}^2 m_T'' + \frac{\rho}{4} n D_P P^2 q_{iP}^2 \end{aligned} \tag{5.33}$$

The first term is due to the waves radiated from the vibrating hull and can be computed in a strip method as for the rigid-body motions. U is the ship speed and q'_i the derivative of the shape function (natural mode) with respect to x. $N(x)$ is the damping constant (per length) for a given strip. N depends on section shape and frequency. For high-frequency

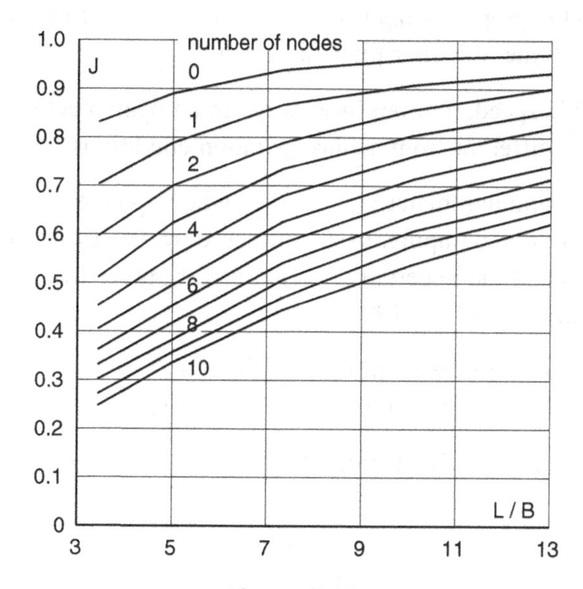

Figure 5.12:
Reduction factor J for hydrodynamic masses for three-dimensional flow in vertical ship vibrations

radiated waves (with wave lengths much shorter than the section width), we can approximate:

$$N(x) \approx \rho \frac{gB(x)\tan\alpha}{2\omega} \tag{5.34}$$

$B(x)$ is the local section width and α the flare angle in the waterline measured against the vertical ($\alpha = 0°$, i.e. $N = 0$, for wall-sided sections).

The second term is due to the pressure resistance of the vertically moving cross-sections. The contribution of the vibrations by themselves is negligibly small, but the interaction with rigid-body motions is considered here by the average vertical velocity $|w|$. c_w is the vertical motion resistance coefficient of the section. Lacking better data, one employs here steady flow resistance values, typically $0.5 < c_w < 1$ for the midbody sections. At the ship ends, where the sections are well-rounded, c_w is negligibly small.

The third and fourth terms consider the effect of the transom stern (thus index T, e.g. $q_{iT} = q_i(x_T)$ is the value of the shape function at the transom). N_T is the damping constant, m''_T the added mass, both for high-frequency and the transom shape. These terms assume a detaching flow at the transom, similar to that in the strip method. Note that the wetted transom stern in operation (with ship wave system and motions in seaways) differs from that at rest. The third term containing N_T is typically much smaller than the fourth term and can usually be neglected.

The last term is due to the propeller. q_{iP} is the value of the shape function at the propeller, D_P the propeller diameter, P the propeller pitch, and n the propeller revolutions (in 1/s).

The terms depend on ship speed, motions in seaways and propeller actions. Therefore vibration damping in port will be different from actual operation conditions.

Besides hydrodynamic damping, material damping and component damping (due to floor and deck coverings) play a role in damping. In the literature, widely different values are stated for damping characteristics and the uncertainty increases for higher frequencies. For simple practical estimates, Asmussen et al. (1998) give:

- for ship in loaded condition: $\vartheta = \min(8;\ 7 \cdot f/20 + 1)\%$
- for ship in ballast condition: $\vartheta = \min(6;\ 5.5 \cdot f/20 + 0.5)\%$

The frequency f is taken in Hz. The degree of damping ϑ is coupled to the logarithmic decrement Λ:

$$\Lambda = \frac{2\pi \cdot \vartheta}{\sqrt{1 - \vartheta^2}} \tag{5.35}$$

The logarithmic decrement describes the ratio of two successive maxima: $e^{\Lambda} = A_1/A_2$.

5.6. Excitation of Vibration

Ship hull vibrations are mainly excited by seaway, propeller, and main engines. In addition there are special cases, where periodic flow separation at structure appendages or torque fluctuations in electric engines excite structural vibrations, for example.

5.6.1. Propellers

Propellers excite vibrations by induced pressure fluctuations on the ship hull and on the rudder. The propeller induces a pressure field, due to the displacement effect of the propeller blades and due to the changing cavitation volume on cavitating propellers. The contribution due to cavitation is often more important for vibration excitation. The exciting frequencies of a propeller are generally rpm × number of blades and higher harmonics. Thus we have, say, for a four-bladed propeller with 120 rpm = 2 Hz: 8, 16, 24, ... Hz as exciting frequencies. In choosing an appropriate number of propeller blades, one tries to place the propeller excitation in an interval between hull natural frequencies. For modern propellers, excited pressure impulses at second, third, or fourth blade frequency are typically higher than at blade frequency. The explanation is that modern numerical propeller design tools (panel methods) allow the cavitation volume on the propeller responsible for pressure fluctuations at blade frequency to be relatively well minimized. Higher-frequency excitations are largely due to tip-vortex cavitation. Scale effects play a more significant role for higher-order excitations. Holden et al. (1980) give empirical formulae to estimate the cavitation-induced pressure on the ship hull of single-screw ships, but numerical methods (panel methods or lifting surface methods) are state-of-the-art to numerically predict induced pressure fluctuations. RANSE methods including cavitation models drift increasingly into industrial applications. Still, numerical simulations in the initial design phase are usually followed by model tests, using facilities that allow high Reynolds numbers. Model tests can measure pressure impulses on the hull. For simpler tests, the ship wake is approximated using a fine grid upstream and the pressure impulses are measured on a flat plate above the propeller approximately at a position where in reality the ship hull is. For tests with complete ship models, a grid of pressure probes is installed in the model above the propeller (typically 10–20 probes). The measured time histories of the pressures are decomposed in a Fourier analysis into the individual exciting orders (z, $2z$, $3z$, etc.) These can be used for comparisons among different design alternatives or as input for FEM simulations of structural vibrations.

Practical experience is that the pressure amplitude above the propeller alone is not adequate to characterize the excitation behavior of a propeller. Therefore, no generally valid limits can be stated for pressure fluctuation amplitudes. These amplitudes depend on propeller tip clearance, transmitted power, cavitation extent, etc. Nevertheless, Asmussen et al. (1998) give some guidelines: '[...] pressure amplitudes at blade frequency of 1 to 2, 2 to 8, and over 8 kPa at a point directly above the propeller can be categorized as "low", "medium" and "high", respectively. Total vertical force fluctuations at blade frequency, integrated from pressure fluctuations, range from about 10 kN for a high-performance special-purpose ship to 1000 kN for a high-performance container vessel. For usual ship types and sizes, corresponding values lie between 100 and 300 kN. Whether these considerable excitation forces result in large vibrations depends on dynamic characteristics of the ship structure and can only be judged rationally on the basis of forced vibration analysis.'

Propellers also excite vibrations by way of unsteady propeller blade forces. These are due to the inhomogeneous wake of the ship and are transmitted via the propeller shaft into the ship. Reducing the inhomogeneity of the ship wake reduces the vibration excitation. The wake is usually determined in experiments in the model basin. In practice, the unsteady propeller forces for given wake are determined in numerical simulations.

5.6.2. Engine

Donath and Bryndum (1988) discuss in detail ship engine vibrations. Engines can be responsible for considerable vibrations. Diesel engines are far more critical in this respect than turbines. The main effects are due to the moving parts and the gas forces between pistons and cylinders. For large main engines, the vibration excitation due to horizontal and vertical total forces is typically almost zero. One often introduces the 'order' of excitation:

$$\text{order} = \frac{\text{exciting frequency}}{\text{rpm}}$$

For slow two-stroke diesels with N cylinders, the first, second,..., N^{th} order may excite vibrations. For four-stroke diesels, half orders may also be excited. A diesel engine located on the centerplane of the ship excites (usually) vertical vibrations of first and second (i.e. with one and two times the frequency of the engine rpm) and horizontal-torsional vibrations of Nth order. (Also, torsional vibrations of second order may be excited.) Engines arranged off-center excite both horizontal and vertical vibrations in all integer orders. However, one can couple two symmetrically arranged engines in rpm and phase such that their excitations in horizontal or vertical vibrations cancel each other.

The natural frequencies of the ship girder change depending on the load condition (ballast, fully loaded). Nevertheless, one tries to exploit remaining off-resonance intervals for the lower natural frequencies. Besides moments around the x axis, moments around the y and z axes are excited, because mass forces of the cylinders act at different x positions. For slender, single-screw ships, the main engine has to be arranged on a slender skeg and the moments around the z axis cannot be as well absorbed as for twin-screw ships where the engines are arranged on a relatively broad flat bottom. For twin-screw ships, the moments around the x axis are more critical. Moments about the x axis appear only at orders N, $2N$, $3N$, ... for two-stroke engines and $0.5N$, N, $1.5N$, $2N$, ... for four-stroke engines. Fundamental natural frequencies of main engine vibrations depend on the distribution of stiffness values and masses of the engine itself, but also to a large extent on the stiffness of adjoining structures. For large engines, particularly slow two-stroke engines, the foundation cannot be regarded as completely stiff and vibration analyses using FEM models of engine and supporting elastic structure show considerably lower natural frequencies than for infinitely rigid support (Asmussen et al. 1998).

Propeller shaft lines have to be considered together with their supporting structures. Torsional vibrations are covered by classification society rules. Axial vibrations are usually calculated by isolated models consisting of masses, springs, and damping elements. For bending vibrations, finite element computations are performed including a simple three-dimensional model of the surrounding aftbody structure. The oil film stiffness in the slide bearings for the propeller shaft and the hydrodynamic (added) mass of the propeller need to be included in the model to get realistic results. As both parameters are difficult to estimate, sensitivity analyses with parametric variations are recommended.

5.6.3. Seaway

Seaway excites a broadband of frequencies, so avoiding resonance is impossible. Ship vibrations are mainly excited by propeller and engines. The seaway excites mainly rigid-body motions such as roll, heave, and pitch. The lowest natural frequencies of vibrations of the ship hull usually lie above the significant frequencies of the seaway and are thus seldom excited. Only for very large ships, the lowest natural frequencies may become less than 1 Hz and seaway may also play a role in exciting continuous vibrations ('springing'). Slamming can induce considerable free vibrations (whipping). Studies of Germanischer Lloyd for a large LNG carrier have shown that whipping may increase bending moments (and thus longitudinal stresses) by 25%.

5.6.4. Vortex-Induced Vibrations

If unpleasant vibrations appear on board ships normally the respective frequency clearly identifies either engine or propeller as exciting source. However, for vortex-induced vibrations, the identification of the exciting sources requires considerably greater effort. In the past, the exciting source was often found only after an extensive trial-and-error approach, starting with modifications of the most likely appendages such as V-brackets, fins, sea chests, etc. This approach is inefficient, time-consuming, and costly.

Modern simulation techniques allow a more detailed insight into the physical mechanisms involving vortex shedding. They also allow rapid assessment of potential design changes. This saves time and cost for shipyard and owner. The simulations used for vortex-induced vibrations combine computational fluid dynamics (CFD) for the vortex generation and the associated vibration excitation and finite element analyses for vibration response. Based on the simulations, many potential excitation sources can be excluded. In the end, a single final sea trial suffices to verify the excitation source in measurements and to quantify the excitation.

Menzel et al. (2008) describe the procedure for an actual case study:

- Sea trials reveal vibration problems. The sea trial measurements determine frequency and amplitudes.
- Comparison with engine and propeller frequencies rules out engine and propeller as exciting source.
- Three-dimensional RANSE model of ship with all appendages (and recesses like sea chests) is created and simulated in the time domain (unsteady RANSE simulation).
- Pressure histories at all appendages and recesses are analyzed to detect unsteady vortex formation with (approximately) the frequency causing the vibration problems.
- If several appendages with such critical frequency are found, finite element analyses can indicate the effect of pressure fluctuations on the structural vibration (e.g. in a given cabin). This typically narrows the source of vibration down to the appendage where large pressure amplitudes are created by the vortices, which have strong effect on a given cabin or structural part of the ship.
- Optional at this point, a dedicated sea trial may be performed with specific and detailed measurement only for this part. Increasingly, the confidence in CFD is there to avoid the added expense and time of sea trials.
- The appendage is then re-designed, typically smoothing transitions, and re-analyzed until the critical vortices disappear in the simulation.
- Then the new design is built and verified in sea trials. There is no case known where sea trials did not confirm the disappearance of the problem.

Industry experience has shown that this approach allows solving complicated vibration problems. The simulation-based approach is time- and cost-efficient, and therefore clearly superior to the traditional trial-and-error approach.

5.7. Effect of Vibrations

Despite careful analysis and design, vibrations on board ships cannot be avoided completely. Suitable upper limits for the effects on ship, cargo, engines, and humans can be found in various regulations and norms. Parameters employed in the evaluation of vibrations are frequency f, displacement s, vibrational velocity v and vibrational acceleration a. For harmonic vibrations, the amplitudes of s, v, and a are coupled to each other by the frequency: $v = 2\pi \cdot f \cdot s$ and $a = 2\pi \cdot f \cdot v$. Displacement and frequency thus determine also velocity and acceleration. Results of experiments or computations for vibrational analyses are often displayed in double-logarithmic form (Fig. 5.13). Above the displayed band, vibrations are no longer acceptable

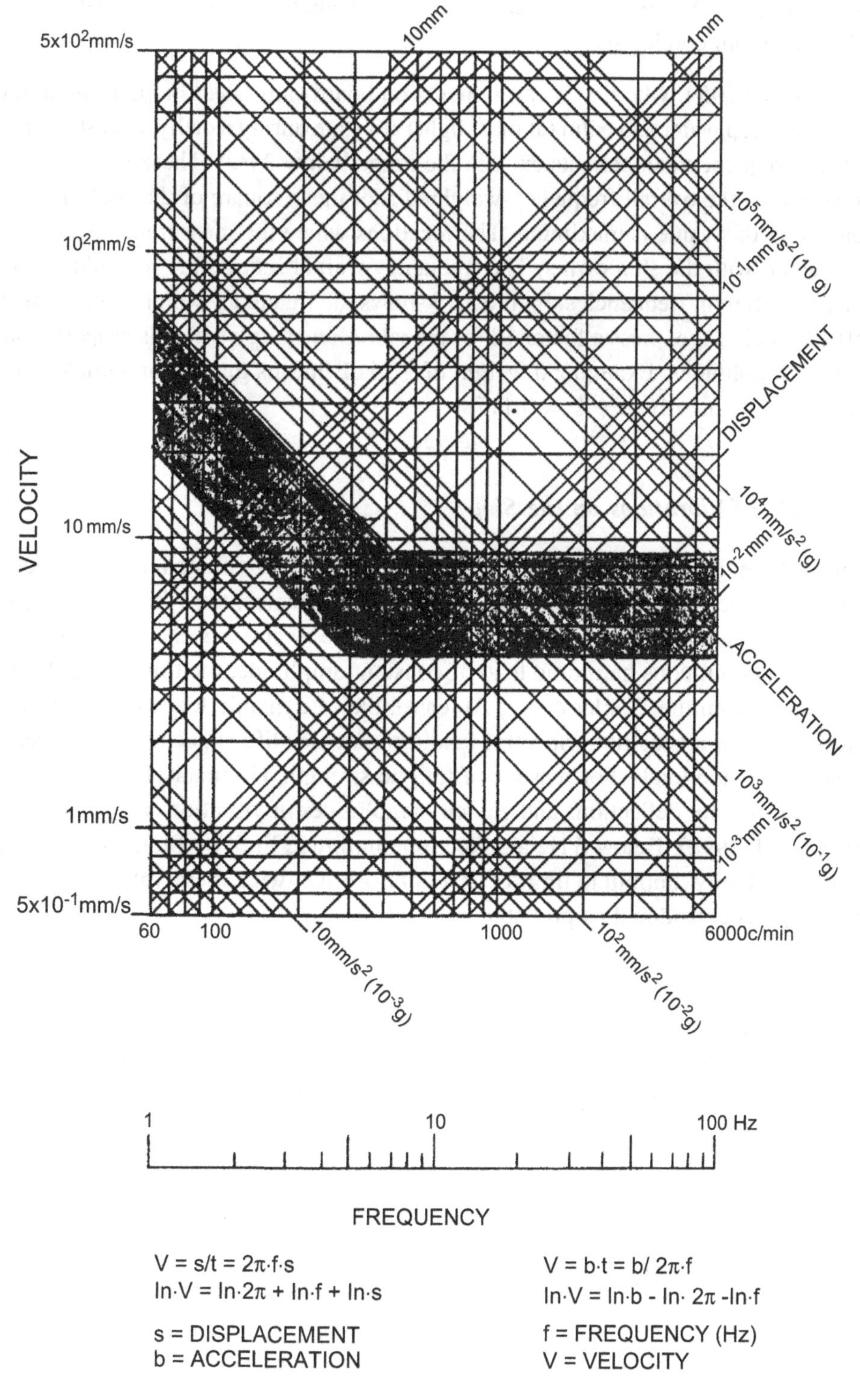

Figure 5.13:
Double-logarithmic vibration diagram, ISO 6954

according to ISO 6954. Below the band, vibrations are uncritical. Inside the band, they are acceptable in certain conditions.

According to ISO 6954, vibration measurements on board ships shall be performed in course straight ahead, deep water and calm (at most light) sea state during usually at least 1 min (2 min if significant frequency components exist in the range below 2 Hz). These measurements are used to determine the overall frequency-weighted root mean square of the acceleration in the direction where this value is maximum. The locations where the measurements are taken are typically agreed with the ship owner. The frequency weight reflects the individual human sensitivity for different frequencies. Figure 5.14 shows the frequency weights according to ISO 6954; 10 dB (decibels) express a factor 10 for the vibration energy which is proportional to the square of the amplitudes. Example: a weight of −15 dB means that the amplitude is to be multiplied by a factor $\sqrt{10^{-15/10}} = 0.1778$.

5.7.1. *Effects of Vibrations on the Ship*

The main problem here is fatigue of the structural design. To estimate the life expectancy of a ship structure under dynamic load, one needs the frequencies, the vibrational amplitudes, the time span of each load group, and the Wöhler (or S−N) curve of the material. For a one-step load, i.e. a continuous harmonic load of constant amplitude, the time to crack initiation follows directly from the Wöhler curve. An ensemble of dynamic loads of different amplitudes is called a spectrum. Spectral loading increases life expectancy considerably compared to one-step loads. For reasons of fatigue strength, Germanischer Lloyd recommends limits for deflections and velocity amplitudes of vibrations in aftbodies of ships (Asmussen et al. 1998). Below 5 Hz a deflection amplitude limit of 1 mm is recommended, above 5 Hz a velocity amplitude limit of 30 mm/s. Above twice these values, premature fatigue damage is considered as probable.

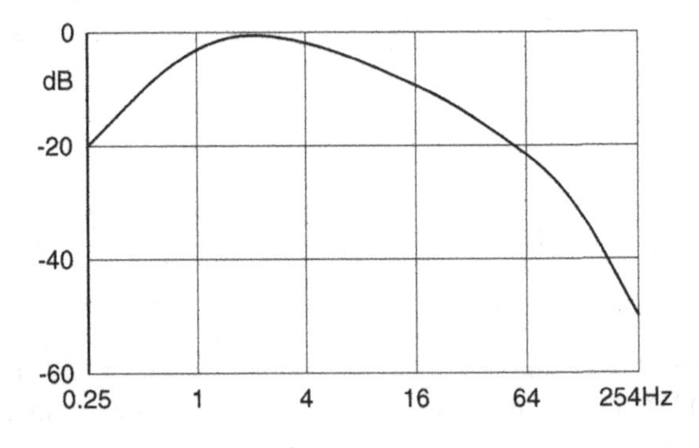

Figure 5.14:
Frequency weight according to ISO 6954

5.7.2. Effects on Engines

ISO 7919 and ISO 10816 specify that engines and connected aggregates should generally not be subject to vibrations exceeding any of the following limits: 0.71 mm deflection amplitude; 14 mm/s velocity amplitude; 0.7 g acceleration amplitude. For rudder gear rooms and bow thruster rooms, we accept velocity amplitudes approximately twice as high and acceleration amplitudes approximately four times as high. Classification societies have also incorporated these limits in their rules.

5.7.3. Effects of Vibrations on Humans

The most frequent cause of re-design due to vibration problems concerns effects on humans. Different standards exist for vibrations in ship rooms. ISO 6954 covers vibrations on merchant ships and its effects on humans. ISO 6954 gives values for the overall frequency-weighted root mean square acceleration and velocity, above which adverse comments are probable (Table 5.5). Below half these values, adverse comments are not probable. The zone between upper and lower values reflects the shipboard vibration environment commonly experienced and accepted. The values are determined over the frequency range from 1 Hz to 80 Hz. The limits are different for categories A (e.g. passenger cabins), B (e.g. crew rooms), and C (e.g. operational rooms).

For harmonic oscillations, root mean square σ and maximum amplitude a are coupled by $a = \sqrt{2} \cdot \sigma$.

Classification societies have introduced 'comfort classification' with respect to vibration levels (e.g. Det norske Veritas, Table 5.6): for comfort rating number 1, 2, or 3, the velocity amplitudes are considered separately for all appearing frequencies. For each individual frequency and each measured location, limit values may not be exceeded. Only for frequencies below 5 Hz can the measured velocity amplitude be reduced by the ratio of actual frequency to 5 Hz.

ISO 2631 also concerns the effect of mechanical vibrations on humans, at low frequencies 0.1–0.5 Hz (ship motion sickness) and higher frequencies 0.5–80 Hz (health, comfort). Important parameters are frequency, direction of vibration, and the form in which the vibrations enter the human body. We thus perceive vibrations differently depending on whether

Table 5.5: Values of overall frequency-weighted r.m.s. values for acceleration and velocity, above which adverse comments are probable, ISO 6954

A		B		C	
143 mm/s^2	4 mm/s	214 mm/s^2	6 mm/s	286 mm/s^2	8 mm/s

Table 5.6: Velocity amplitude limits (mm/s) for three comfort classes following Det norske Veritas

Comfort rating number	1	2	3
High speed and light craft			
– passenger localities, navigation bridge, offices	2.0	3.5	5.0
– control rooms	3.0	4.5	6.0
Passenger ships			
– top-grade cabins	1.5	2.0	2.5
– standard cabins, public spaces	1.5	2.5	4.0
– open deck recreation	2.5	3.5	5.0
Yacht (owner and guest areas)			
– accommodation on sea/in port	1.0/0.5	2.0/1.0	3.0/2.0
– outdoor recreation areas	2.0/0.5	3.0/1.0	4.0/2.0
– navigation bridge	1.5	2.5	4.0
Cargo ships			
– cabins, mess, recreation rooms, offices, bridge	2.5	3.5	5.0
– control rooms, work places	3.5	4.5	6.0

they enter the body via our feet, our hands, or our buttocks. The relevant frequency range lies approximately between 1 and 100 Hz. For vibrations below 1 Hz, the body reacts with motion sickness (nausea). Objective criteria cannot be formulated as individual people react very differently to vibrations. This may be due to different natural frequencies of individual body parts. Subjective criteria are comfort (feeling well), ability to perform, and health. ISO 2631 classifies the direction of vibrations in a body-fixed coordinate system (Fig. 5.15), and gives admissible vibrational accelerations (e.g. Fig. 5.16). The curves have the duration of the effect as a parameter.

Table 5.7 lists typical natural frequencies of body parts and symptoms for vertical vibrational exciting of a sitting human with amplitude at tolerance threshold.

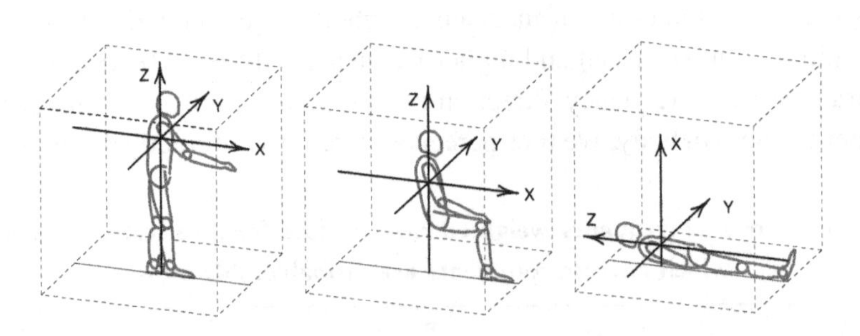

Figure 5.15:
Vibration directions with reference to human body

To obtain

"exposure limits" :multiply acceleration values by 2 (6 dB higher) ,
"reduced comfort boundary " :divide acceleration values by 3,15 (10 dB lower)

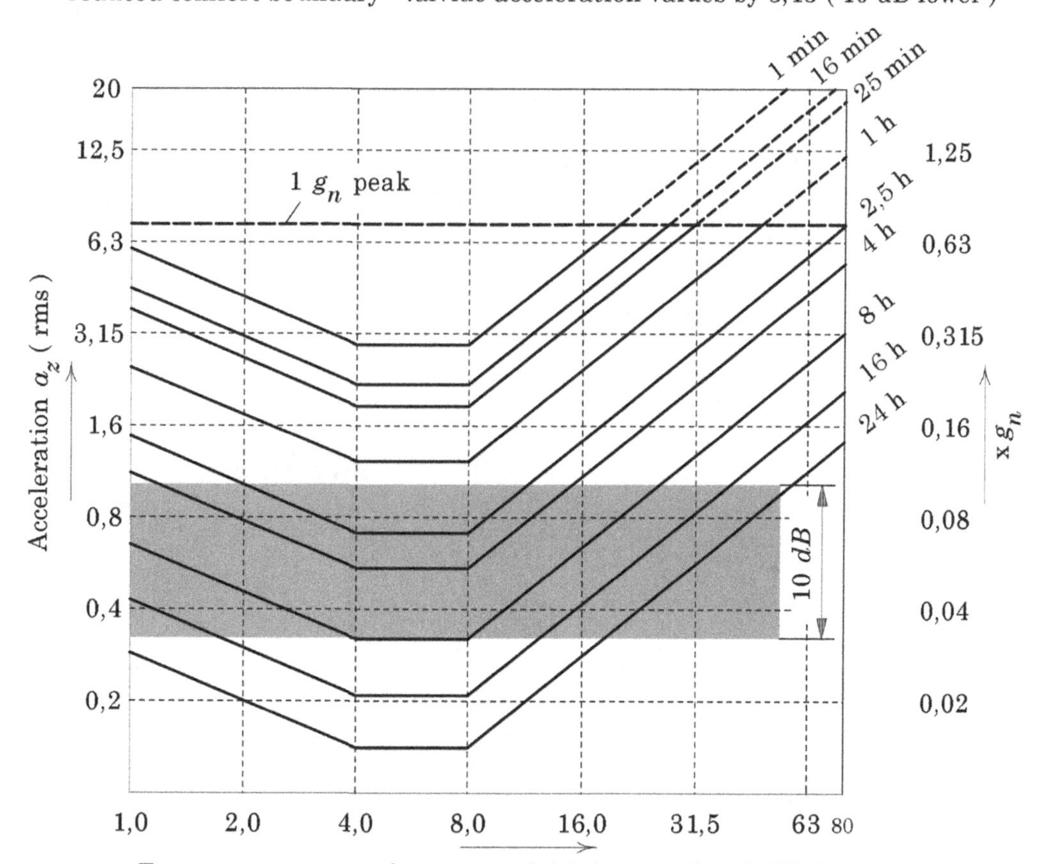

Figure 5.16:
Threshold values in the direction of the body's longitudinal axis, ISO 2631

Table 5.7: Effect of vibrations on sitting human

Body part	Effect of threshold vibration	Natural frequency
Brain	General discomfort, nausea	4.5–9 Hz
Head	Difficulty to speak	13–20 Hz
Chest	Pain in chest, breathing difficulties	4–8 Hz
Back	Back pain	8–12 Hz
Intestines	Urge to release feces	10.5–16 Hz
Bladder	Urge to release urine	10–18 Hz
Legs	Increased muscle tension	13–20 Hz
Abdomen	Abdominal pain	4.5–10 Hz

CHAPTER 6

Ship Maneuvering

Chapter Outline

6.1. Introduction

Ship maneuvering comprises:

- course keeping (this concerns only the direction of the ship's longitudinal axis);
- course changing;
- track keeping (important in restricted waters);
- speed changing (especially stopping).

Maneuvering requirements are a standard part of the contract between shipyard and shipowner. IMO regulations specify minimum requirements for all ships, but shipowners may introduce

Practical Ship Hydrodynamics. DOI: 10.1016/B978-0-08-097150-6.10006-5

additional or more severe requirements for certain ship types, e.g. tugs, ferries, dredgers, exploration ships. Important questions for the specification of ship maneuverability may include:

- Does the ship keep a reasonably straight course (in autopilot or manual mode)?
- Under what conditions (current, wind) can the ship berth without tug assistance?
- Up to what ratio of wind speed to ship speed can the ship still be kept on all courses?
- Can the ship lay rudder in acceptable time from one side to the other?

Ship maneuverability is described by the following main characteristics:

- initial turning ability: ability to initiate a turn (rather quickly);
- sustained turning ability: ability for sustained (rather high) turning speed;
- yaw checking ability: ability to stop turning motion (rather quickly);
- stopping ability: ability to stop (in rather short distance and time);
- yaw stability: ability to move straight ahead in the absence of external disturbances (e.g. wind) at one rudder angle (so-called neutral rudder angle).

The sustained turning ability appears to be the least important, since it describes the ship behavior only for a time long after initiating a maneuver. The stopping ability is of interest only for slow speeds. For avoiding obstacles at high ship speed, it is far more effective to change course than to stop. (Course changes require less distance than stopping maneuvers for full speed.)

Understanding ship maneuvering and the related numerical and experimental tools is important for the designer for the choice of maneuvering equipment of a ship. Items of the maneuvering equipment may be:

- rudders;
- fixed fins (e.g. above the rudder; skeg);
- jet thrusters;
- propellers (including fixed pitch, controllable pitch, slewable, and cycloidal (e.g. Voith–Schneider propellers);
- adjustable ducts for propellers, steering nozzles;
- waterjets.

Both maneuvering and seakeeping of ships concern time-dependent ship motions, albeit with some differences:

- The main difficulty in both fields is to determine the fluid forces on the hull (including propeller and rudder) due to ship motions (and possibly waves).
- At least a primitive model of the maneuvering forces and motions should be part of any seakeeping simulation in oblique waves.

- Contrary to seakeeping, maneuvering is often investigated in shallow (and usually calm) water and sometimes in channels.
- Linear relations between velocities and forces are reasonable approximations for many applications in seakeeping; in maneuvering they are applicable only for rudder angles of a few degrees. This is one reason for the following differences.
- Seakeeping is mostly investigated in the frequency domain; maneuvering investigations employ time-domain simulations.
- In seakeeping, motion equations are written in an inertial coordinate system; in maneuvering simulations a ship-fixed system is applied. (This system, however, typically does not follow heel motions.)
- For fluid forces, viscosity is usually assumed to be of minor importance in seakeeping computations. In maneuvering simulations, the free surface is often neglected. Ideally, both free surface and viscous effects should be considered for both seakeeping and maneuvering.

Here we will focus on the most common computational methods for manoeuvring flows. Far more details, especially on maneuvering devices, can be found in Brix (1993).

6.2. Simulation of Maneuvering with Known Coefficients

6.2.1. Introduction and Definitions

The hydrodynamic forces of main interest in maneuvering are:

- the longitudinal force (resistance) X;
- the transverse force Y;
- the yaw moment N;

depending primarily on:

- the longitudinal speed u and acceleration $\dot{u}$;
- transverse speed v at midship section and acceleration $\dot{v}$;
- yaw rate (rate of rotation) $r = \dot{\psi}$ (rad/time) and yaw acceleration $\dot{r} = \ddot{\psi}$, where ψ is the yaw angle;
- the rudder angle δ (positive to port).

For heel angles exceeding approximately 10°, these relations are influenced substantially by heel. The heel may be caused by wind or, for Froude numbers exceeding approximately 0.25, by the maneuvering motions themselves. Thus at least for fast ships we are also interested in:

- the heeling moment K;
- the heel angle ϕ.

For scaling these forces and moments from model to full scale, or for estimating them from results in similar ships, X, Y, K, and N are made non-dimensional in one of the following ways:

$$\begin{Bmatrix} X' \\ Y' \\ K' \\ N' \end{Bmatrix} = \frac{1}{q \cdot L^2} \begin{Bmatrix} X \\ Y \\ K/L \\ N/L \end{Bmatrix} \quad \text{or} \quad \begin{Bmatrix} C_X \\ C_Y \\ C_K \\ C_N \end{Bmatrix} = \frac{1}{q \cdot L \cdot T} \begin{Bmatrix} X \\ Y \\ K/L \\ N/L \end{Bmatrix} \tag{6.1}$$

with $q = \rho \cdot u^2/2$ and ρ water density. Note that here we use the instantaneous longitudinal speed u (for $u \neq 0$) as reference speed. Alternatively, the ship speed at the beginning of the maneuver may be used as reference speed. L is the length between perpendiculars. The term 'forces' will from now on include both forces and moments, unless otherwise stated.

The motion velocities and accelerations are also made non-dimensional by suitable powers of u and L:

$$v' = v/u\,; \quad r' = r \cdot L/u\,; \quad \dot{u}' = \dot{u} \cdot L/u^2\,; \quad \dot{v}' = \dot{v} \cdot L/u^2\,; \quad \dot{r}' = \dot{r} \cdot L^2/u^2 \tag{6.2}$$

6.2.2. Force Coefficients

CFD or model tests may be used to determine the force coefficients. Then the body forces may be approximated by expressions like:

$$Y' = Y_{\dot{v}}' \cdot \dot{v}' + Y_{\dot{r}}' \cdot \dot{r}' + Y_v' \cdot v' + Y_{v^3}' \cdot (v')^3 + Y_{vr^2}' \cdot v'(r')^2 + Y_{v\delta^2}' \cdot v'\delta^2 + Y_r' \cdot r' + Y_{r^3}' \cdot (r')^3 + \ldots \tag{6.3}$$

$Y_{\dot{v}}'$... are non-dimensional coefficients. Unlike the above formula, such expressions may also involve terms like $Y_{ru}' \cdot r' \cdot \Delta u'$, where $\Delta u' = (u - V)/u$. V is a reference speed, normally the speed at the beginning of the maneuver. Comprehensive tables of such coefficients have been published, e.g. Wolff (1981) for models of five ship types (tanker, Series 60 $C_B = 0.7$, mariner, container ship, ferry) (Tables 6.1 and 6.2). The coefficients for u are based on $\Delta u = u - V$ in these tables. Corresponding to the small Froude numbers, the values do not contain heeling moments and the dependency of coefficients on heel angle. Such tables, together with the formulae for X, Y, and N as given above, may be used for time simulations of motions of such ships for an arbitrary time history of the rudder angle.

Wolff's results are deemed to be more reliable than other experimental results because they were obtained in large-amplitude, long-period motions of relatively large models (L between 6.4 and 8.7 m). Good accuracy in predicting the maneuvers of sharp single-screw ships in full scale from coefficients obtained from experiments with such models has been demonstrated. For full ships, for twin-screw ships, and for small models, substantial differences between model and full-scale maneuvering motions are observed. Correction methods from model to full scale need still further improvement.

Table 6.1: Data of four models used in maneuvering experiments (Wolff 1981)

	Tanker	Series 60	Container	Ferry
Scale	1:35	1:26	1:34	1:16
L_{pp}	8.286 m	7.034 m	8.029 m	8.725 m
B	1.357 m	1.005 m	0.947 m	1.048 m
T_{fp}	0.463 m	0.402 m	0.359 m	0.369 m
T_m	0.459 m	0.402 m	0.359 m	0.369 m
T_{ap}	0.456 m	0.402 m	0.359 m	0.369 m
C_B	0.805	0.700	0.604	0.644
Coord. origin aft of FP	4.143 m	3.517 m	4.014 m	4.362 m
LCG (x_G)	−0.270 m	0.035 m	−0.160 m	−0.149 m
Radius of gyration i_z	1.900 m	1.580 m	1.820 m	1.890 m
No. of propellers	1	1	2	2
Propeller turning	Right	Right	Outward	Outward
Propeller diameter	0.226 m	0.279 m	0.181 m	0.215 m
Propeller P/D	0.745	1.012	1.200	1.135
Propeller A_E/A_0	0.60	0.50	0.86	0.52
No. of blades	5	4	5	4

For small deviations of the ship from a straight path, only linear terms in the expressions for the forces need to be retained. In addition we neglect heel and all those terms that vanish for symmetrical ships to obtain the equations of motion:

$$(X'_{\dot{u}} - m')\dot{u}' + X'_u \Delta u' + X'_n \Delta n' = 0 \tag{6.4}$$

$$(Y'_{\dot{v}} - m')\dot{v}' + (Y'_{\dot{r}} - m'x'_G)\dot{r}' + Y'_v v' + (Y'_r - m')r' = -Y'_\delta \delta \tag{6.5}$$

$$(N'_{\dot{v}} - m'x'_G)\dot{v}' + (N'_{\dot{r}} - I'_{zz})\dot{r}' + N'_v v' + N'_v v' + (N'_r - m'x'_G)r = -N'_\delta \delta \tag{6.6}$$

I_{zz} is the moment of inertia with respect to the z-axis:

$$I_{zz} = \int (x^2 + y^2)\,\mathrm{d}m \tag{6.7}$$

$m' = m/(\frac{1}{2}\rho L^3)$ is the non-dimensional mass, $I'_{zz} = I_{zz}/(1/2 \rho L^5)$ the non-dimensional moment of inertia coefficient.

If we just consider the linearized equations for side forces and yaw moments, we may write:

$$M'\vec{\dot{u}}' + D'\vec{u}' = \vec{r}'\Delta \tag{6.8}$$

with:

$$M' = \begin{bmatrix} -Y'_{\dot{v}} + m' & -Y'_{\dot{r}} + m'x'_G \\ -N'_{\dot{v}} + m'x'_G & -N'_{\dot{r}} + I'_{zz} \end{bmatrix}; \quad \vec{u}' = \begin{Bmatrix} v' \\ r' \end{Bmatrix} \tag{6.9}$$

$$D' = \begin{bmatrix} -Y'_v & -Y'_r + m' \\ -N'_v & -N'_r + m'x'_G \end{bmatrix}; \; \vec{r}' = \begin{Bmatrix} Y'_\delta \\ N'_\delta \end{Bmatrix} \tag{6.10}$$

M' is the mass matrix, D' the damping matrix, $\vec{r}'$ the rudder effectiveness vector, and $\vec{u}'$ the motion vector. The terms on the right-hand side thus describe the steering action of the rudder. Some modifications of the above equation of motion are of interest:

1. If in addition a side thruster at location x_t is active with thrust T, the (non-dimensional) equation of motion modifies to:

$$M'\dot{\vec{u}}' + D'\vec{u}' = \vec{r}'\Delta + \begin{Bmatrix} T' \\ T'x'_t \end{Bmatrix} \tag{6.11}$$

Table 6.2: Non-dimensional hydrodynamic coefficients of four ship models (Wolff 1981); values to be multiplied by 10^{-6}

Model of	*Tanker*	*Series 60*	*Container*	*Ferry*
Initial F_n	*0.145*	*0.200*	*0.159*	*0.278*
m'	14622	11432	6399	6765
$x'_G m'$	365	57	−127	−116
I'_{zz}	766	573	329	319
$X'_{\dot{u}}$	−1077	−1064	0	0
$X'_{\dot{u}u^2}$	−5284	0	0	0
X'_u	−2217	−2559	−1320	−4336
X'_{u^2}	1510	0	1179	−2355
X'_{u^3}	0	−2851	0	−2594
X'_{v^2}	−889	−3908	−1355	−3279
X'_{r^2}	237	−838	−151	−571
X'_{δ^2}	−1598	−1346	−696	−2879
X'_{v^2u}	0	−1833	−2463	−2559
X'_{δ^2u}	2001	2536	0	3425
X'_{r^2u}	0	0	−470	−734
X'_{vr}	9478	7170	3175	4627
$X'_{v\delta}$	1017	942	611	877
$X'_{r\delta}$	−482	−372	−340	−351
X'_{vu}	745	0	0	0
X'_{vu^2}	0	0	−207	0
X'_{ru}	0	−270	0	0
X'_r	48	0	0	−19
X'_δ	166	0	0	0
$X'_{\delta u^2}$	0	150	0	0
$X'_{v^2\delta}$	−4717	0	0	0
$X'_{r^2\delta}$	−365	0	0	0
X'_{v^3}	1164	2143	0	0
X'_{r^3}	−118	0	0	0
X'_{δ^3u}	−278	0	0	0
X'_{δ^4}	0	621	213	2185
X'_{v^3u}	0	0	−3865	0
X'_{r^3u}	0	0	−447	0

Model of	*Tanker*	*Series 60*	*Container*	*Ferry*
$Y'_{\dot{v}}$	−11420	−12608	−6755	−7396
$Y'_{\dot{v}v^2}$	−21560	−34899	−10301	0
$Y'_{\dot{r}}$	−714	−771	−222	−600
$Y'_{\dot{r}r^2}$	−468	166	−63	0
Y'_0	−244	26	0	0
Y'_u	263	−69	−33	57
Y'_v	−15338	−16630	−8470	−12095
Y'_{v^3}	−36832	−45034	0	−137302
Y'_{vr^2}	−19040	−37169	−31214	−44365
$Y'_{v\delta^2}$	0	0	−4668	2199
Y'_r	4842	4330	2840	1901
Y'_{r^2}	0	152	85	0
Y'_{r^3}	1989	2423	−1945	−1361
Y'_{ru}	0	−1305	2430	−1297
Y'_{ru^2}	0	0	4769	0
Y'_{rv^2}	22878	10230	−33237	−36490
$Y'_{r\delta^2}$	1492	0	0	−2752
Y'_δ	3168	2959	1660	3587
Y'_{δ^2}	0	0	0	98
Y'_{δ^3}	3621	−7494	0	0
Y'_{δ^4}	1552	613	−99	0
Y'_{δ^5}	−5526	4344	−1277	−6262
$Y'_{\delta v^2}$	0	0	13962	0
$Y'_{\delta r^2}$	1637	0	2438	0
$Y'_{\delta u}$	−4562	−4096	0	−5096
$Y'_{\delta u^2}$	0	974	0	0
Y'_{δ^3u}	2640	4001	0	3192
$Y'_{v\|v\|}$	−11513	−19989	−47566	0
$Y'_{r\|r\|}$	−351	0	1731	0
$Y'_{\delta\|\delta\|}$	−889	2029	0	0
Y'_{v^3r}	12398	0	0	0
Y'_{r^3u}	0	2070	0	0

TABLE 6.2 continued

Model of	*Tanker*	*Series 60*	*Container*	*Ferry*
$N'_{\dot{v}}$	−523	326	239	426
$N'_{\dot{v}v^2}$	2311	1945	5025	10049
$N'_{\dot{r}}$	−576	−461	−401	−231
$N'_{\dot{r}r^2}$	−130	−250	132	0
N'_0	67	9	0	0
N'_u	−144	37	8	−36
N'_v	−5544	−6570	−3800	−3919
N'_{v^2}	−132	0	0	0
N'_{v^3}	−2718	−16602	−23865	−33857
N'_{vu}	0	−1146	−2179	−3666
N'_{vr^2}	3448	4421	−4586	0
$N'_{v\delta^2}$	2317	0	1418	570
N'_r	−3074	−2900	−1960	−2579
N'_{r^2}	0	−45	0	0
N'_{r^3}	−865	−1919	−729	−2253
N'_{ru}	0	0	−473	0
N'_{ru^2}	913	0	0	0
N'_{rv^2}	−16196	−20530	−27858	−60110

Model of	*Tanker*	*Series 60*	*Container*	*Ferry*
$N'_{r\delta^2}$	−324	0	−404	237
N'_{δ}	−1402	−1435	−793	−1621
N'_{δ^2}	0	−138	0	−73
N'_{δ^3}	−1641	3907	0	0
N'_{δ^4}	−536	0	0	0
N'_{δ^5}	2220	−2622	652	2886
$N'_{\delta v^2}$	0	0	−6918	−2950
$N'_{\delta r^2}$	−855	0	−1096	−329
$N'_{\delta u}$	2321	1856	0	2259
$N'_{\delta u^2}$	0	−568	0	0
$N'_{\delta^2 u}$	316	0	0	0
$N'_{\delta^3 u}$	−1538	−1964	0	−1382
$N'_{v\|v\|}$	0	5328	8103	0
$N'_{r\|r\|}$	0	0	−1784	0
N'_{vr}	−394	0	0	0
$N'_{\delta\|\delta\|}$	384	−1030	0	0
$N'_{v^3 u}$	−27133	−13452	0	0
$N'_{r^3 u}$	0	−476	0	−1322

2. For steady turning motion ($\dot{\vec{u}}' = 0$), the original linearized equation of motion simplifies to:

$$D'\vec{u}' = \vec{r}'\delta \tag{6.12}$$

Solving this equation for r' yields:

$$r = \frac{Y'_\delta N_v - Y_v N_\delta}{Y'_v(Y'_r - m')C'}\delta \tag{6.13}$$

C' is the yaw stability index:

$$C' = \frac{N'_r - m'x'_G}{-m'} - \frac{N'_v}{Y'_v} \tag{6.14}$$

$Y'_v(Y'_r - m')$ is positive, the nominator (almost) always negative. Thus C' determines the sign of r'. Positive C' indicates yaw stability, negative C' yaw instability. Yaw instability is the tendency of the ship to increase the absolute value of an existing drift angle. However, the formula is numerically very sensitive and measured coefficients are often too inaccurate for predictions. Therefore, usually more complicated analyses are necessary to determine yaw stability.

3. If the transverse velocity in the equation of motion is eliminated, we obtain a differential equation of second order of the form:

$$T_1T_2\ddot{r} + (T_1 + T_2)\cdot\dot{r} + r = -K(\delta + T_2\dot{\delta}) \tag{6.15}$$

The T_i are time constants. $|T_2|$ is much smaller than $|T_1|$ and thus may be neglected, especially since linearized equations are anyway a (too) strong simplification of the problem, yielding the simple 'Nomoto' equation:

$$T\dot{r} + r = -K\delta \tag{6.16}$$

T and K here denote time constants. K is sometimes called rudder effectiveness. This simplified equation neglects not only all non-linear effects, but also the influence of transverse speed, longitudinal speed, and heel. As a result, the predictions are too inaccurate for most practical purposes. The Nomoto equation allows, however, a quick estimate of rudder effects on course changes. A slightly better approximation is the 'Norrbin' equation:

$$T\dot{r} + r + \alpha r^3 = -K\delta \tag{6.17}$$

α here is a non-linear 'damping' factor of the turning motions. The constants are determined by matching measured or computed motions to fit the equations best. The Norrbin equation still does not contain any unsymmetrical terms, but for single-screw ships the turning direction of the propeller introduces an asymmetry, making the Norrbin equation questionable.

4. The stability index is difficult to compute due to the numeric sensitivity. Subsequently, the slope of the spiral curve in the origin (for three degrees of freedom) has gained popularity as a single, relatively simple indicator for ships that handle 'correctly':

$$\frac{N'_{v'}\cdot Y'_\delta - Y'_{v'}\cdot N'_\delta}{Y'_{v'}(N'_{r'} - m'x'_g) - N'_{v'}(Y'_{r'} - m')} < 0 \tag{6.18}$$

The following regression formulae for linear velocity and acceleration coefficients have been proposed (Clarke et al. 1983):

$$Y'_{\dot{v}} = -\pi(T/L)^2\cdot(1 + 0.16C_B\cdot B/T - 5.1(B/L)^2) \tag{6.19}$$

$$Y'_{\dot{r}} = -\pi(T/L)^2\cdot(0.67B/L - 0.0033(B/T)^2) \tag{6.20}$$

$$N'_{\dot{v}} = -\pi(T/L)^2\cdot(1.1B/L - 0.041B/T) \tag{6.21}$$

$$N'_{\dot{r}} = -\pi(T/L)^2 \cdot (1/12 + 0.017C_B \cdot B/T - 0.33B/L) \quad (6.22)$$

$$Y'_v = -\pi(T/L)^2 \cdot (1 + 0.40C_B \cdot B/T) \quad (6.23)$$

$$Y'_r = -\pi(T/L)^2 \cdot (-0.5 + 2.2B/L - 0.08B/T) \quad (6.24)$$

$$N'_v = -\pi(T/L)^2 \cdot (0.5 + 2.4T/L) \quad (6.25)$$

$$N'_r = -\pi(T/L)^2 \cdot (0.25 + 0.039B/T - 0.56B/L) \quad (6.26)$$

T is the mean draft. These formulae apply to ships on even keel. For ships with draft difference $t = T_{ap} - T_{fp}$, correction factors may be applied to the linear even-keel velocity coefficients (Inoue and Kijima 1978):

$$Y'_v(t) = Y'_v(0) \cdot (1 + 0.67t/T) \quad (6.27)$$

$$Y'_r(t) = Y'_r(0) \cdot (1 + 0.80t/T) \quad (6.28)$$

$$N'_v(t) = N'_v(0) \cdot (1 - 0.27t/T \cdot Y'_v(0)/N'_v(0)) \quad (6.29)$$

$$N'_r(t) = N'_r(0) \cdot (1 + 0.30t/T) \quad (6.30)$$

These formulae are based both on theoretical considerations and on model experiments with four 2.5 m models of the Series 60 with different block coefficients for $0.2 < t/T < 0.6$.

In cases where u and/or the propeller turning rate n vary strongly during a maneuver or even change sign as in a stopping maneuver, the above coefficients will vary widely. Therefore, the so-called four-quadrant equations (e.g. Sharma 1986) are better suited to represent the forces. These equations are based on a physical explanation of the forces due to hull, rudder, and propeller, combined with coefficients to be determined in experiments.

6.2.3. Physical Explanation and Force Estimation

In the following, forces due to non-zero rudder angles are not considered. If the rudder at the midship position is treated as part of the ship's body, only the difference between rudder forces at the actual rudder angle δ and those at $\delta = 0°$ have to be added to the body forces treated here. The gap between ship stern and rudder may be disregarded in this case. Propeller forces and hull resistance in straightforward motion are neglected here.

We use a coordinate system with origin fixed at the midship section on the ship's center plane at the height of the center of gravity (Fig. 6.1). The x-axis points forward, y to starboard, z vertically downward. Thus the system participates in the motions u, v, and r of the ship, but does not follow the ship's heeling motion. This simplifies the integration in time (e.g. by

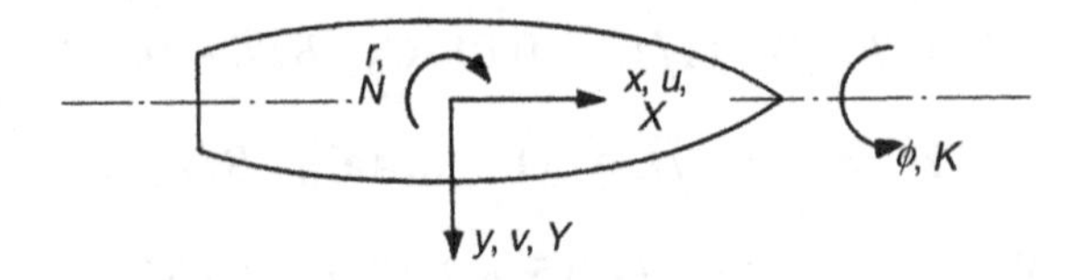

Figure 6.1:
Coordinates x, y; direction of velocities u, v, r, forces X, Y, and moments K, N

a Runge–Kutta scheme) of the ship's position from the velocities u, v, r and eliminates several terms in the force formulae.

Hydrodynamic body forces can be imagined to result from the change of momentum (= mass · velocity) of the water near to the ship. Most important in maneuvering is the transverse force acting upon the hull per unit length (e.g. meter) in the x-direction. According to the slender-body theory, this force is equal to the time rate of change of the transverse momentum of the water in a 'strip' between two transverse planes spaced one unit length. In such a 'strip' the water near to the ship's side mostly follows the transverse motion of the respective ship section, whereas water farther from the hull is less influenced by transverse ship motions. The total effect of this water motion on the transverse force is the same as if a certain 'added mass' per length m' moved exactly like the ship section in transverse direction. (This approach is thus similar to the strip method approach in ship seakeeping.)

The added mass m' may be determined for any ship section as:

$$m' = \frac{1}{2}\pi \cdot \rho \cdot T_x^2 \cdot c_y \tag{6.31}$$

T_x is the section draft and c_y a coefficient. c_y may be calculated:

- analytically if we approximate the actual ship section by a 'Lewis section' (conformal mapping of a semicircle); Figure 6.2 shows such solutions for parameters (T_x/B) and β = immersed section area/($B \cdot T_x$);
- for arbitrary shape by a close-fit boundary element method as for 'strips' in seakeeping strip methods, but for maneuvering the free surface is generally neglected;
- by field methods including viscosity effects.

Neglecting influences due to heel velocity $\dot{\phi}$ and heel acceleration $\ddot{\phi}$, the time rate of change of the transverse momentum of the 'added mass' per length is:

$$\left(\frac{\partial}{\partial t} - u \cdot \frac{\partial}{\partial x}\right)[m'(v + x \cdot r)] \tag{6.32}$$

$\partial/\partial t$ takes account of the local change of momentum (for fixed x) with time t. The term $u \cdot \partial/\partial x$ results from the convective change of momentum due to the longitudinal motion of the water 'strip' along the hull with appropriate velocity $-u$ (i.e. from bow to stern). $v + x \cdot r$ is the

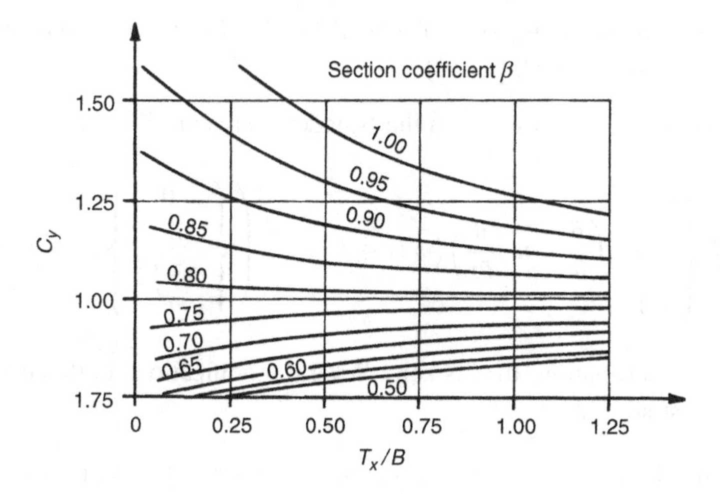

Figure 6.2:
Section added mass coefficient c_y for low-frequency, low-speed horizontal acceleration

transverse velocity of the section in the y-direction resulting from both transverse speed v at midship section and the yaw rate r. The total transverse force is obtained by integrating the above expression over the underwater ship length L. The yaw moment is obtained by multiplying each force element with the respective lever x, and the heel moment is obtained by using the vertical moment $z_y m'$ instead of m', where z_y is the depth coordinate of the center of gravity of the added mass. For Lewis sections, this quantity can be calculated theoretically (Fig. 6.3). For CFD approaches the corresponding vertical moment is computed directly as part

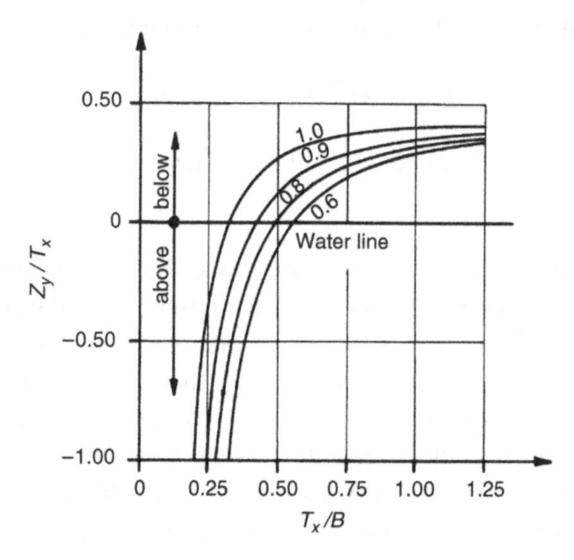

Figure 6.3:
Height coordinate z_y of section added mass m'

of the numerical solution. Söding gives a short FORTRAN subroutine to determine c_y and z_y for Lewis sections in Brix (1993, p. 252).

Based on these considerations we obtain the 'slender-body contribution' to the forces as:

$$\begin{Bmatrix} X \\ Y \\ K \\ N \end{Bmatrix} = \int_L \begin{Bmatrix} 0 \\ 1 \\ 1 \\ x \end{Bmatrix} \cdot \left(\frac{\partial}{\partial t} - u \cdot \frac{\partial}{\partial x}\right)\left(m'(v + x \cdot r)\right) \cdot \left(\begin{Bmatrix} 0 \\ m' \\ -z_y m' \\ m' \end{Bmatrix}(v + x \cdot r)\right) \mathrm{d}x \quad (6.33)$$

The 'slender-body contribution' to X is zero. Several modifications to this basic formula are necessary or at least advisable:

1. For terms involving $-\partial/\partial t$, i.e. for the acceleration-dependent parts of the forces, correction factors k_1, k_2 should be applied. They consider the lengthwise flow of water around bow and stern which is initially disregarded in determining the sectional added mass m'. The acceleration part of the above basic formula then becomes:

$$\begin{Bmatrix} X_1 \\ Y_1 \\ K_1 \\ N_1 \end{Bmatrix} = \int_L \begin{Bmatrix} 0 \\ -k_1 m' \\ z_y m' \\ -k_2 x m' \end{Bmatrix} \cdot (k_1 \dot{v} + k_2 x \cdot \dot{r})\, \mathrm{d}x \quad (6.34)$$

k_1 and k_2 are approximated here by regression formulae which were derived from the results of three-dimensional flow calculations for accelerated ellipsoids:

$$k_1 = \sqrt{1 - 0.245\varepsilon - 1.68\varepsilon^2} \quad (6.35)$$

$$k_2 = \sqrt{1 - 0.76\varepsilon - 4.41\varepsilon^2} \quad (6.36)$$

with $\varepsilon = 2T_x/L$.

2. For parts in the basic formula due to $u \cdot \partial/\partial x$, one should distinguish terms where $\partial/\partial x$ is applied to the first factor containing m' from terms where the second factor $v + x \cdot r$ is differentiated with respect to x (which results in r). For the former terms, it was found by comparison with experimental values that the integral should be extended only over the region where $\mathrm{d}m'/\mathrm{d}x$ is negative, i.e. over the forebody. This may be understood as the effect of flow separation in the aftbody. The flow separation causes the water to retain most of its transverse momentum behind the position of maximum added mass, which for ships without trim may be taken to be the midship section. The latter terms, however, should be integrated over the full length of the ship. This results in:

$$\begin{Bmatrix} X_2 \\ Y_2 \\ K_2 \\ N_2 \end{Bmatrix} = u \begin{Bmatrix} 0 \\ -m'_m \\ z_{ym}m'_m \\ -x_m m'_m \end{Bmatrix} \cdot (v + x_m \cdot r) + u \cdot r \int_{x_a}^{x_m} \begin{Bmatrix} 0 \\ m' \\ -z_{ym}m' \\ xm' \end{Bmatrix} \mathrm{d}x - u \int_{x_m}^{x_f} \begin{Bmatrix} 0 \\ 0 \\ 0 \\ m' \end{Bmatrix} (v + x \cdot r)\,\mathrm{d}x \tag{6.37}$$

x_a is the x coordinate of the aft end, x_f of the forward end of the ship. The index m refers to the x coordinate where m' is maximum. For negative u the differences in treating the fore- and aftbody are interchanged.

3. The slender-body theory disregards longitudinal forces associated with the added mass of the ship in the longitudinal direction. These additional terms are taken from potential-flow theory without flow separation (Newman 1977):

$$\begin{Bmatrix} X_3 \\ Y_3 \\ K_3 \\ N_3 \end{Bmatrix} = \begin{Bmatrix} -m_x \cdot \dot{u} \\ -m_x \cdot u \cdot r \\ 0 \\ -m_x \cdot u \cdot v \end{Bmatrix} \tag{6.38}$$

m_x is the added mass for longitudinal motion; it may be approximated by a formula which was also fitted to theoretical values for ellipsoids:

$$m_x = \frac{m}{\pi\sqrt{L^3/\nabla} - 14} \tag{6.39}$$

∇ denotes here the volume displacement. Theoretically additional terms proportional to $r \cdot v$ and r^2 should appear in the formula for X. According to experiments with ship models, however, the $r \cdot v$ term is much smaller and the r^2 term may even have a different sign than the theoretical expression. Therefore these terms, which are influenced substantially by flow separation, have been omitted. Further, some theoretical terms of small magnitude involving heeling motion or referring to the heeling moment have also been omitted in the above formula for X_3, etc.

4. Because slender-body theory neglects flow separation in transverse flow around ship sections (only longitudinal flow separation is roughly taken into account), an additional 'cross-flow resistance' of the ship sections has to be added. The absolute value of this resistance per unit length is:

$$\frac{1}{2}\rho \cdot T_x \cdot v_x^2 \cdot C_D \tag{6.40}$$

$v_x = v + x \cdot r$ is the transverse velocity of the section. C_D is a cross-flow resistance coefficient. The direction of the resistance is opposite to the direction of v_x. Thus for arbitrary direction of motion, the term $-v_x|v_x|$ is required instead of v_x^2. Therefore the cross-flow resistance adds the following contributions to the body forces:

$$\begin{Bmatrix} X_4 \\ Y_4 \\ K_4 \\ N_4 \end{Bmatrix} = \frac{1}{2}\rho \int_L \begin{Bmatrix} 0 \\ -1 \\ z_D \\ -x \end{Bmatrix} (v + x \cdot r) \cdot |v + x \cdot r| \cdot T_x C_D \, \mathrm{d}x \tag{6.41}$$

z_D is the z coordinate (measured downward from the center of gravity G of ship's mass m) of the action line of the cross-flow resistance. For typical cargo ship hull forms, this force acts on about 65% of the draft above the keel line. Thus a constant (mean) value over ship length of:

$$z_D = \overline{KG} - 0.65T \tag{6.42}$$

may be applied to the formula for X_4, etc. For tug models, values of 1.0 ± 0.1 instead of the above 0.65 were found.

C_D is estimated as 1.0 averaged over the whole ship length for cargo vessels like container ships with bilge keels. For fuller hulls values between 0.5 and 0.7 may be suitable. The C_D values are generally higher in the aftbody than in the forebody due to stronger flow separation in the aftbody.

Results of transverse towing tests (at zero speed) with and without heel with large models are presented in Table 6.3. These results differ from the situation at considerable forward speed.

The sum of contributions 1 to 4 constitutes the total body force:

$$\begin{Bmatrix} X \\ Y \\ K \\ N \end{Bmatrix} = \begin{Bmatrix} X_1 \\ Y_1 \\ K_1 \\ N_1 \end{Bmatrix} + \begin{Bmatrix} X_2 \\ Y_2 \\ K_2 \\ N_2 \end{Bmatrix} + \begin{Bmatrix} X_3 \\ Y_3 \\ K_3 \\ N_3 \end{Bmatrix} + \begin{Bmatrix} X_4 \\ Y_4 \\ K_4 \\ N_4 \end{Bmatrix} \tag{6.43}$$

Table 6.3: Results of transverse towing tests with large models upright and with 10° heel; models were equipped with rudder and propeller but without bilge keels

	Cargo ship	Tanker	Tanker	Container ship	Twin-screw salvage tug
L/B	6.66	5.83	6.11	7.61	5.21
B/T	2.46	2.43	2.96	2.93	2.25
C_B	0.66	0.84	0.81	0.58	0.58
C_D	0.562	0.983	0.594	0.791	0.826
$C_{D10°}$	0.511	1.151	–	1.014	–

For steady traversing or pure yaw motion without forward speed, only terms listed under 4 above are relevant.

The yaw stability is very sensitive to small changes in the body forces. Therefore a reliable prediction of yaw stability based on the slender-body theory or regression analysis of model tests is not possible. Substantial improvements of theoretical calculations seem possible only if the flow separation around the hull is determined in detail by computational simulation of the viscous, turbulent flow in a RANSE code (with appropriate turbulence model) or even LES simulations. However, the slender-body approach described here appears to be useful and sufficient in most cases in deep water. Extensions of the theory to shallow water exist.

6.2.4. Influence of Heel

For exact motion predictions including the coupling of maneuvering motions with heel one should take account of:

- the heeling moments due to weight and mass moments, hydrostatic and hydrodynamic moments on hull, rudder, and propeller, and possibly wind heeling moments or other external influences;
- the dependence of X, Y, and N on heel angle, heel velocity, and heel acceleration.

Details may be drawn from Bohlmann's (1989, 1990) work on submarine maneuvering.

By choosing our coordinate origin at the height of the ship's center of gravity, many of these influences are zero, others are small in cargo ships. For example, the dependence of X, Y, and N on the heel rate and heel acceleration can be neglected if the interest is not in the rolling motions themselves, but only in their influence on maneuvering motions. In this case, the heel angle may be determined by the equilibrium resulting from the maneuvering heel moment K as stated before, the hydrostatic righting moment, and possibly the wind- and propeller-induced moments. However, the dependence of X, Y, and N on the heel angle may be substantial.

The following procedure is recommended to evaluate the influence of heel:

1. In the equations for X_1 etc., m' is determined taking into account the heel angle. This leads to larger m' values in the midship range due to increase of draft with heel for the full midship sections. Capturing the influence of heel in the computations of m' is straightforward in CFD computations, but also a Lewis transformation approach can be extended to include heel (Söding 1984).
2. In the equations for X_1, instead of v the expression $v - u \cdot \partial y_B/\partial x$ should be used, where y_B is the y coordinate of the center of gravity of the immersed section area due to heel. The term takes account of the curvature of the 'centerline' of the heeled hull.
3. The cross-flow resistance coefficient C_D depends on the heel angle. C_D may decrease by 1–3% per degree of heel in the direction of drift motion. Due to the increase of section

draft (at least in the midship region) with heel, however, the actual cross-flow resistance may increase with heel angle.

4. For larger heel angles exceeding approximately 25°, the cross-flow velocity in the equation for X_4, etc. should be determined with respect to the curved line being composed of the points of maximum draft of the ship sections. If this line has transverse coordinate $y_T(x)$, instead of $v + x \cdot r$ the expression $v + x \cdot r - \partial y_T/\partial x \cdot u$ has to be used in the equation for X_4, etc. for heel angles exceeding 25°. For smaller heel, a linear interpolation of the correction term $\partial y_T/\partial x \cdot u$ over heel is recommended.

This procedure improves predictions, but still has substantial deficiencies for larger heel angles.

6.2.5. Shallow Water and Other Influences

Body forces depend not only on the actual acceleration, speed, and (in case of heel) state of the vessel, but also on the previous time history of body motion. This is due to vortex shedding and waves generated by the ship. However, these 'memory effects' are very small in ordinary maneuvering motions. Exceptions where memory effects may be important are:

- PMM experiments with the usual too small amplitudes;
- self-induced motions of a moored ship.

Shallow water, non-uniform current and interactions with other ships may substantially influence the body forces as discussed in detail in Brix (1993). The influence of shallow water can be roughly described as follows. If the ship keel is just touching the sea bottom, the effective side ratio of the ship hull is increased from approximately 0.1 (namely $2T/L$; factor 2 due to mirror image at waterplane) to ∞. This increases the transverse forces by approximately a factor of 40 (following the simple estimate formula for rudder lift in Section 6.4). The rudder itself increases its effective side ratio from approximately 2 for deep water to ∞ for extremely shallow water. The rudder forces are then increased by a factor of approximately 2.6. The hull forces for a yaw stable ship decrease the course-changing ability, the rudder forces increase the course-changing ability. Since the hull forces increase more than the rudder forces on shallow water, the net result for yaw stable ships is:

- increased radius of turning circle;
- increased turning time;
- increased yaw checking time.

For yaw instable ships, this may be different, especially if the yaw stability changes drastically. Shallow water may increase or decrease yaw stability. One of several effects is the change of trim. Boundary element methods, namely 'wave resistance' codes, may be used to predict trim and sinkage of real ship geometries, usually with good accuracy on shallow water. The results of such computations have been used to estimate the amount of yaw instability.

6.2.6. Stopping

The rudder behind a reverse turning propeller is almost without effect. The track of a stopping ship is thus largely determined by the maneuvering forces of the propeller(s) and wind. For yaw instable ships, the track can be largely influenced by small initial port or starboard motions. For sister ships (large tankers) under 'same' conditions, stopping times vary between 12 and 22 minutes with largely differing tracks. The differences are attributed to such small (random) initial differences in yaw motions.

For low speeds, the stopping times and distances can be determined as follows. One assumes that between two points in time t_1 and t_2 the reverse thrust (minus thrust deduction) T is approximately constant and that the resistance R is proportional to speed u:

$$\frac{R}{u^2} = \frac{R_0}{U_0^2} = k \tag{6.44}$$

k is the stopping constant. The index 0 denotes the values at the beginning of the maneuver. If we assume a straight stopping track, the fundamental equation of motion is:

$$m \cdot \dot{u} = -k(u^2 + u_T^2) \tag{6.45}$$

The mass m includes the hydrodynamic added mass m'' for longitudinal motion which may be estimated by Eq. (6.39). $u_T = U_0\sqrt{T/R_0}$ is the speed the ship would have after a long time if the thrust T were directed forward. The above differential equation can be solved (by separation of variables) to yield:

$$\Delta t = t_2 - t_1 = \frac{m}{ku_T}\left[\arctan\frac{u_1}{u_T} - \arctan\frac{u_2}{u_T}\right] \tag{6.46}$$

The distance is given by multiplying the above differential equation by $u = ds/dt$ and solving again (by separation of variables) to yield:

$$\Delta s = s_2 - s_1 = \frac{m}{2k}\ln\left(\frac{u_1^2 + u_T^2}{u_2^2 + u_T^2}\right) \tag{6.47}$$

These two equations for Δt and Δs can be used to compute stepwise the stopping process by splitting the process into time intervals where the thrust T can be assumed to be constant. The reverse thrust is best determined by using four-quadrant diagrams for the propellers, if these diagrams are available.

6.2.7. Jet Thrusters

Transverse jet thrusters consist of a transverse pipe through the ship hull usually located at the bow or at the stern. The pipe contains a screw propeller which pumps water either to port or starboard, thus creating a side thrust (and moment). The purpose of a jet thruster is to increase maneuverability at low speeds, allowing the ship to maneuver with no or less tug assistance. As the rudder astern already supplies maneuvering forces, jet thrusters are more effective at the bow and usually placed there ('bow thrusters'). Also, stern jets need to cope with potential collision problems in arranging both jet pipe and propeller shaft. Jet thrusters may also serve as an emergency backup for the main rudder. Backups for rudders are needed for ships with dangerous cargo. Jet thrusters are less attractive for ships on long-distance routes calling at few ports. The savings in tug fees may be less than the additional expense for fuel.

For ocean-going ships, side thrusts of 0.08−0.12 kN per square meter underwater lateral area are typical values. These values relate to zero forward speed of the ship. Installed power P, cross-section area of the pipe A, and flow velocity v in the jet thruster are related by:

$$T = \rho \cdot A \cdot v^2 \tag{6.48}$$

$$P = \frac{1}{\eta} \cdot \frac{1}{2} \rho \cdot A \cdot v^3 \tag{6.49}$$

η is here the efficiency of the thruster propeller. These equations yield:

$$\frac{P}{T} = \frac{v}{2\eta} \tag{6.50}$$

$$\frac{T}{A} = \rho v^2 \tag{6.51}$$

$\eta = 0.8$ and $v = 11$ m/s yield typical relations: approximately 120 kN/m^2 thrust per thruster cross-section area and 7 kW power per kN thrust.

With increasing speed, jet thrusters become less effective and rudders become more effective. The reason is that the jet is bent backwards and may reattach to the ship hull. The thrust is then partially compensated by an opposite suction force. This effect may be reduced by installing a second (passive) pipe without a propeller downstream of the thruster (Brix 1993).

6.2.8. CFD for Ship Maneuvering

For most ships, the linear system of equations determining the drift and yaw velocity in steady turning motion is nearly singular. This produces large relative errors in the predicted steady turning rate, especially for small rudder angles and turning rates. For large rudder angle and turning rate, non-linear forces alleviate these problems somewhat. But non-linear hull forces depend crucially on the cross-flow resistance or the direction of the longitudinal vortices, i.e. on quantities which are determined empirically and which vary widely. In addition, extreme rudder forces depend strongly on the rudder stall angle which − for a rudder behind the hull and propeller − requires at least two-dimensional RANSE simulations. Thus large errors are frequently made in predicting both the ship's path in hard maneuvers and the course-keeping qualities. (The prediction of the full ship is fortunately easier as at the higher Reynolds numbers stall rarely occurs.) In spite of that, published comparisons between predictions and measurements based on inviscid, classical approaches have indicated almost always excellent accuracy; a notable exception is Söding (1993). The difference is that Söding avoids all information which would not be available had the respective model not been tested previously. The typical very good agreement published by others is then suspected to be either chosen as best results from a larger set of predictions or due to empirical corrections of the calculation method based on experiments which include the ship used for demonstrating the attained accuracy. Naturally, these tricks are not possible for a practical prediction where no previous test results for the ship design can be used. Thus accurate maneuvering predictions require RANSE approaches, and even then care has to be taken in grid generation and turbulence modeling. It may also be possible to predict full-scale ship motions with sufficient accuracy, but the experience published so far is insufficient to establish this as state of the art. However, differences between alternative designs and totally unacceptable designs may be detected using simper methods for maneuvering prediction.

The simplest approach to body force computations is the use of regression formulae based on slender-body theory, but with empirical coefficients found from analyzing various model experiments, e.g. Clarke et al. (1983). The next more sophisticated approach would be to apply slender-body methods directly, deriving the added mass terms for each strip from analytical (Lewis form) or BEM computations. The application of three-dimensional RANSE methods yields the best results, but only a few industry applications had appeared by 2010. The main individual approaches are ranked in increasing complexity:

- *Lifting surface methods*
  An alternative to slender-body theory, applicable to rudder and hull (separately or in combination), is the lifting surface model. It models the inviscid flow about a plate (center plane), satisfying the Kutta condition (smooth flow at the trailing edge) and usually the free-surface condition for zero F_n (double-body flow). The flow is determined as a superposition of horseshoe vortices which are symmetrical with respect to the water surface (mirror plane). The strength of each horseshoe vortex is determined by a collocation method from Biot–Savart's law. For stationary flow conditions, in the ship's wake there are no vertical vortex lines, whereas in instationary flow vertical vortex lines are required also in the wake. The vortex strength in the wake follows from three conditions:
  1. Vortex lines in the wake flow backwards with the surrounding fluid velocity, approximately with the ship speed u.
  2. If the sum of vertical ('bound') vortex strength increases over time within the body (due to larger angles of attack), a corresponding negative vorticity leaves the trailing edge, entering into the wake.
  3. The vertical vortex density is continuous at the trailing edge.

Except for a ship in waves, it seems accurate enough to use the stationary vortex model for maneuvering investigations.

Vortex strengths within the body are determined from the condition that the flow is parallel to the midship (or rudder) plane at a number of collocation points. The vortices are located at one-quarter of the chord length from the bow, the collocation points at three-quarters of the chord length from the bow. This gives a system of linear equations to determine the vortex strengths. Transverse forces on the body may then be determined from the law of Kutta–Joukowski, i.e. the body force is the force exerted on all 'bound' (vertical) vortices by the surrounding flow.

Alternatively, one can smooth the bound vorticity over the plate length, determine the pressure difference between port and starboard of the plate, and integrate this pressure difference. For shallow water, reflections of the vortices are necessary both at the water surface and at the bottom. This produces an infinite number of reflections, a subset of which is used in numerical approximations. If the horizontal vortex lines are arranged in the ship's center plane, only transverse forces depending linearly on v and r are generated. The equivalent to the non-linear cross-flow forces in slender-body theory is found in the vortex models if the horizontal vortex lines are oblique to the center plane. Theoretically the position of the vortex lines could be determined iteratively to ensure that they move with the surrounding fluid flow which is influenced by all other vortices. But practically this procedure is usually not applied because of the high computing effort and convergence problems. According to classical foil theory, the direction of the horizontal vortices should be halfway between the ship longitudinal direction and the motion direction in deep water. More modern procedures arrange the vortices in the longitudinal direction within the ship length, but in an oblique plane (for steady motion at a constant yaw angle) or on a circular cylinder (for steady turning motion) in the wake. The exact direction of the vortices is determined depending on water depth. Also important is the arrangement of vortex lines and collocation points on the material plate. Collocation points should be about halfway between vortex lines in both longitudinal and vertical directions. High accuracy with few vortex lines is attained if the distance between vertical vortices is smaller at both ends of the body, and if the distance between horizontal vortices is small at the keel and large at the waterline.

- *Lifting body methods*
  A body with finite thickness generates larger lift forces than a plate. This can be taken into account in different ways:

  1. by arranging horseshoe vortices (or dipoles) on the hull on both sides;
  2. by arranging a source and a vortex distribution on the center plane;
  3. by arranging source distributions on the hull and a vortex distribution on the center plane.

  In the third case, the longitudinal distribution of bound vorticity can be prescribed arbitrarily, whereas the vertical distribution has to be determined from the Kutta condition along the trailing edge. The Kutta condition can be approximated in different ways. One suitable formulation is: the pressure (determined from the Bernoulli equation) should be equal at port and starboard along the trailing edge. If the ship has no sharp edge at the stern (e.g. below the propeller's axis for a stern bulb), it is not clear where the flow separation (and thus the Kutta condition) should be assumed. This may cause large errors for transverse forces for the hull alone, but when the rudder is modeled together with the hull, the uncertainty is much smaller. Forces can be determined by integrating the pressure over the hull surface. For a very thin body, the lifting surface and lifting body models should result in similar forces. In practice, however, large differences are found. The lifting-body model with source distributions on both sides of the body has difficulties if the body has a sharp bow. Assuming a small radius at the bow waterline produces much better results. For a ship hull it seems difficult to obtain more accurate results from lifting-body theory than using slender-body theory. For the rudder and for the interaction between rudder and hull, however, lifting-surface or lifting-body theory is the method of choice for angles of attack where no stall is expected to occur. Beyond the stall angle, only RANSE methods (or even more sophisticated viscous flow computations) may be used.

By the early 1990s research applications for lifting-body computations including free-surface effects appeared for steady drift motions. The approach of Zou (1990) is typical. First the wave resistance problem is solved including dynamic trim and sinkage. Assuming small asymmetry, the difference between symmetrical and asymmetrical flow is linearized. The asymmetrical flow is then determined by a lifting-body method with an additional source distribution above the free surface.

- *Field methods*

  In spite of the importance of viscosity for maneuvering, viscous hull force calculations appeared in the 1990s only as research applications and were mostly limited to steady flow computations around a ship with a constant yaw angle. By 2010, simulations for freely maneuvering ships, even in waves, were presented in research applications. However, industry applications still relied largely on inviscid approaches with semi-empirical corrections, due to the large required resources for RANSE maneuvering simulations. Difficulties in RANSE computations for maneuvering are:

  - The number of computational cells is much higher than for resistance computations, because both port and starboard sides must be discretized and because vortices are shed over nearly the full ship length.
  - The large-scale flow separation requires advanced turbulence models, adding to the computational effort.

  State-of-the-art computations for ship hulls at model scale Reynolds numbers were capable of predicting transverse forces and moments well, even for large yaw angles, but predicted the longitudinal force (resistance) with large relative errors. Computations have included dynamic trim and sinkage, which play an important role in shallow-water maneuvering.

RANSE computations including free-surface effects will grow in importance and have started to drift into practical applications. They are expected to substantially improve the accuracy of maneuvering force predictions over the next decade.

6.3. Experimental Approaches

6.3.1. Maneuvering Tests for Full-Scale Ships in Sea Trials

The main maneuvering characteristics as listed in the introduction to maneuvering are quantified in sea trials with the full-scale ship. Usually the design speed is chosen as initial speed in the maneuver. Trial conditions should feature deep water (water depth > 2.5 ship draft), little wind (less than Beaufort 4) and 'calm' water to ensure comparability to other ships. Trim influences the initial turning ability and yaw stability more than draft. For comparison with other ships, the results are made non-dimensional with ship length and ship length travel time (L/V).

The main maneuvers used in sea trials follow recommendations of the Manoeuvring Trial Code of ITTC (1975) and the IMO circular MSC 389 (1985). IMO also specifies the display of some of the results in bridge posters and a maneuvering booklet on board ships in the IMO resolution A.601(15) (1987) (Provision and display of maneuvering information on board ships).

The main maneuvers in sea trials are:

1. *Turning circle test*
   Starting from straight motion at constant speed, the rudder is turned at maximum speed to an angle υ (usually maximum rudder angle) and kept at this angle, until the ship has performed a turning circle of at least 540°. The trial is performed for both port and starboard sides. The essential information obtained from this maneuver (usually by GPS) consists of (Fig. 6.4):
   - tactical diameter;
   - maximum advance;
   - transfer at 90° change of heading;
   - times to change heading 90° and 180°;
   - transfer loss of steady speed.

   Typical values are tactical diameter of 4.5–7L for slender, 2.4–4 L for short and full ships. Decisive is the slenderness ratio $L/\sqrt[3]{\nabla}$, where ∇ is the volume displacement.

   Fast displacement ships with $F_n > 0.25$ may feature dangerously large heel angles in turning circles. The heel is always outwards, i.e. away from the center of the turning circle. (Submarines and boats with dynamic lift like hydrofoils are exceptional in that they may heel inwards.) The heel is induced by the centrifugal force $m \cdot u \cdot r$ acting outwards on the ship's center of gravity, the hull force $Y_v v + Y_r r$ acting inwards, and the much smaller

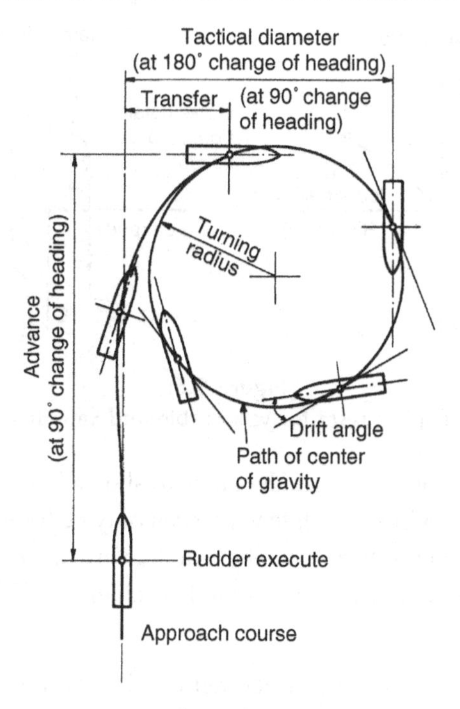

Figure 6.4:
Definitions used on turning circle

rudder force $Y_\delta \delta$ acting outwards. For maneuvering predictions it is important to consider that the ship is faster at the beginning of the turning circle and slower at sustained turning. The heeling angle exceeds dynamically the static heel angle due to forces listed above.

The turning circle test is used to evaluate the turning ability of the ship.

2. *Spiral maneuvers*

   We distinguish between:

   - *'Direct' spiral maneuver (Dieudonne)* – With the ship on an initial straight course, the rudder is put hard to one side until the ship has reached a constant rate of change of heading. The rudder angle is then decreased in steps (typically 5°, but preferably less near zero rudder angle) and again held until a steady condition is reached. This process is repeated until the rudder has covered the whole range to the maximum rudder angle on the other side. The rate of turn is noted for each rudder angle. The test should be performed at least for yaw unstable ships going both from port to starboard and from starboard to port.
   - *'Indirect' (reverse) spiral maneuver (Bech)* – The ship is steered at a constant rate of turn and the mean rudder angle required to produce this yaw rate is measured. This way, points on the curve rate of turn vs. rudder angle may be taken in any order.

The spiral test results in a curve as shown in Fig. 6.5. The spiral test is used to evaluate the turning ability and the yaw stability of the ship. For yaw unstable ships, there may be three

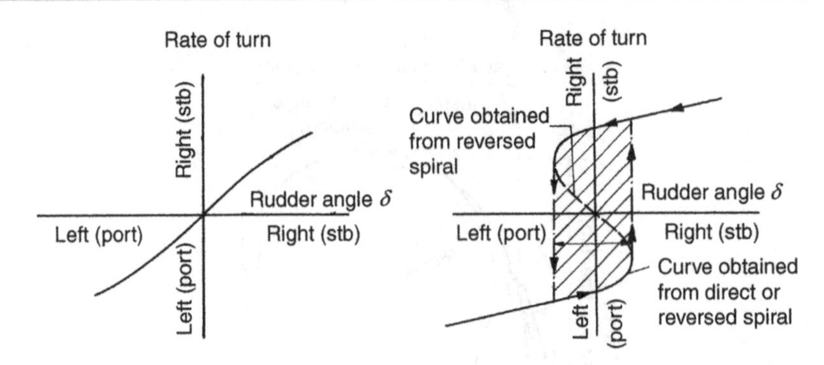

Figure 6.5:
Results of spiral tests for yaw stable and yaw unstable ship

possible rates of turn for one given rudder angle as shown in Fig. 6.5. The one in the middle (dotted line) represents an instable state which can only be found by the indirect method. In the direct method, the rate of turn 'switches' at the vertical sections of the curve suddenly to the other part of the curve if the rudder angle is changed. This is indicated by the dotted arrows in Fig. 6.5.

The spiral test, especially with the direct method, is time-consuming and sensitive to external influences. The results show that a linearization of the body force equations is acceptable only for small $|r|$ (Fig. 6.5). For yaw stable ships, the bandwidth of acceptable rudder angles to give small $|r|$ is small, e.g. $\pm 5°$. For yaw unstable ships, large $|r|$ may result for any δ.

3. *Pull-out maneuver*

After a turning circle with steady rate of turn the rudder is returned to midship. If the ship is yaw stable, the rate of turn will decay to zero for turns both port and starboard. If the ship is yaw unstable, the rate of turn will reduce to some residual rate of turn (Fig. 6.6).

The pull-out maneuver is a simple test to give a quick indication of a ship's yaw stability, but requires very calm weather. If the yaw rate in a pull-out maneuver tends towards a finite value

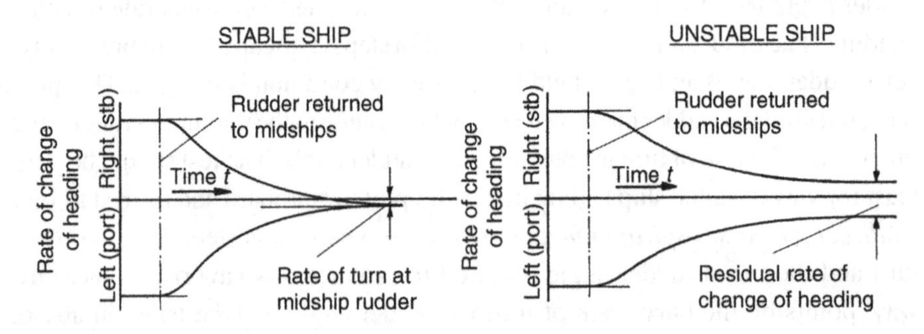

Figure 6.6:
Results of pull-out maneuver

in single-screw ships, this is often interpreted as yaw instability, but it may be at least partially due to the influence of asymmetry induced by the propeller in single-screw ships or wind.

4. *Zigzag maneuver*
   The rudder is reversed alternately by a rudder angle δ to either side at a deviation ψ_e from the initial course. After a steady approach the rudder is put over to starboard (first execute). When the heading is ψ_e off the initial course, the rudder is reversed to the same rudder angle to port at maximum rudder speed (second execute). After counter rudder has been applied, the ship continues turning in the original direction (overshoot) with decreasing turning speed until the yaw motion changes direction. In response to the rudder the ship turns to port. When the heading is ψ_e off the initial course to port, the rudder is reversed again at maximum rudder speed to starboard (third execute). This process continues until a total of, say, five rudder executes have been completed. Typical values for ψ_e are 10° and 20°. The test was especially developed for towing tank tests, but it is also popular for sea trials. The test yields initial turning time, yaw checking time, and overshoot angle (Fig. 6.7).

For the determination of body force coefficients a modification of the zigzag maneuver is better suited: the incremental zigzag test. Here, after each period other angles δ and ψ_e are chosen to cover the whole range of rudder angles. If the incremental zigzag test is properly executed it may substitute for all other tests as the measured coefficients should be sufficient for an appropriate computer simulation of all other required maneuvering quantities. Figure 6.8 shows results of many model zigzag tests as given by Brix (1993). These yield the following typical values:

- initial turning time t_a: 1–1.5 ship length travel time;
- time to check starboard yaw t_s: 0.5–2 ship travel length time (more for fast ships);
- starboard overshoot angle α_s: 5–15°;
- turning speed to port r (yaw rate): 0.2–0.4 per ship travel length time.

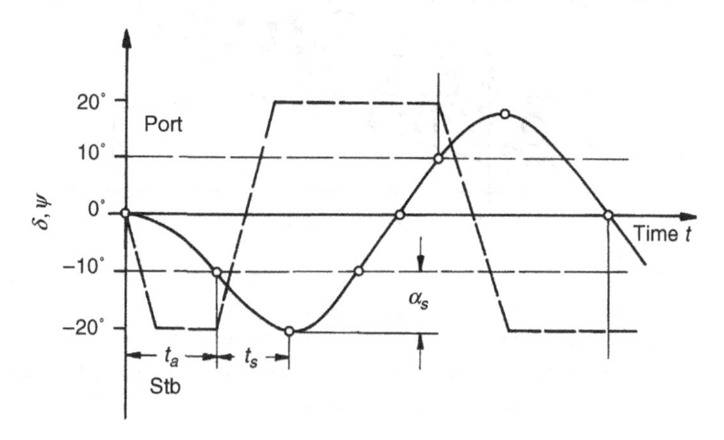

Figure 6.7:
Scheme of zigzag maneuver; t_a initial turning time, t_s yaw checking time, α_s overshoot angle; – – – – rudder angle δ, —— —— course angle ψ

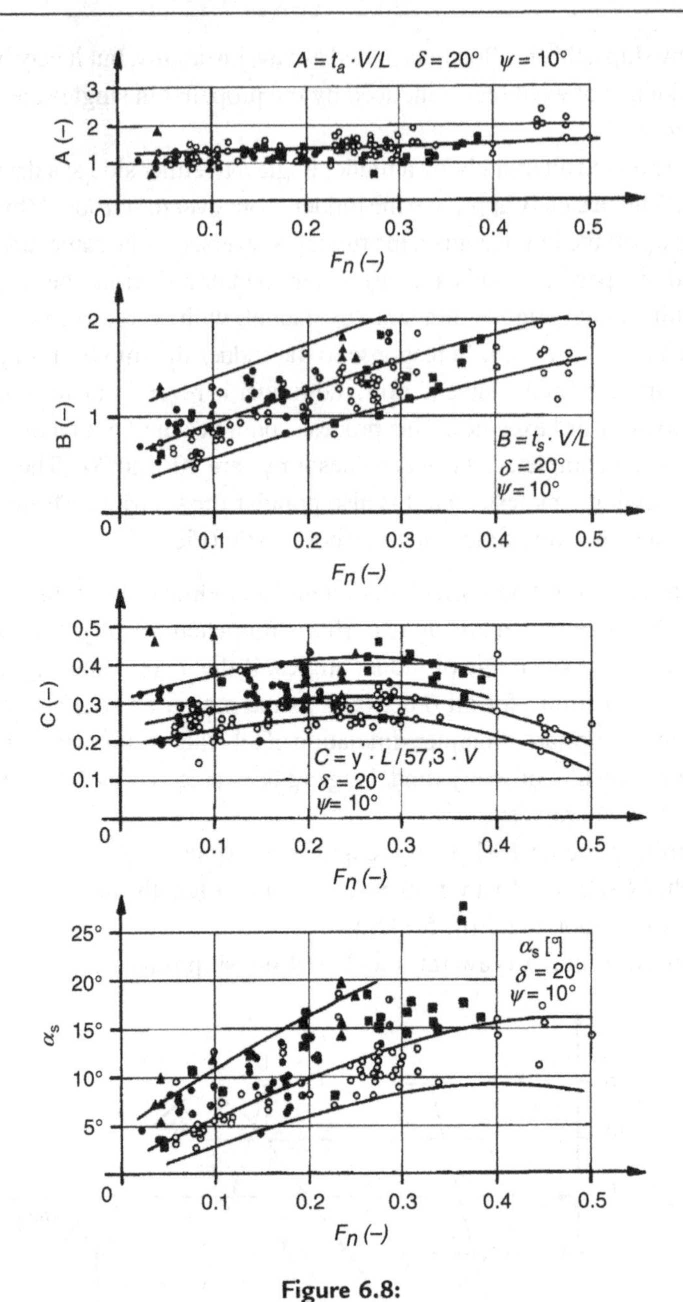

Figure 6.8:
Non-dimensional data obtained from zigzag model tests (Brix 1993); A = non-dimensional initial turning times; $\beta = B$ = non-dimensional times to check starboard yaw; C = non-dimensional turning speed to port; α_s starboard overshoot angle

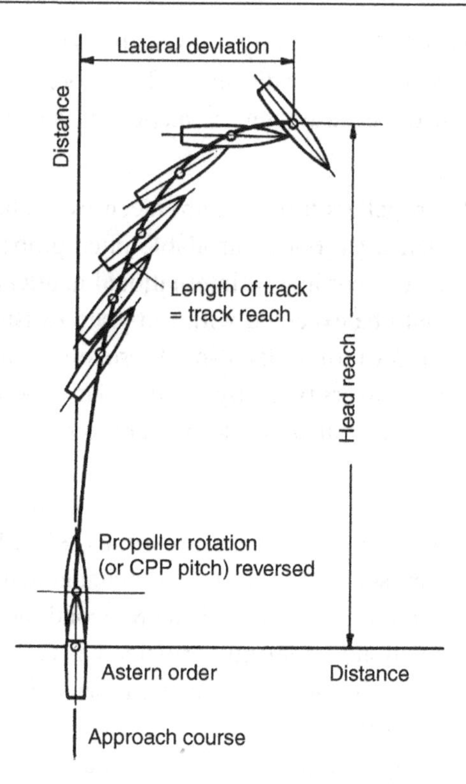

Figure 6.9:
Results of stopping trial

5. *Stopping trial*

   The most common maneuver in stopping trials is the crash-stop from full-ahead speed. For ships equipped with fixed-pitch propellers, the engine is stopped and then as soon as possible reversed at full astern. Controllable-pitch propellers (CPP) allow a direct reversion of the propeller pitch. Sometimes the rudder is kept a midships, sometimes one tries to keep the ship on a straight course, which is difficult as the rudder effectiveness usually decreases drastically during the stopping maneuver and because the reversing propeller induces substantial transverse forces on the aftbody. The reaction to stopping maneuvers is strongly non-linear. Thus environmental influences (e.g. wind) and slight changes in the initial conditions (e.g. slight deviation of the heading to either port or starboard) may change the resulting stopping track considerably.

   The maneuver ends when $u = 0$. Results of the stopping maneuver are (Fig. 6.9):
   - head reach (distance traveled in the direction of the ship's initial course);
   - lateral deviation (distance to port or starboard measured normal to the ship's initial course);
   - stopping time.

Crash-stops from full speed are nautically not sensible as turning usually offers better avoidance strategies involving shorter distances. Therefore stopping maneuvers are also recommended at low speed, because then the maneuver is of practical interest for navigation purposes.

Single-screw ships with propellers turning right (seen from abaft clockwise) will turn to starboard in a stopping maneuver. For controllable-pitch propellers, the propeller pitch is reversed for stopping. Since according to international nautical conventions, collision avoidance maneuvers should be executed with starboard evasion, single-screw ships should be equipped with right-turning fixed-pitch propellers or left-turning CPPs. Simulations of stopping maneuvers typically use the four-quadrant diagrams for propellers to determine the propeller thrust also in astern operation (see Section 2.2, Chapter 2).

6. *Hard rudder test*
   With the ship on an initially straight course, the rudder is put hard to 35° port. As soon as this rudder angle is reached (i.e. without waiting for a specific heading or rate of turning), the rudder is reversed to hard starboard. The time for changing the rudder angle from 35° on one side to 30° on the other side must not exceed 28 seconds according to IMO regulations (SOLAS 1960). This regulation is rightfully criticized as the time limit is independent of ship size. The IMO regulation is intended to avoid under-dimensioning of the rudder gear.
7. *Man-overboard maneuver (Williamson turn)*
   This maneuver brings the ship in minimum time on opposite heading and same track as at the beginning of the maneuver, e.g. to search for a man overboard. The rudder is laid initially hard starboard, at, say, 60° (relative to the initial heading) hard port, and at, say, −130° to midship position again (Fig. 6.10). The appropriate angles (60°, −130°) vary with each ship and loading condition and have to be determined individually such that at the end of the maneuver the deviation in heading is approximately 180° and in track approximately zero. This is determined in trial-and-error tests during ship trials. However, an approximate starting point is determined in computational simulations beforehand.

6.3.2. Model Tests

Model tests to evaluate maneuverability are usually performed with models ranging between 2.5 m and 9 m in length. The models are usually equipped with propeller(s) and rudder(s), electrical motor, and rudder gear. Small models are subject to considerable scaling errors and usually do not yield satisfactory agreement with the full-scale ship, because the too small model Reynolds number leads to different flow separation at model hull and rudder and thus different non-dimensional forces and moments, especially the stall angle (angle of maximum lift force shortly before the flow separates completely on the suction side), which will be much smaller in models (15° to 25°) than in the full-scale ship (>35°). Another scaling error also contaminates tests with larger models: the flow velocity at the rudder outside the propeller

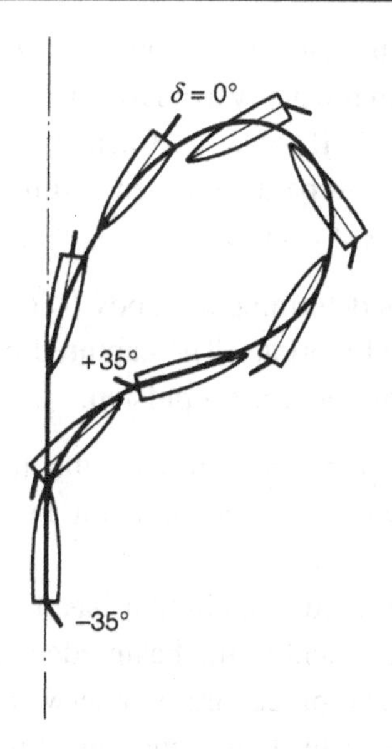

Figure 6.10:
Man-overboard maneuver (Williamson turn)

slipstream is too small (due to a too large wake fraction in model scale) and the flow velocity inside the propeller slipstream is too large (because the too large model resistance requires a larger propeller thrust). The effects cancel each other partially for single-screw ships, but usually the propeller effect is stronger. This is also the case for twin-screw twin-rudder ships, but for twin-screw midship-rudder ships the wake effect dominates for free-running models. For a captured model, propeller thrust minus thrust deduction does not have to equal resistance. Then the propeller loading may be chosen lower such that scale effects are minimized. However, the necessary propeller loading can only be estimated.

Model tests are usually performed at Froude similarity. For small Froude numbers, hardly any waves are created and the non-dimensional maneuvering parameters become virtually independent of the Froude number. For $F_n < 0.3$, for example, the body forces Y and N may vary with speed only by less than 10% for deep water. For higher speeds the wave resistance changes noticeably and the propeller loading increases, as does the rudder effectiveness if the rudder is placed in the propeller slipstream. Also, in shallow water, trim and sinkage change with F_n, influencing Y and N. If the rudder pierces the free surface or is close enough for ventilation to occur, the Froude number is always important.

Model tests with free-running models are usually performed indoors to avoid wind effects. The track of the models is recorded either by cameras (two or more) or from a carriage following the model in longitudinal and transverse directions. Turning circle tests can only be performed in broad basins and even then usually only with rather small models. Often, turning circle tests are also performed in towing tanks with an adjacent round basin at one end. The maneuver is then initiated in the towing tank and ends in the round basin. Spiral tests and pull-out maneuvers require more space than is usually available in towing tanks. However, towing tanks are well suited for zigzag maneuvers. If the ship's track is precisely measured in these tests, all necessary body force coefficients can be determined and the other maneuvers can be numerically simulated with sufficient accuracy.

Model tests with captured models determine the body force coefficients by measuring the forces and moments for prescribed motions. The captured models are also equipped with rudders, propellers, and electric motors for propulsion.

- Oblique towing tests can be performed in a regular towing tank. For various yaw and rudder angles, resistance, transverse force, and yaw moment are measured, sometimes also the heel moment.
- Rotating arm tests were performed in a circular basin. The carriage was supported by an island in the center of the basin and at the basin edge. The carriage rotates around the center of the circular basin. The procedure is otherwise similar to oblique towing tests. Due to the disturbance of the water by the moving ship, only the first revolution could be used to measure the desired coefficients. Large non-dimensional radii of the turning circle are only achieved for small models (inaccurate) or large basins (expensive). Today, this technology is obsolete and replaced by planar motion mechanisms which can also generate accelerations, not just velocities.
- Planar motion mechanisms (PMMs) are installed on a towing carriage. They superimpose sinusoidal transverse or yawing motions (sometimes also sinusoidal longitudinal motions) to the constant longitudinal speed of the towing carriage. The periodic motion may be produced mechanically from a circular motion via a crankshaft or by computer-controlled electric motors (computerized planar motion carriage (CPMC)). The CPMC is far more expensive and complicated, but allows the extension of model motions over the full width of the towing tank, arbitrary motions and a precise measuring of the track of a free-running model.

6.4. Rudders

6.4.1. General Remarks and Definitions

Rudders are hydrofoils pivoting on a vertical or nearly vertical axis. They are normally placed at the ship's stern behind the propeller(s) to produce a transverse force and a steering moment about the ship's center of gravity by deflecting the water flow to a direction of the foil plane. Table 6.4 gives offsets of several profiles used for rudders depicted in Fig. 6.11. Other profile shapes and hydrodynamic properties are available from Abbott and Doenhoff (1959) and Whicker and Fehlner (1958).

Table 6.4: Offsets of rudder profiles

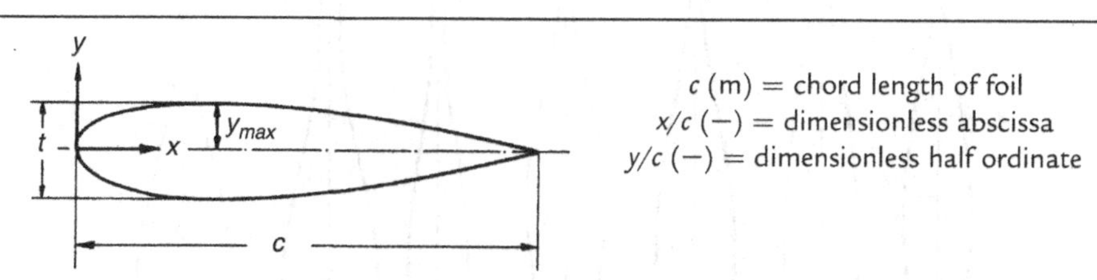

Note: last digits of profile designation correspond to the thickness form, e.g. 25 for $t/c = 0.25$.
For differing thickness t_0 the half ordinates y_0 to be obtained by multiplication

$$\frac{y'}{c} = \frac{t'}{t} \cdot \frac{y}{c}$$

	IFS62-TR 25	IFS61-TR 25	IFS58-TR 15	HSVA-MP71-20	HSVA-MP73-20	NACA 00-20	NACA 643-018
x/c	y/c	y/c	y/c	y/c	y/c	y/c	y/c
0.0000	0.0000	0.0000	0.0000	0.0000	0.04420*	0.04420*	0.02208*
0.0125	0.0553	0.0553	0.0306	0.0230	0.03156	0.03156	0.02177
0.0250	0.0732	0.0732	0.0409	0.0306	0.04356	0.04356	0.03005
0.0500	0.0946	0.0946	0.0530	0.0419	0.05924	0.05924	0.04186
0.1000	0.1142	0.1142	0.0655	0.0583	0.07804	0.07804	0.05803
0.1500	0.1226	0.1226	0.0715	0.0706	0.08910	0.08910	0.06942
0.2000	**0.1250**	**0.1250**	0.0743	0.0801	0.09562	0.09562	0.07782
0.2500	0.1234	0.1226	**0.0750**	0.0881	0.09902	0.09902	0.08391
0.3000	0.1175	0.1176	0.0740	0.0939	**0.10000**	**0.10000**	0.08789
0.4000	0.0993	0.1002	0.0669	0.0996	0.09600	0.09672	**0.08952**
0.4500	—	—	—	**0.1000**	—	—	0.08630
0.5000	0.0742	0.0766	0.0536	0.0965	0.08300	0.08824	0.08114
0.6000	0.0480	0.0533	0.0377	0.0766	0.06340	0.07606	0.06658
0.7000	0.0263	0.0357	0.0239	0.0546	0.04500	0.06106	0.04842
0.8000	0.0123	0.0271	0.0168	0.0335	0.02740	0.04372	0.02888
0.9000	0.0080	0.0250	0.0150	0.0140	0.01200	0.02414	0.01101
1.0000	0.0075	0.0250	0.0150	0.0054	0.00540	0.00210	0.00000

*radius.

Rudders are placed at the ship's stern for the following reasons:

- The rudder moment turning the ship is created by the transverse force on the rudder and an oppositely acting transverse force on the ship hull acting near the bow. This moment increases with distance between the rudder force and the hull force.
- Rudders outside of the propeller slipstream are ineffective at small or zero ship speed (e.g. during berthing). In usual operation at forward speed, rudders outside of the propeller slipstream are far less effective. Insufficient rudder effectiveness at slow ship speed can be temporarily increased by increasing the propeller rpm, e.g. when passing other ships. During stopping, rudders in the propeller slipstream are ineffective.
- Bow rudders not exceeding the draft of the hull are ineffective in ahead motion, because the oblique water flow generated by the turned rudder is redirected longitudinally by the

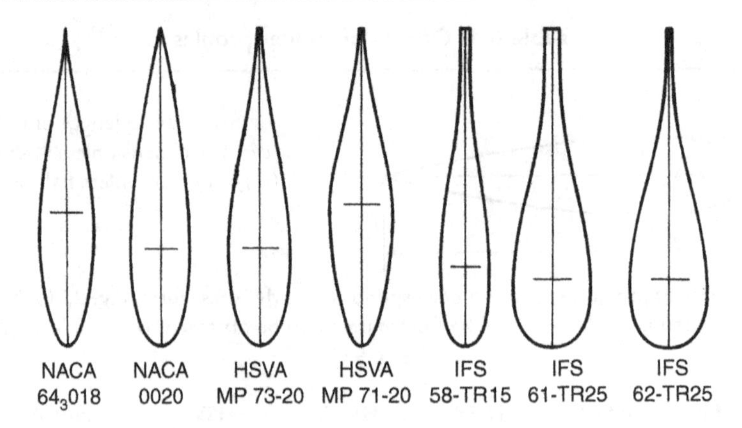

Figure 6.11:
Some rudder profiles, offsets given in Table 6.4

hull. Thus, transverse forces on a bow rudder and on the forward moving hull cancel each other. The same generally applies to stern rudders in backward ship motion. The yaw instability of the backward-moving ship in one example could not be compensated by rudder actions if the drift angle exceeded $\beta = 1.5°$. To improve the maneuverability of ships which frequently have to move astern (e.g. car ferries), bow rudders may be advantageous. In reverse flow, maximum lift coefficients of rudders range between 70% and 100% of those in forward flow. This force is generally not effective for steering the ship astern with a stern rudder, but depending on the maximum astern speed it may cause substantial loads on the rudder stock and steering gear due to the unsuitable balance of normal rudders for this condition.

The rudder effectiveness in maneuvering is mainly determined by the maximum transverse force acting on the rudder (within the range of rudder angles achievable by the rudder gear). Rudder effectiveness can be improved by:

- rudder arrangement in the propeller slipstream (especially for twin-screw ships);
- increasing the rudder area;
- better rudder type (e.g. spade rudder instead of semi-balanced rudder);
- rudder engine which allows larger rudder angles than the customary 35°;
- shorter rudder steering time (more powerful hydraulic pumps in rudder engine).

Figure 6.12 defines the parameters of main influence on rudder forces and moments generated by the dynamic pressure distribution on the rudder surface. The force components in the flow direction α and perpendicular to it are termed drag D and lift L, respectively. The moment about a vertical axis through the leading edge (nose) of the rudder (positive clockwise) is termed Q_N. If the leading edge is not vertical, the position at the mean height of the rudder is used as a reference point.

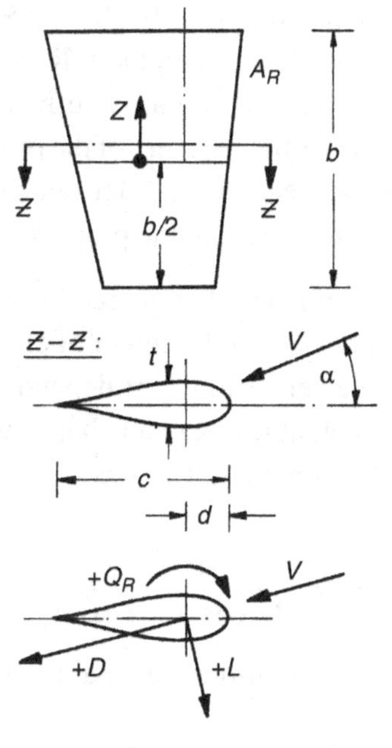

Figure 6.12:
Definition sketch of rudder geometry and rudder forces; A_R = rudder area; b = rudder height; c = chord length; d = rudder stock position; D = drag; L = lift; Q_R = rudder stock torque; t = rudder thickness; v = flow velocity; z = vertical rudder coordinate at $b/2$; α = angle of attack; δ = rudder angle; $\Lambda = b^2/A_R$ = aspect ratio

The moment about the rudder stock at a distance d behind the leading edge (nose) is:

$$Q_R = Q_N + L \cdot d \cdot \cos\alpha + D \cdot d \cdot \sin\alpha \tag{6.52}$$

The stagnation pressure $q = ½\,\rho V^2$ and the mean chord length $c_m = A_R/b$ are used to define the following non-dimensional force and moment coefficients:

$$\text{lift coefficient} \quad C_L = L/(q \cdot A_R) \tag{6.53}$$

$$\text{drag coefficient} \quad C_D = D/(q \cdot A_R) \tag{6.54}$$

$$\text{nose moment coefficient} \quad C_{QN} = Q_N/(q \cdot A_R \cdot c_m) \tag{6.55}$$

$$\text{stock moment coefficient} \quad C_{QR} = Q_R/(q \cdot A_R \cdot c_m) \tag{6.56}$$

The stock moment coefficient is coupled to the other coefficients by:

$$C_{QR} = C_{QN} + \frac{d}{c_m}(C_L \cdot \cos\alpha + C_D \cdot \sin\alpha) \tag{6.57}$$

For low fuel consumption of the ship (for constant rudder effectiveness), we want to minimize the ratio C_L/C_D for typical small angle of attacks as encountered in usual course-keeping mode. Due to the propeller slipstream, angles of attack of typically 10° to 15° (with opposing sign below and above the propeller shaft) occur for zero-deflected rudders. Reducing the rudder resistance by 10% in this range of angles of attack improves the propulsive efficiency by more than 1%. Various devices to improve ship propulsion by partial recovery of the propeller's rotative energy have been proposed in the course of time, e.g. Schneekluth and Bertram (1998). However, the major part of this energy is recovered anyhow by the rudder in the propeller slipstream.

Size, and thus cost, of the rudder engine are determined by the necessary maximum torque at the rudder stock. The stock moment is zero if the center of effort for the transverse rudder force lies on the rudder stock axis. As the center of effort depends on the angle of attack, this is impossible to achieve for all angles of attack. Rudder shapes with strongly changing centers of effort therefore require larger rudder engines. The position of the center of effort behind the leading edge (nose) is:

$$c_s = \frac{c \cdot C_{QN}}{C_L \cos\alpha + C_D \sin\alpha} \tag{6.58}$$

The denominator in this formula is the non-dimensional force coefficient for the normal force on the rudder.

6.4.2. Fundamental Hydrodynamic Aspects of Rudders and Simple Estimates

C_L, C_D, and C_{QN} can be determined in wind tunnel tests or computations. Extensive wind tunnel measurements have been published by Thieme (1992) and Whicker and Fehlner (1958).

Figure 6.13 shows an example. Practically these data allow rough estimates only of rudder forces and moments of ships, because in reality the flow to the rudder is irregular and highly turbulent and has a higher Reynolds number than the experiments, and because interactions with the ship's hull influence the rudder forces. For angles of attack smaller than stall angle α_s (i.e. the angle of maximum C_L) the force coefficients may be approximated with good accuracy by the following formulae:

$$C_L = C_{L1} + C_{L2} = 2\pi \frac{\Lambda \cdot (\Lambda + 0.7)}{(\Lambda + 1.7)^2} \cdot \sin\alpha + C_Q \cdot \sin\alpha \cdot |\sin\alpha| \cdot \cos\alpha \tag{6.59}$$

$$C_D = C_{D1} + C_{D2} + C_{D0} = \frac{C_L^2}{\pi \cdot \Lambda} + C_Q \cdot |\sin\alpha|^3 + C_{D0} \tag{6.60}$$

$$C_{QN} = -(C_{L1} \cdot \cos\alpha + C_{D1} \cdot \sin\alpha) \cdot \left(0.47 - \frac{\Lambda + 2}{4(\Lambda + 1)}\right) - 0.75 \cdot (C_{L2} \cdot \cos\alpha + C_{D2} \cdot \sin\alpha) \tag{6.61}$$

Figure 6.14 illustrates the C_L formula. The first term in the C_L formula follows from potential thin-foil theory for the limiting aspect ratios 0 and $\Lambda = 1$. For other aspect ratios it is an

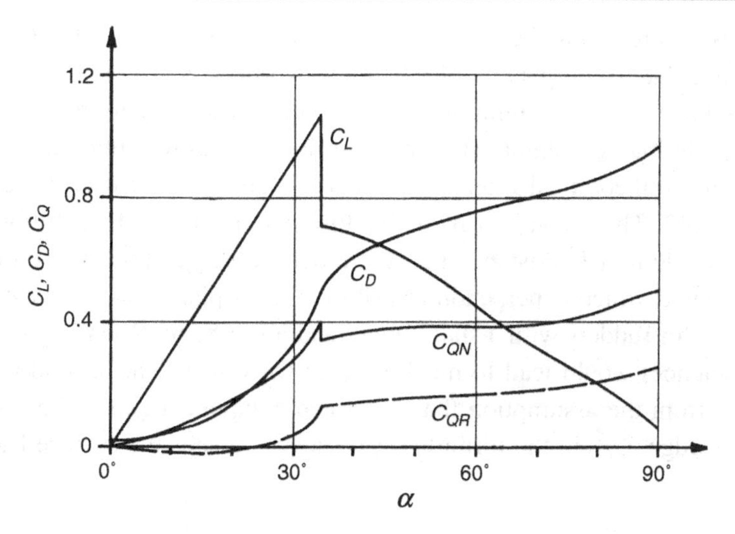

Figure 6.13:
Force and moment coefficients of a hydrofoil $\Lambda = 1$; rudder stock position $d/c_m = 0.25$; NACA-0015; $R_n = 0.79 \cdot 10^6$; Q_N = nose moment; Q_R = rudder stock torque

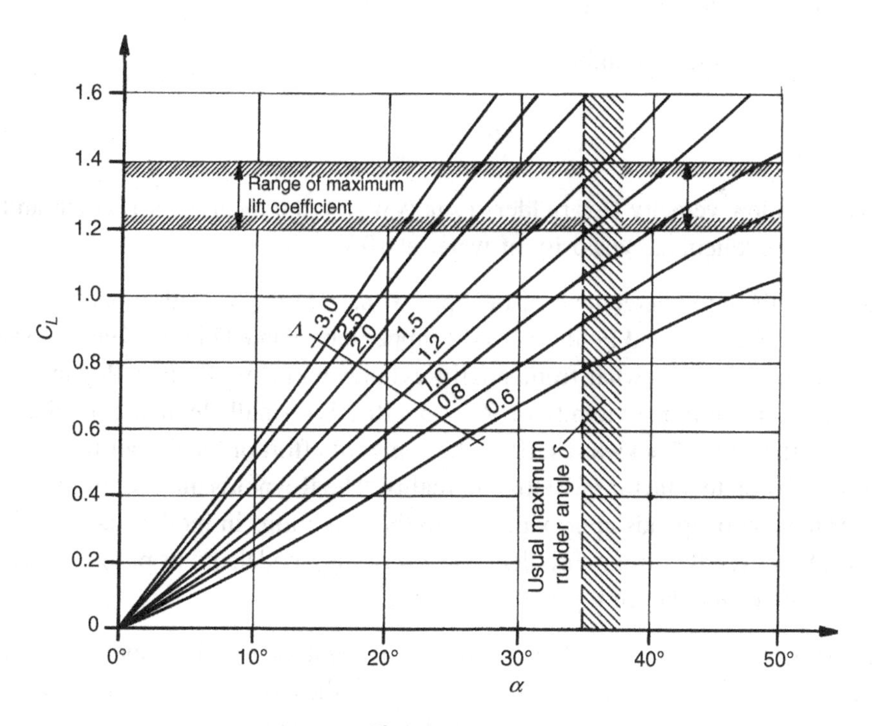

Figure 6.14:
Lift coefficient C_L versus angle of attack α with the aspect ratio Λ as parameter

approximation to theoretical and experimental results. The first term in the C_D formula is the induced resistance due to the generation of trailing vortices behind the foil. The equation includes a 10% increase in the minimum induced drag which occurs for an elliptical load distribution over the rudder height. The first term in the C_{QN} formula would be a good approximation of the theoretical moment in ideal fluid if 0.5 were used instead of the empirical value 0.47. The second terms in the formulae for C_L and C_D follow from the assumption of an additional resistance-like force depending quadratically on the velocity component $V \cdot \sin \alpha$ which is perpendicular to the rudder plane. A resistance coefficient $C_Q \approx 1$ may be used for rudders with a sharp upper and lower edge. Rounding the edges (which is difficult in practice) would lead to much smaller C_Q values. The second term in the C_{QN} formula follows from the assumption that this force component acts at 75% chord length from the leading edge. C_{D0} in the formula for C_D approximates the surface friction. We may approximate:

$$C_{D0} = 2.5 \cdot \frac{0.075}{(\log R_n - 2)^2} \tag{6.62}$$

This is 2.5 times the frictional resistance coefficient following the ITTC 1957 formula. C_{D0} refers to the rudder area which is about half the wetted area of the rudder. In addition a form factor has to be taken into account to yield the factor 2.5.

For hydrofoils the Reynolds number is defined as:

$$R_n = \frac{V \cdot c}{\nu} \tag{6.63}$$

where V is the inflow velocity (for rudders usually V_A), c the mean chord length and $\nu \approx 1.35 \cdot 10^6$ m^2/s the kinematic viscosity of water at 10°C.

Table 6.5 shows the good agreement of the approximate formulae with model test measurements of Whicker and Fehlner (1958) (upper table) and Thieme (1962) (lower table). Thieme's results suffer somewhat from small Reynolds numbers. Rudder Reynolds numbers behind a large ship are in the vicinity of $R_n = 5 \cdot 10^7$. Too small Reynolds numbers result in larger drag coefficients, a backward shift of the center of effort of the rudder force and smaller stall angles α_s (by up to a factor of 2) than in reality. The Reynolds number of the last column in the lower table corresponds approximately to the conditions in model maneuvering experiments. However, the strong turbulence behind a ship model and its propeller act similarly to a larger Reynolds number in these experiments.

The formulae for C_L, C_D, and C_{QN} do not take into account the profile shape. The profile shape affects mainly the stall angle α_s, the maximum lift and the cavitation properties of the rudder, but hardly the lift at a given angle of attack below α_s. Table 6.5 shows that, compared to the 'standard' NACA profiles, concave profiles with thickness maximum in front of the

Table 6.5: Measured (M) and computed (C) force and moment coefficients of different profiles (Thieme 1964, Whicker and Fehlner 1958); +: independent from profile shape; *: uncertain values, probably due to experimental technique

Profile	+)	NACA 0015	+)	NACA 0015	+)	NACA 0015
Λ	1	1	2	2	3	3
$(t/c)_{max}$	+)	15	+)	15	+)	15
at x/c		30		30		30
$R_n/10^6$	2.7	2.7	2.7	2.7	2.7	2.7
Source	C	M	C	M	C	M
C_L at $\alpha = 10°$	0.27	0.27	0.44	0.44	0.55	0.55
C_L at $\alpha = 20°$	0.59	0.60	0.92	0.93	1.14	1.10
C_L at $\alpha = \alpha_s$	1.17	1.26	1.33	1.33	1.32	1.25
α_s (°)	38.5	38.5	28.7	28.7	23.0	23.0
C_L/C_D at $\alpha = 10°$	8.11	7.26	10.45	10.35	12.28	12.40
C_L/C_D at $\alpha = 20°$	4.62	4.25	5.70	5.79	6.63	7.05
C_L/C_D at $\alpha = \alpha_s$	2.28	2.20	3.88	4.00	5.76	6.00
c_s/c at $\alpha = 10°$	0.17	0.16	0.18	0.19	0.19	0.18
c_s/c at $\alpha = \alpha_s$	0.30	0.31	0.24	0.25	0.23	0.23

Profile	NACA 0015	NACA 0025	IFS62 TR 25	IFS61 TR 25	IFS58 TR 15	Plate $t/c =$ 0.03	NACA 0015
Λ	1	1	1	1	1	1	1
$(t/c)_{max}$	15	25	25	25	15	3	15
at x/c	30	30	20	20	25	–	30
$R_n/10^6$	0.79	0.78	0.78	0.79	0.79	0.71	0.20
Source	M	M	M	M	M	M	M
C_L at $\alpha = 10°$	0.29	0.27	0.33	0.32	0.32	0.34	0.35
C_L at $\alpha = 20°$	0.62	0.59	0.71	0.69	0.67	0.72	0.55
C_L at $\alpha = \alpha_s$	1.06	1.34	1.48	1.34	1.18	1.14	0.72
α_s (°)	33.8	46.0*	46.0*	41.0*	33.5	40*	35.0*
C_L/C_D at $\alpha = 10°$	7.20	5.40	4.70	4.00	6.40	3.80	2.80
C_L/C_D at $\alpha = 20°$	4.40	4.20	3.60	3.60	3.90	2.50	1.75
C_L/C_D at $\alpha = \alpha_s$	2.30	1.70	1.50	1.80	2.40	1.30	1.19
c_s/c at $\alpha = 10°$	0.18	0.20	0.27	0.26	0.25	0.28	0.28
c_s/c at $\alpha = \alpha_s$	0.35	0.35	0.36	0.25	0.33	0.41	0.43

standard value of 30% of the chord length (measured from the leading edge) produce larger maximum lift and less change of the center of effort of the rudder force. The latter fact allows a smaller steering gear if the rudder is properly balanced. On the other hand, these profiles have higher drag coefficients, thus requiring more propulsive power for the same ship speed. (For a rudder behind a propeller, the slipstream rotation causes angles of attack of typically 10° to 15°.) A 10% increase of the rudder resistance in this angle-of-attack range accounts for approximately 1% increase in the necessary propulsion power. For ship speeds exceeding 22 knots and the rudder in the propeller slipstream, profiles with the opposite tendency (backward-shifted maximum thickness) are preferred because they are less prone to cavitation.

Greater profile thickness produces greater maximum lift at the (correspondingly greater) stall angle α_s, but it increases the rudder drag, and in most cases the danger of cavitation in high-speed ships. Thus, the smallest thickness possible to accommodate the rudder post and bearing is normally used. For rudders of small aspect ratio, the greater maximum lift of thick rudders is realized only in yaw checking, but not at all if the steering gear allows the normal $\delta = 35°$ rudder angle only (Fig. 6.14). A trailing edge of substantial thickness decreases the change of the center of effort c_s with angle of attack α, but it causes substantially increased drag; thus, because of too large drag, the application of these profiles should be avoided, at least in long-range vessels.

The approximate formulae for the force coefficients are only valid for angles of attack $\alpha < \alpha_s$. Beyond the stall angle α_s the flow separates near the profile leading edge (nose) on the suction side of the profile without reattachment. Then the lift decreases strongly and the drag increases (Fig. 6.13). The sudden drop in lift beyond the stall angle α_s is not found for certain other profiles and in rudders behind a propeller.

The stall angle α_s depends primarily on:

- the aspect ratio Λ;
- the profile shape and thickness;
- the Reynolds number;
- probably the surface roughness;
- the turbulence of the inflow;
- the spatial distribution of the inflow velocity.

Because of the last four parameters, an exact prediction of maximum rudder lift from wind tunnel or towing tank experiments is impossible. Whereas a greater aspect ratio Λ (height-to-chord ratio b/c_m) increases the lift for a given angle of attack $\alpha < \alpha_s$, the maximum lift coefficient is practically independent of the aspect ratio (Fig. 6.14). Thus increasing the rudder area by increasing the chord length is of equal effect as by increasing the rudder height with respect to the maximum rudder force if the stall angle is reached by the steering gear; otherwise, an increase in rudder height is much more effective than a corresponding increase in chord length. (A rudder angle $\delta = 35°$ relative to the ship's longitudinal axis corresponds to angles of attack α_s, of nearly the same size in initial turning, of smaller size during steady turning and of larger size in the yaw checking phase.)

Because of the different stall angles α_s and lift curve slopes of rudders of different aspect ratios it would be advantageous to use an effective rudder angle δ_{eff} instead of the geometrical rudder angle δ for rules, e.g. about the maximum rudder angle and the rudder rate of the steering gear, as well as for nautical use. This would be 'fairer' for rudders of different aspect ratio; it would also make better use of rudders of smaller aspect ratio (today their greater stall angle α_s is frequently not realized because of a too small maximum rudder angle δ) and would lead to more equal response of different ships on (effective) rudder angles. If geometric and effective rudder angles are defined to coincide for a normal aspect ratio of $\Lambda = 2$, their relationship is (Fig. 6.15):

$$\delta_{\text{eff}} = \frac{2.2 \cdot \Lambda}{\Lambda + 2.4} \cdot \delta \qquad (6.64)$$

For aspect ratios $\Lambda < 3$, which are typical for ship rudders, the vertical distribution of the lift force in homogeneous, unbounded flow is practically elliptic:

$$\text{lift per length} = \frac{L}{b} \cdot \frac{4}{\pi} \cdot \sqrt{1 - \left(\frac{z}{b/2}\right)^2} \qquad (6.65)$$

Here z is the vertical distance from the mean height between the lower and upper edges of the rudder. The distribution is hardly influenced by the rudder shape for the usual trapezoidal shape with a taper ratio $0.5 < c_{\text{min}}/c_{\text{max}} < 1.0$. Thus, for a free-running rudder of trapezoidal shape the lift center is nearly at half the rudder height, not at the center of gravity of the shape. This effect is even more pronounced for lower aspect ratios. If there is only a small gap between the upper edge of the rudder and fixed parts of the hull (at the rudder angles concerned), the center of effort moves up a little, but never more than 7.5% of b, the value for a rudder without a gap at its upper edge.

Air ventilation may occur on the suction side of the rudder if the rudder pierces or comes close to the water surface. The extent of the ventilation may cover a large part of the rudder (even the

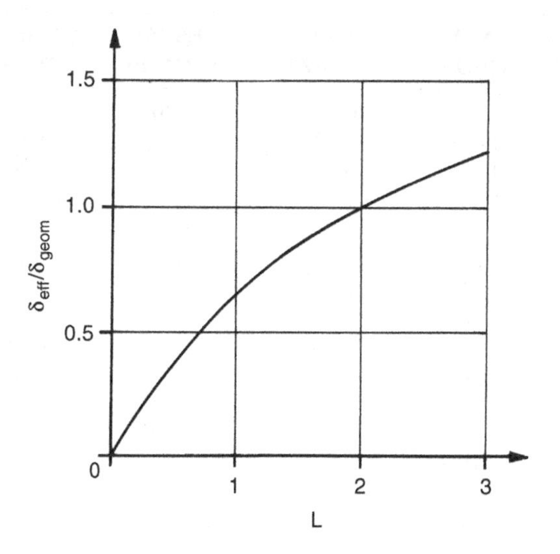

Figure 6.15:
Ratio between effective and geometrical angles of attack

whole rudder height), decreasing the rudder effectiveness drastically. This is important for maneuvers at ballast draft for full speed, e.g. at ship trials.

The dynamic pressure along the profile of a rudder depends on the local velocity v according to Bernoulli's law:

$$p_{\mathrm{dyn}} = \frac{\rho}{2} \cdot (V^2 - v^2) = q \cdot \left(1 - \frac{v^2}{V^2}\right) \tag{6.66}$$

For the usual straight profiles v/V is decomposed into two components:

1. Component v_t/V due to the profile thickness t. This component is equal on both sides of the profile. v_t/V may be taken from Table 6.6. For different profile thickness t, the velocity ratio v_t/V must be corrected by

$$\left[\left(\frac{v_t}{V}\right)_{\mathrm{actual}} - 1\right] = \left[\left(\frac{v_t}{V}\right)_{\mathrm{table}} - 1\right] \cdot \frac{t_{\mathrm{actual}}}{t_{\mathrm{table}}} \tag{6.67}$$

   Information on other profiles may be found in Abbott and Doenhoff (1959) or computed by CFD (e.g. boundary element method).
2. Component v_a/V due to the angle of attack α_s. This component has opposite sign on both sides of the profile. It is practically independent of the profile shape. Only in the front part

Table 6.6: $v_t = V$; flow speed v_t along the profile over inflow velocity V as a function of the profile abscissa x, $\alpha = 0°$

x/c (%)	NACA 643-018	NACA 0020	HSVA MP73-20	HSVA MP71-20	IFS58 TR15	IFS61 TR25	IFS62 TR25
0	0.00	0.00	0.00	0.00	0.00	0.00	0.00
0.75	0.77	0.69	0.69	0.57	0.79	0.67	0.68
1.25	0.96	0.91	0.91	0.88	1.06	0.95	0.94
2.5	1.05	1.03	1.08	1.00	1.20	1.09	1.18
6.0	1.11	1.17	1.22	1.10	1.29	1.47	1.48
7.5	1.15	1.25	1.27	1.12	1.30	1.52	1.53
10	1.17	1.27	1.29	1.14	1.28	1.50	1.52
15	1.20	1.30	1.31	1.18	1.26	1.47	1.48
20	1.22	1.29	1.30	1.20	1.23	1.43	1.44
30	1.25	1.26	1.27	1.24	1.20	1.31	1.33
40	1.26	1.21	1.24	1.28	1.16	1.18	1.21
50	1.20	1.17	1.17	1.30	1.08	1.06	1.08
60	1.13	1.13	1.07	1.14	1.00	0.96	0.97
70	1.06	1.08	1.01	1.04	0.94	0.90	0.90
80	0.98	1.03	0.95	0.96	0.93	0.90	0.87
90	0.89	0.96	0.88	0.87	0.96	0.94	0.90
95	0.87	0.91	0.89	0.87	0.97	0.95	0.93

does it depend on the profile nose radius. Figure 6.16 illustrates this for a lift coefficient $C_{Ll} \approx 1$. The values given in Fig. 6.16 have to be multiplied by the actual local lift coefficient:

3.

$$C_{Ll} = \frac{\text{lift per length}}{c \cdot q} = C_L \cdot \frac{4}{\pi} \cdot \sqrt{1 - \left(\frac{z}{b/2}\right)^2} \tag{6.68}$$

where C_L (Λ) is estimated by Eq. (6.59).
The dynamic pressure is then:

$$p_{\text{dyn}} = \left[1 - \left(\frac{v_t}{V} \pm \frac{v_a}{V} \cdot C_{Ll}\right)^2\right] \cdot q \tag{6.69}$$

Due to the quadratic relationship in this equation, the pressure distribution will generate the given C_{Ll} only approximately. For better accuracy, the resulting local lift coefficient should be integrated from the pressure difference between both sides of the profile. If it differs substantially from the given value, the pressure distribution is corrected by superimposing the v_a/V distribution in Eq. (6.69) with a factor different from C_{Ll} such that the correct C_{Ll} is attained by the integration of the pressure difference.

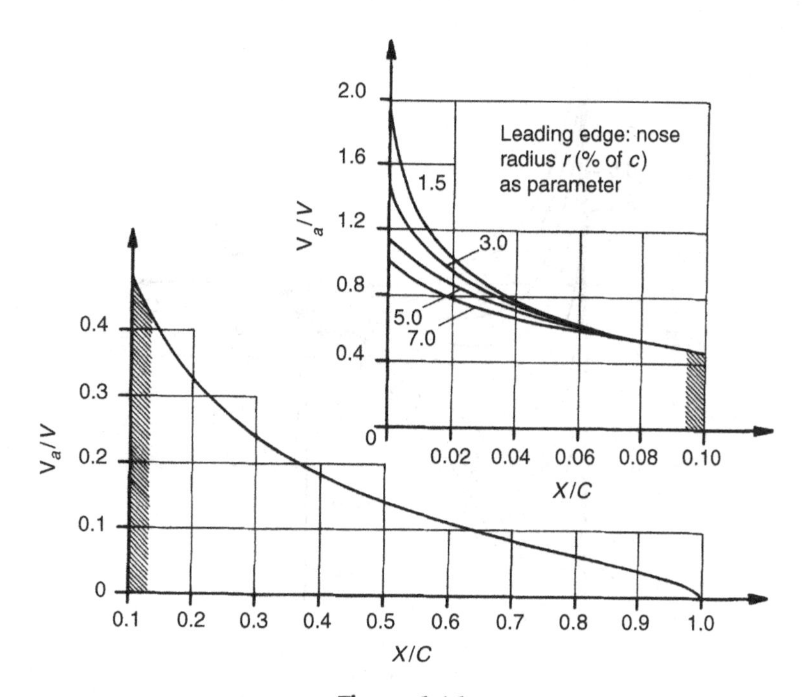

Figure 6.16:
Relative change of velocity v_a/V on the surface due to the angle of attack α_s for $C_{Ll} \approx 1$

The dynamic pressure is negative over most of the profile length, for moderate lift coefficients even on the pressure side of the rudder. This is illustrated in Fig. 6.17 for an NACA0021 profile. The curve for $C_{LI} = 0$ corresponds to the component due to the profile thickness alone. For other C_{LI} values, the upper and lower curves refer to the pressure and suction sides, respectively. For profiles with a curved mean line, an additional velocity component has to be added. It may be taken from Abbott and Doenhoff (1959, pp. 77ff and App. II), or it may be determined by a two-dimensional potential-flow calculation for which various methods and codes are available. Brix (1993, p. 84) gives a sample calculation for the NACA643-018 profile for $\alpha = 15°$.

6.4.3. Rudder Types

Various rudder types have been developed over the years (Fig. 6.18):

- *Rudder with heel bearing (simplex)*
  The most common rudder type formerly built was a rectangular profile rudder with heel bearing. The heel has to have considerable width to withstand the horizontal forces. Flow separation at the heel increases resistance and the non-homogeneity of the wake field at the propeller plane, which in turn increases propeller-induced vibrations. Therefore modern single-screw ships are usually equipped with other rudder types, but the rudder with heel bearing is still popular for small craft and some fishery vessels, because it is cheap.

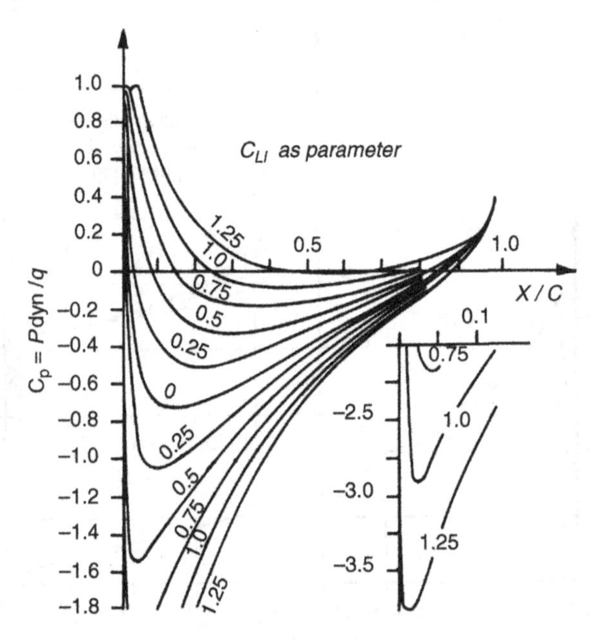

Figure 6.17:
Distribution of the non-dimensional dynamic pressure along an NACA0021 profile as a function of the local lift coefficient C_{LI} (Riegels 1958)

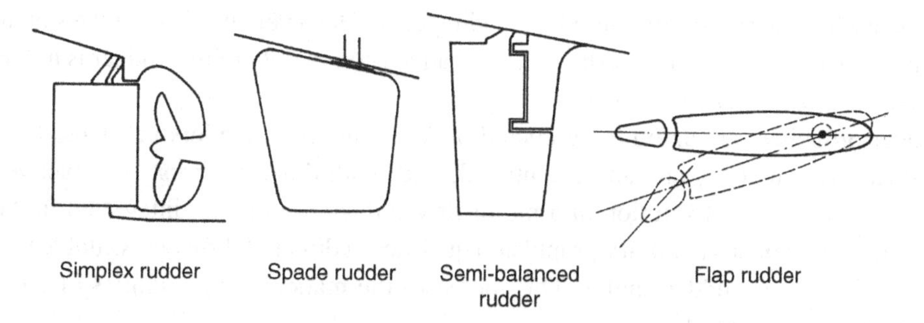

Figure 6.18:
Various rudder types

- *Spade rudder*
  This type of rudder is commonly applied, especially on ferries, ro-ro ships, and special craft. The rudder stock is subject to high bending moments, especially for high ship speed and large rudder height. For stock diameters larger than 1 m, this design becomes impractical.
- *Semi-balanced rudders*
  For semi-balanced rudders, a fixed guide-head (sometimes called rudder horn) extends over the upper part of the rudder. This type of rudder has the following properties:
  - Decreased rudder bending moment compared to spade rudders.
  - Reduced rudder effectiveness compared to spade rudders. For steady turning circles, the semi-balanced rudder produces only approximately half the transverse force as a spade rudder of the same area (including the area of the rudder horn). The reasons for the reduced transverse force are:
    - The horizontal gap between horn and rudder opens wide for large rudder angles. Sometimes horizontal plates are attached at the horn along the gap as a remedy for this problem (rudder scissors).
    - Unfavorable angle of attack for the rudder horn (fixed guide-head).
    - Drag/lift ratio of the rudder about twice as high as for spade rudders.
- *Flap rudders*
  Flap rudders (e.g. Becker rudders) consist of a movable rudder with a trailing edge flap activated by a mechanical or hydraulical system, thus producing a variable flap angle as a function of the rudder angle. This system works like an airfoil with a flap. Flap rudders give a much higher lift per rudder angle and a 60–70% higher maximum lift compared to a conventional rudder of the same shape, size, and area.

Less frequently, the following rudder types are employed:

- *Rudders with rotating cylinders*
  These rudders have a rotating cylinder at the leading edge. Whereas the freely arranged rotating cylinder works according to the Magnus effect, the combination of a rotating cylinder and a conventional rudder shifts the stall to a much higher angle of attack by eliminating the

boundary layer at the separation-prone leading edge. However, at full scale the stall angle of conventional rudders is often so high that the added complexity of this rudder is not justified.

- *Active rudder/rudder propellers*
  Rudder propellers are azimuthing ducted or free-running propellers in a fixed or hinged vertical position. They are active control devices with directed thrust. The 'active propeller' is a special solution of a motor-driven ducted propeller integrated in the main rudder. Thus, besides auxiliary propulsion qualities, a directed thrust is available within the range of the main rudder angles. This increases the maneuvering qualities of the ship, especially at low speeds.
- *Steering nozzle with rudder*
  Steering nozzle may be fitted with flapped or unflapped rudders. This highly effective steering device is sometimes fitted to tugs, or research or fishery ships.

A fixed fin above the rudder improves the yaw checking stability as much as if the area of the fixed fin were included in the movable rudder. However, for course-changing ability only the movable rudder is decisive. In fact, a fixed fin has a course-stabilizing property and increases, for example, the turning circle diameter. A gap between the rudder top and the hull increases the rudder resistance at center position due to the induced resistance of the oblique inflow of the propeller slipstream and the resistance of the rudder.

Twin-screw ships may be fitted with spade or semi-balanced rudders, either behind the propellers or as midship rudders. For fast ships with a rudder arrangement on the centerplane cavitation problems are avoided, but this arrangement is less effective than rudders in the propeller slipstream, especially on shallow water.

6.4.4. Interaction of Rudder and Propeller

Rudders are normally placed in the propeller slipstream for the following reasons:

- A profiled rudder increases the propulsive efficiency by utilizing a part of the rotational energy contained in the propeller slipstream.
- In steady ahead motion, the rudder forces are typically more than twice those of a rudder outside of the propeller slipstream.
- Even for a stationary or slowly moving ship, substantial rudder forces may be generated by increasing the propeller rpm (especially to provide increased rudder effectiveness during transient maneuvers).

Because the rudder forces are proportional to the flow speed squared at the rudder, an accurate determination of the speed in the propeller slipstream at the rudder position is required to correctly predict rudder forces. According to the momentum theory of the propeller, the mean axial speed of the slipstream far behind the propeller is:

$$V_\infty = V_A\sqrt{1 + C_{Th}} \tag{6.70}$$

where C_{Th} is the thrust loading coefficient according to Eq. (2.5). V_A is the mean axial speed of inflow to the propeller, A_P the propeller area. The theoretical slipstream radius r_∞ far behind the propeller flows from the law of continuity, assuming that the mean axial speed at the propeller is the average between V_A and V_∞:

$$r_\infty = r_0\sqrt{\frac{1}{2}\left(1 + \frac{V_A}{V_\infty}\right)} \tag{6.71}$$

Here r_0 is half the propeller diameter D.

Normally the rudder is in a position where the slipstream contraction is not yet completed. The slipstream radius and axial velocity may be approximated by (Söding 1982):

$$r = r_0 \cdot \frac{0.14(r_\infty/r_0)^3 + r_\infty/r_0 \cdot (x/r_0)^{1.5}}{0.14(r_\infty/r_0)^3 + (x/r_0)^{1.5}} \tag{6.72}$$

and

$$V_x = V_\infty \cdot \left(\frac{r_\infty}{r}\right)^2 \tag{6.73}$$

Here x is the distance of the respective position behind the propeller plane. To determine rudder force and moment, it is recommended to use the position of the center of gravity of the rudder area within the propeller slipstream.

The above expression for r is an approximation of a potential-flow calculation. Compared to the potential flow result, the slipstream will increase in diameter with increasing distance x from the propeller plane due to turbulent mixing with the surrounding fluid. This may be approximated (Söding 1986) by adding:

$$\Delta r = 0.15x \cdot \frac{V_x - V_A}{V_x + V_A} \tag{6.74}$$

to the potential slipstream radius and correcting the slipstream speed according to the momentum theorem:

$$V_{\text{corr}} = (V_x - V_A)\left(\frac{r}{r + \Delta r}\right)^2 + V_A \tag{6.75}$$

The results of applying this procedure are shown in Fig. 6.19. V_{corr} is the mean value of the axial speed component over the slipstream cross-section.

The rudder generates a lift force by deflecting the water flow up to considerable lateral distances from the rudder. Therefore the finite lateral extent of the propeller slipstream

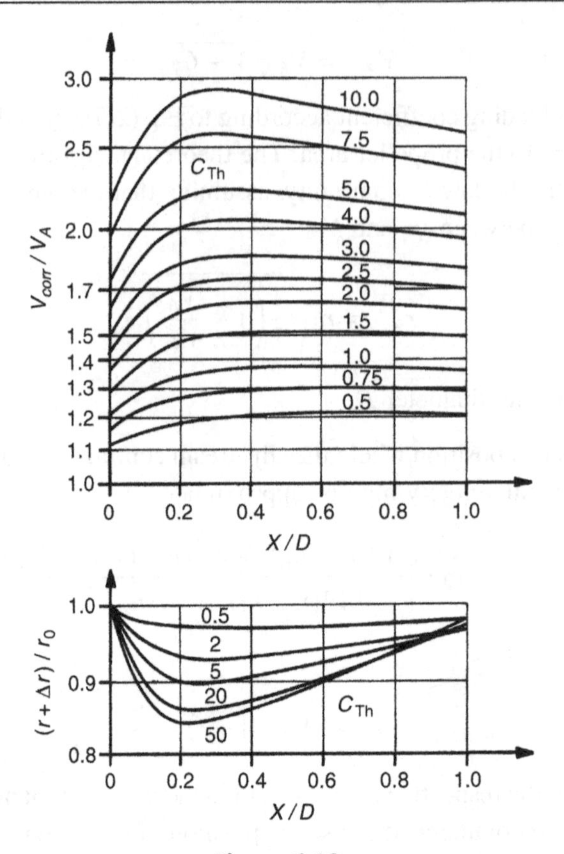

Figure 6.19:
Mean axial slipstream speed as a function of propeller approach speed V_A and slipstream radius $(r+\Delta r)/r_0$ due to potential flow and turbulent mixing at different positions x/D behind the propeller

diminishes the rudder lift compared to a uniform inflow velocity. This is approximated (Söding 1982) (based on two-dimensional potential flow computations for small angles of attack) by multiplying the rudder lift determined from the velocity within the rudder plane by the correction factor λ determined from:

$$\lambda = \left(\frac{V_A}{V_{\text{corr}}}\right)^f \quad \text{with} \quad f = 2 \cdot \left(\frac{2}{2 + d/c}\right)^8 \tag{6.76}$$

Here V_A is the speed outside of the propeller slipstream laterally from the rudder. d is the half-width of the slipstream. For practical applications, it is recommended to transform the circular cross-section (radius $r + \Delta r$) of the propeller slipstream to a quadratic one (edge length $2d$) of equal area. This leads to the relation:

$$d = \sqrt{\frac{\pi}{4}}(r + \Delta r) = 0.886 \cdot (r + \Delta r) \tag{6.77}$$

The inflow velocity in the rudder plane varies along the rudder height due to the wake distribution and the propeller slipstream. The effect of this variation may be approximated by using the mean squared velocity:

$$\overline{V^2} = \frac{1}{A_R} \int_0^b V^2 \cdot c\, dz \tag{6.78}$$

for the determination of the rudder lift.

Lifting-surface calculations show that, compared to a uniform distribution, the lift coefficient (defined with reference to $\overline{V^2}$) is some 5% higher for rudders extending downward beyond the lower edge of the propeller slipstream (Fig. 6.20). Hence it is recommended to extend the rudder as far to the baseline of the ship as possible.

A simple global correction for the lift force of a rudder behind a propeller (to be added to the lift computed by the usual empirical formulae for rudders in free stream) is (Söding 1998a, b):

$$\Delta L = T \cdot \left(1 + \frac{1}{\sqrt{1 + C_{Th}}}\right) \cdot \sin\delta \tag{6.79}$$

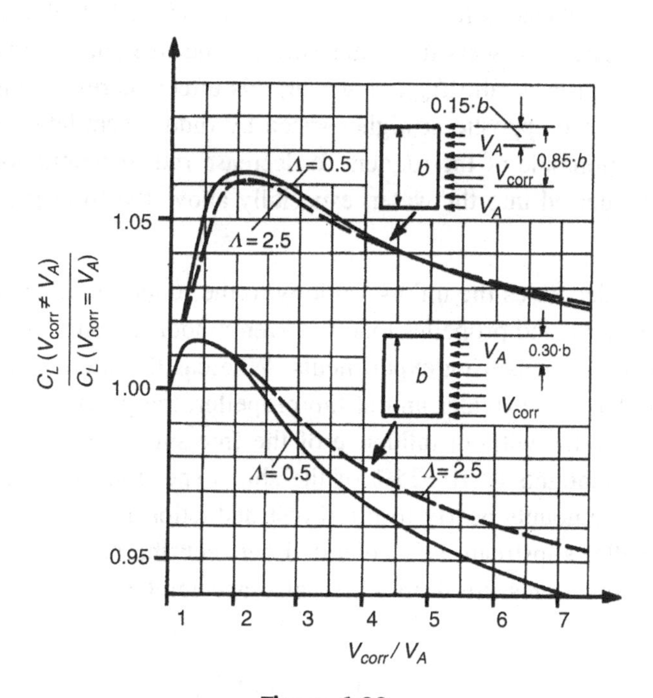

Figure 6.20:
Lift coefficients as a function of the vertical distribution of flow and the aspect ratio Λ

The additional drag (or decrease in propeller thrust) is:

$$\Delta D = T \cdot \left(1 + \frac{1}{\sqrt{1 + C_{Th}}}\right) \cdot (1 - \cos \delta) \tag{6.80}$$

In these formulae, T denotes the propeller thrust.

6.4.5. Interaction of Rudder and Ship Hull

If the hull above the rudder is immersed, it suppresses the flow from the pressure to the suction side around the upper edge of the rudder. This has effects similar to an increase of the rudder aspect ratio Λ:

- It decreases the induced drag.
- It increases the slope of the lift curve versus angle of attack α.
- It hardly influences the maximum lift at the stall angle α_s.

The magnitude of this effect depends on the size of the gap between the upper edge of the rudder and the hull. For very small gaps, the aspect ratio Λ_{eff} is theoretically twice the nominal value, in practice $\Lambda_{eff} \approx 1.6 \cdot \Lambda_{geom}$. To close the gap between hull and rudder at least for small rudder angles δ – and thus increasing the rudder effectiveness – a fixed fin above the rudder is advantageous for small rudder angles. If the hull above the rudder is not immersed or if the rudder intersects the water surface, the free surface may also increase somewhat the effective aspect ratio Λ_{eff}. However, this effect decreases with increasing ship speed and may turn to the opposite at higher speed by rudder ventilation drawn from the surface along the suction side of the rudder. To decrease rudder ventilation, a broad stern shape sufficiently immersed into the water, especially above the front part of the rudder, is advantageous.

The wake of the hull decreases the inflow velocity to the rudder and increases the propeller load. Differences in wake and propeller load between model and ship are the main cause of scale effects in model maneuvering experiments. Whereas the wake due to hull surface friction will be similar at the rudder and at the propeller, the potential wake – at least for small Froude numbers, i.e. without influence of the free surface – is nearly zero at the rudder, but typically amounts to 10–25% of the ship's speed at the propeller of usual single-screw ships. It amounts nearly to the thrust deduction fraction t. Thus the flow outside of the propeller slipstream is accelerated between the propeller and the rudder by about $t \cdot V$. This causes a pressure drop which also accelerates the propeller slipstream to approximately:

$$V_x = \frac{V_{corr}^2 + t \cdot V^2}{V_{corr}} \tag{6.81}$$

The corresponding slipstream contraction is:

$$r_x = (r + \Delta r) \cdot \sqrt{\frac{V_{\text{corr}}}{V_x}} \tag{6.82}$$

For non-zero rudder angle and forward ship speed, an interaction between the flow around rudder and hull occurs which decreases the lift force at the rudder; however, an additional transverse force of equal direction is generated at the aftbody. Compared to the rudder lift without hull interaction, the total transverse force is increased by the factor $(1 + a_H)$. The term a_H may be approximated (Söding 1982):

$$a_H = \frac{1}{1 + (4.9 \cdot e/T + 3c/T)^2} \tag{6.83}$$

Here T is the draft of the ship, e the mean distance between the front edge of the rudder and the aft end of the hull, and c the mean rudder chord length. Compared to the free-running rudder, the center of effort of the total transverse force is shifted forward by approximately:

$$\Delta x = \frac{0.3T}{e/T + 0.46} \tag{6.84}$$

Potential flow computations show that Δx may increase to up half the ship's length in shallow water if the gap length e between rudder and hull tends to zero, as may be the case for twin-screw ships with a center rudder. This would decrease the ship's turning ability on shallow water. For a non-zero drift velocity v (positive to starboard, measured amidships) and a non-zero yaw rate r (positive clockwise if seen from above) of the ship, the hull in front of the rudder influences the flow direction at the rudder position. Without hull influence, the transverse flow velocity v relative to the hull at the rudder position x_R is:

$$v_R = -(v + x_R \cdot r) \tag{6.85}$$

x_R is the distance between rudder and midship section, negative for stern rudders. However, experiments of Kose (1982) with a freely rotating, unbalanced rudder behind a ship model without propeller indicated a mean transverse velocity at the rudder's position of only:

$$v_R = -(0.36 \cdot v + 0.66 \cdot x_R \cdot r) \tag{6.86}$$

From the rudder angle δ (positive to port side), the mean longitudinal flow speed V_x (positive backward) and the mean transverse flow speed v_R at the rudder position, the angle of attack follows:

$$\alpha = \delta + \arctan \frac{v_R}{V_{\text{corr}}} \tag{6.87}$$

6.4.6. Rudder Cavitation

Rudder cavitation may occur even at small rudder angles for ship speeds exceeding 22 knots with rudder(s) in the propeller slipstream and $P/A_P > 700\,\text{kW/m}^2$. Here P is the delivered power, A_P the propeller disk area.

Rudder cavitation – as with propeller cavitation – is caused by water evaporation where at points of high flow velocity the pressure locally drops below the vapor pressure of the water. Cavitation erosion (loss of material by mechanical action) occurs when small bubbles filled with vapor collapse on or near to the surface of the body. During the collapse a microscopic high-velocity jet forms, driven by surface tension and directed onto the body surface. It causes small cracks and erosion, which in seawater may be magnified by corrosion (galvanic loss of material). Paint systems, rubber coatings, enamel, etc. offer no substantial resistance to cavitation, but austenitic steel and some types of bronze seem to retard the erosion compared to the mild steel normally used for rudders.

The cavitation number σ (Fig. 6.21) is a non-dimensional characteristic value for studying cavitation problems in model experiments:

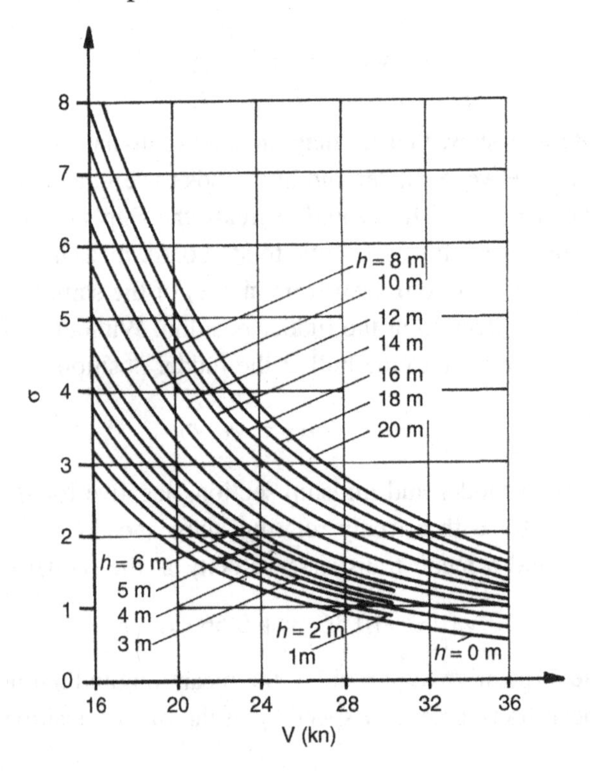

Figure 6.21:
Cavitation number σ as a function of the ship speed V with the submersion h (depth below water surface) as parameter

$$\sigma = \frac{p - p_v}{1/2 \rho V^2} \tag{6.88}$$

p is the pressure in undisturbed flow, i.e. atmospheric pressure plus hydrostatic pressure, p_v vaporization pressure, V ship speed, ρ density of water.

There are different types of rudder cavitation:

1. *Bubble cavitation on the rudder side plating*
   For large rudder angles, cavitation is unavoidable in ships of more than about 10 knots. It will decrease the rudder lift substantially if the cavitation causes a complete separation of flow from the suction side. Otherwise its influence on rudder forces is small (Kracht 1987). Cavitation erosion is of interest only if it occurs within the range of rudder angles $\delta = \pm 5°$ used for course keeping. Evaluation of model experiments shows that the onset of cavitation is indeed observed if the pressure determined by potential-flow theory is smaller than the water vaporization pressure p_v. p_v lies typically between 1% and 3% of the atmospheric pressure. It may therefore (not in model tests, but for full-scale ships) simply be taken as zero. Thus, to test for blade side cavitation in the design stage of ships, one may proceed as follows:
   - Determine the slipstream radius $r + \Delta r$ and the inflow speed to the rudder V_{corr} from Fig. 6.19 or the corresponding formulae at about 80% of the propeller tip radius above and below the propeller axis.
   - Correct these values to obtain V_x and r_x by Eqs (6.81) and (6.82).
   - Because of non-uniform distribution of the slipstream velocity, add 12% of V to obtain the maximum axial speed at the rudder:

   $$V_{max} = V_x + 0.12 \cdot (V_{corr} - V_A) \tag{6.89}$$

   - Estimate the inflow angle α to the rudder due to the rotation of the propeller slipstream by:

   $$\alpha = \arctan\left(4.3 \cdot \frac{K_Q}{J^2} \cdot \sqrt{\frac{1 - \overline{w}}{1 - w_{local}}} \cdot \frac{V_A}{V_{max}}\right) \tag{6.90}$$

     $\overline{w}$ is the mean wake number and w_{local} the wake number at the respective position. The equation is derived from the momentum theorem with an empirical correction for the local wake. It is meant to apply about 0.7 to 1.0 propeller diameter behind the propeller plane. The position relevant to the pressure distribution is about one-half chord length behind the leading edge of the rudder.
   - Add $\delta = 3° = 0.052$ rad as an allowance for steering rudder angles.
   - Determine the maximum local lift coefficient C_{Llmax} from Fig. 6.22, where $\alpha + \delta$ are to be measured in radians. c is the chord length of the rudder at the respective height,

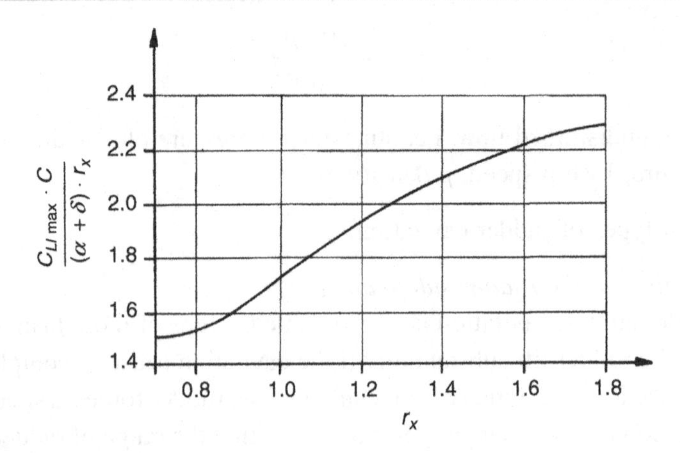

Figure 6.22:
Diagram for determining the local values of maximum lift coefficient C_{Llmax}

r_x the propeller slipstream radius following from Eq. (6.70) with r_x in the place of r_∞. Figure 6.22 is based on lifting-line calculations of a rudder in the propeller slipstream. It takes into account the dependence of the local lift coefficient on the vertical variation of inflow velocity and direction.

- Determine the extreme negative non-dimensional pressure on the suction side depending on profile and local lift coefficient C_{Llmax}. For this we use Fig. 6.23 derived from potential flow calculations.
- Add to p_{dyn} (negative) the static pressure $p_{stat} = 103\,\text{kPa} + \rho \cdot g \cdot h$. h is the distance between the respective point on the rudder and the water surface, e.g. 80% of the propeller radius above the propeller axis.

If the resulting minimum pressure on the suction side is negative or slightly positive (less than 3 kPa), the side plating of the rudder is prone to cavitation. For a right-turning propeller (turning clockwise looking forward) cavitation will occur:

- on the starboard side in the upper part of the rudder relative to the propeller axis;
- on the port side in the lower part of the rudder relative to the propeller axis.

Brix (1993, pp. 91–92) gives an example for such a computation. Measures to decrease rudder side cavitation follow from the above prediction method:

- Use profiles with small p_{dyn} at the respective local lift coefficient. These profiles have their maximum thickness at approximately 40% behind the leading edge.
- Use profiles with an inclined (relative to the mean rudder plane) or curved mean line to decrease the angle of attack (Brix et al. 1971). For a right-turning propeller, the rudder nose should be on the port side above the propeller axis, on the starboard side below it.

2. *Rudder sole cavitation*

Due to the pressure difference between both sides of the rudder caused, say, by the rotation of the propeller slipstream, a flow component around the rudder sole from the pressure to

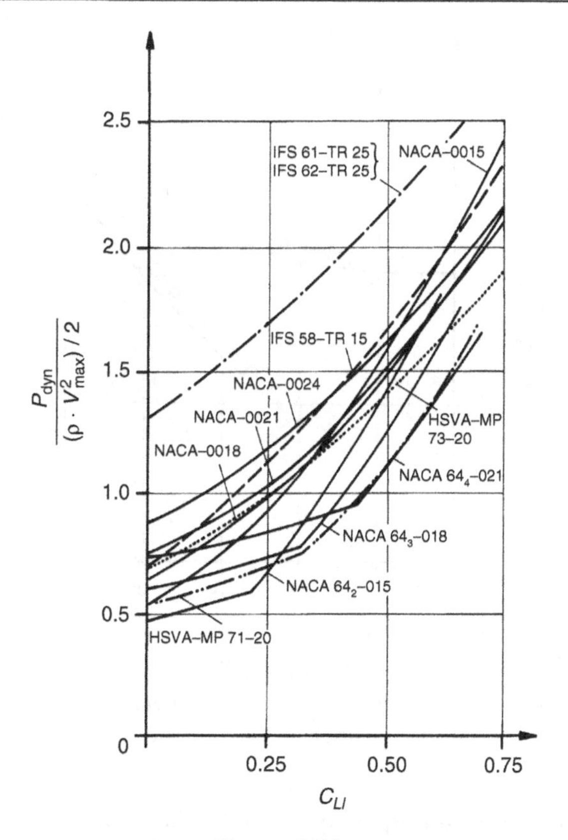

Figure 6.23:
Extreme negative dynamic pressure of the suction side as a function of the local lift coefficient C_{LI} and the profile

the suction side occurs. It causes a rudder tip vortex (similar to propeller tip vortices) which may be filled by a cavitation tube. This may cause damage if it attaches to the side of the rudder. However, conditions for this are not clear at present. If the rudder has a sharp corner at the front lower edge, even for vanishing angles of attack the flow cannot follow the sharp bend from the leading edge to the base plate, causing cavitation in the front part of the rudder sole. As a precaution the base plate is bent upward at its front end (Brix et al. 1971). This lowers the cavitation number below which sole cavitation occurs (Fig. 6.24). For high ship speeds exceeding, say, 26 knots cavitation has still been reported. However, it is expected that a further improvement could be obtained by using a smoothly rounded lower face or a baffle plate at the lower front end (Kracht 1987). No difficulties have been reported at the rudder top plate due to the much lower inflow velocity.

3. *Propeller tip vortex cavitation*

Every propeller causes tip vortices. These are regions of low pressure, often filled with cavitation tubes. Behind the propeller they form spirals which are intersected by the rudder.

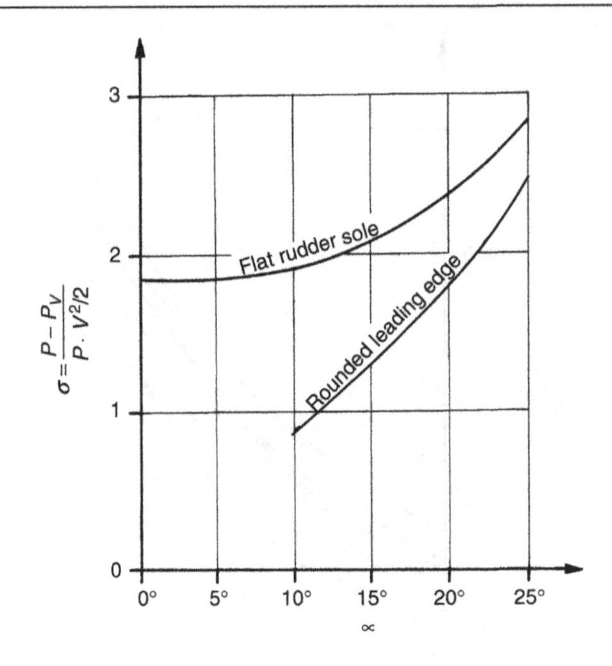

Figure 6.24:
Cavitation number B below which rudder sole cavitation occurs as a function of the angle of attack and the rudder sole construction

Therefore, cavitation erosion frequently occurs at the rudder at the upper and sometimes lower slipstream boundaries, mainly (for right-turning propellers) on the upper starboard side of the rudder. This problem is not confined to high-speed ships. The best means to reduce these effects is to decrease gradually the propeller loading to the blade tips by appropriately reduced pitch, and to use a high propeller skew. These methods also reduce propeller-induced vibrations.

4. *Propeller hub cavitation*
   Behind the propeller hub a vortex is formed which is often filled by a cavitation tube. However, it seems to cause fewer problems at the rudder than the tip vortices, possibly due to the lower axial velocity behind the propeller hub.
5. *Cavitation at surface irregularities*
   Surface irregularities disturbing the smooth flow cause high flow velocities at convex surfaces and edges, correspondingly low pressures, and frequently cavitation erosion. At the rudder, such irregularities may be zinc anodes and shaft couplings. It is reported that also behind scoops, propeller bossing, etc. cavitation erosion occurred, possibly due to increased turbulence of the flow. Gaps between the horn and the rudder blade in semi-balanced rudders are especially prone to cavitation, leading to erosion of structurally important parts of the rudder. For horizontal and vertical gaps (also in flap rudders) the rounding of edges of the part behind the gap is recommended.

6.4.7. Rudder Design

There are no regulations for the rating of the rudder area. The known recommendations give the rudder area as a percentage of the underwater lateral area $L \cdot T$. Det Norske Veritas recommends:

$$\frac{A_R}{L \cdot T} \geq 0.01 \cdot \left(1 + 25\left(\frac{B}{L}\right)^2\right) \tag{6.91}$$

This gives a rudder area of approximately 1.5% of the underwater lateral area for ships of usual width; for unusually broad ships (large mass, low yaw stability) a somewhat larger value is given. This corresponds to typical rudder designs and can serve as a starting point for further analyses of the steering qualities of a ship.

Recommended minimum criteria for the steering qualities of a ship are:

- Non-dimensional initial turning time in Z 20°/10° maneuvers: $t'_a = 1 + 1.73F_n$.
- Non-dimensional yaw checking time in Z 20°/10° maneuvers: $t'_s = 0.78 + 2.12F_n$.
- The rudder should be able to keep the ship on a straight course with a rudder angle of maximum 20° for wind from arbitrary direction and $v_w/V = 5$. v_w is the wind speed, V the ship speed.
- The ship must be able to achieve a turning circle of less than $5 \cdot L$ at the same v_w/V for maximum rudder angle.

The criteria for initial turning time and yaw checking time were derived by Brix using regression analysis for 20°/10° zigzag test results for many ships (Fig. 6.8). The time criteria are critical for large ships (bulkers, tankers), while the wind criteria are critical for ships with a large lateral area above the water (ferries, combatants, container ships). An additional criterion concerning yaw stability would make sense, but this would be difficult to check computationally.

The rudder design can be checked against the above criteria using computations (less accurate) or model tests (more expensive). A third option would be the systematically varied computations of Wagner, described in Brix (1993, pp. 95–102). This approach yields a coefficient $C_{Y\delta}$ for rudder effectiveness which inherently meets the above criteria. The method described in Brix (1993) uses design diagrams. For computer calculations, empirical formulae also derived by Wagner exist.

6.4.8. CFD for Rudder Flows and Conclusions for Rudder Design

The determination of forces on the rudder is important for practical design purposes:

- The transverse force is needed to evaluate the maneuverability of ships already in the design stage as required by IMO.
- The longitudinal force influences noticeably the propulsive efficiency.
- The shaft torsional moment is decisive for selecting a suitable rudder gear.

In principle, there are three sources of information for these forces:

- Model experiments which produce accurate forces at model Reynolds numbers, but suffer from severe scale effects when predicting the maximum lift at full scale.
- RANSE computations appear to be the most reliable source of information and should gain in importance also for practical design of rudders.
- BEM computations can often give sufficiently accurate results with a minimum of effort if some empirical relationships and corrections are applied.

Söding (1998a, b) described the state of the art for BEM comparing the results to RANSE and experimental results. The RANSE computations used for comparison were finite-volume methods employing a standard k–ε turbulence model with wall function.

Söding's BEM approach for rudder flows introduces some special features:

- Special adaptations of the BEM take the irrotational inflow to the rudder induced by hull and propeller into account.
- The propeller slipstream is averaged in the circumferential direction. The radial thrust distribution is assumed such that it approaches gradually zero at the outer limit and is zero in the hub region.
- The ship hull above the rudder can either be modeled as horizontal mirror plane or as a separate body discretized by boundary elements.

The BEM results were compared to RANSE and experimental results for various rudders. According to potential theory, a thin foil in two-dimensional flow (i.e. for aspect ratio $\Lambda = \infty$) at a small angle of attack α produces lift nearly linearly increasing with α corresponding to:

$$\frac{dC_L}{d\alpha} = 2\pi \tag{6.92}$$

In three-dimensional flow, the lift gradient is decreased by a reduction factor $r(\Lambda)$ which is well approximated by:

$$r(\Lambda) = \Lambda\frac{\Lambda + 0.7}{(\Lambda + 1.7)^2} \tag{6.93}$$

Except for Λ, details of the rudder shape in side view (e.g. rectangular or trapezoidal) have hardly any influence on $dC_L/d\alpha$. However, the profile thickness and shape have some influence. Computations and measurements of the lift coefficient corrected for infinite aspect ratio by the formula above yield the following conclusions:

- All values differ from the theoretical value 2π by less than $\pm 17\%$.
- For the same profile, measurements and computations by any method differ generally by only a few per cent, except for NACA profiles with thickness ratio greater than 25%.
- Two-dimensional and three-dimensional RANSE computations hardly differ from each other except for thick NACA profiles.

- The Reynolds number based on axial inflow velocity and mean rudder chord length has relatively little effect on the lift gradient.
- The BEM fails to predict the low lift gradient of profiles with large opening angle of the trailing edge. For such profiles, the Kutta condition used in potential flow is a poor approximation.
- Substantial thickness at the trailing edge increases the lift slope.

Further detailed investigations based on RANSE computations produced the following insight into the effect of profile thickness:

- Thick profiles produce more lift than thinner ones if they have a sharp end (concave sides), and a lower lift if they end in a larger angle (convex or flat sides).
- The mostly used NACA00 profiles are worse than the other profiles investigated, both with respect to lift slope and to the ratio between lift and drag.
- For all profiles, the lift/drag ratio decreases with increasing thickness. Therefore, for a good propulsive efficiency, one should use the thinnest possible profile.
- The IFS profile generates the largest lift. However, when compared to the HSVA MP73-25 profile the difference is small and the lift/drag ratio is worse than for the HSVA profile. The IFS profile is also more liable to suffer from cavitation due to its very uneven pressure distribution on the suction side.

BEM is not capable of predicting the stall angle because stall is inherently a viscous phenomenon. For hard-over maneuvers, the stall angle and its associated maximum lift may be more important than $dC_L/d\alpha$. RANSE computations show that higher Reynolds numbers produce larger maximum C_L. Thus experimental values without extrapolation to actual Reynolds numbers are misleading with respect to maximum lift forces. Other conclusions for the maximum lift at stall angle from RANSE computations are:

- The maximum C_L ranges between 1.2 and more than 2. This upper limit is substantially larger than assumed in classification rules.
- The aspect ratio Λ is of minor influence only. Larger aspect ratio produces somewhat smaller $C_{L,\max}$.
- Small Λ yield large stall angles. (They also yield small $dC_L/d\alpha$, hence little change in the maximum C_L.)
- The taper ratio of the rudder has practically no influence on the maximum C_L.
- Profiles with concave sides produce larger $C_{L,\max}$ than those with flat or convex sides.

Three-dimensional RANSE computations give slightly lower maximum C_L than two-dimensional RANSE computations. The relation between the two-dimensional and three-dimensional values approximately determines the maximum lift while avoiding the complexities (and cost) of three-dimensional RANSE computations, especially for complex configurations and non-uniform inflow.

The recommended procedure is then:

- Perform a two-dimensional RANSE computation for the actual profile and Reynolds number in uniform flow to determine the maximum C_L.
- Perform a panel calculation for the three-dimensional arrangement.
- Convert the computed lift to C_L using an average inflow velocity. The averaged velocity is the root mean square axial velocity averaged over the rudder height.
- Determine the approximate stall angle as that where the three-dimensional C_L in potential flow amounts to 95% of the maximum $C_{L,2d}$ in the two-dimensional RANSE computation.
- Truncate the computed lift forces at that angle, but not drag and stock moment.

In practice, the aftbody arrangement with propeller and rudder is rather more complicated and may even involve additional complexities such as nozzles, fins, and bulbs. These make grid generation for field methods (and even BEM) complicated. However, by 2010 RANSE simulations for hull–propeller–rudder interaction at full-scale Reynolds numbers started to drift into industry applications.

APPENDIX A

Boundary Element Methods

Introduction

The Laplace equation is a *linear* differential equation, i.e. arbitrary linear combinations (superpositions) of elementary solutions of the Laplace equation will again form a possible solution. This chapter is devoted to various elementary solutions used in the computation of ship flows. It is not really necessary to understand the given formulae, but the concepts should be understood. Fortran subroutines for elements are in the public domain and may be obtained on the internet (see Preface).

Consider the case if still water is seen from a passing airplane with speed V, or from a razor blade ship not disturbing the flow. Here the water appears to flow uniformly in the negative direction x with the speed V. The water has no velocity component in the y or z direction, and everywhere uniformly the velocity is $-V$ in the x direction. The corresponding potential is:

$$\phi = -Vx$$

Another elementary potential is that of an undisturbed incident wave as given in Section 4.3.1, Chapter 4.

Various elements (elementary solutions) exist to approximate the disturbance effect of the ship. These more or less complicated mathematical expressions are useful to model displacement ('sources') or lift ('vortices', 'dipoles'). The common names indicate a graphical physical interpretation of the abstract mathematical formulae and will be discussed in the following.

The basic idea of all the related boundary element methods is to superimpose elements in an unbounded fluid. Since the flow does not cross a streamline just as it does not cross a real fluid boundary (such as the hull), any unbounded flow field in which a streamline coincides with the actual flow boundaries in the bounded problem can be interpreted as a solution for the bounded flow problem in the limited fluid domain. Outside this fluid domain, the flow cannot (and should not) be interpreted as a physical flow, even though the computation can give velocities, pressures, etc. everywhere.

The velocity at a field point $\vec{x}$ induced by some typical panel types and some related formula work is given in the following. Expressions are often derived in a local coordinate system. The derivatives of the potential are transformed from the local x–y–z system to a global $\bar{x}$–$\bar{y}$–$\bar{z}$

system. In two dimensions, we limit ourselves to x and z as coordinates, as these are the typical coordinates for a strip in a strip method. $\vec{n} = (n_x, n_z)$ is the outward unit normal in global coordinates, coinciding with the local z vector. $\vec{t}$ and $\vec{s}$ are unit tangential vectors, coinciding with the local x and y vectors, respectively. The transformation from the local to the global system is as follows:

1. *Two-dimensional case*

$$\phi_{\bar{x}} = n_z \cdot \phi_x + n_x \cdot \phi_z$$

$$\phi_{\bar{z}} = -n_x \cdot \phi_x + n_z \cdot \phi_z$$

$$\phi_{\overline{xx}} = \left(n_z^2 - n_x^2\right) \cdot \phi_{xx} + (2n_x n_z) \cdot \phi_{xz}$$

$$\phi_{\overline{xz}} = \left(n_z^2 - n_x^2\right) \cdot \phi_{xz} - (2n_x n_z) \cdot \phi_{xx}$$

$$\phi_{\overline{xxz}} = n_z\left(1 - 4n_x^2\right) \cdot \phi_{xxz} - n_x\left(1 - 4n_z^2\right) \cdot \phi_{xzz}$$

$$\phi_{\overline{xzz}} = n_x\left(1 - 4n_z^2\right) \cdot \phi_{xxz} + n_z\left(1 - 4n_x^2\right) \cdot \phi_{xzz}$$

2. *Three-dimensional case*

$$\phi_{\bar{x}} = t_1 \cdot \phi_x + s_1 \cdot \phi_y + n_1 \cdot \phi_z$$

$$\phi_{\bar{y}} = t_2 \cdot \phi_x + s_2 \cdot \phi_y + n_2 \cdot \phi_z$$

$$\phi_{\bar{z}} = t_3 \cdot \phi_x + s_3 \cdot \phi_y + n_3 \cdot \phi_z$$

$$\phi_{\bar{x}\bar{x}} = t_1^2 \phi_{xx} + s_1^2 \phi_{yy} + n_1^2 \phi_{zz} + 2(s_1 t_1 \phi_{xy} + n_1 t_1 \phi_{xz} + n_1 s_1 \phi_{yz})$$

$$\phi_{\bar{x}\bar{y}} = t_1 t_2 \phi_{xx} + t_1 s_2 \phi_{xy} + t_1 n_2 \phi_{xz} + s_1 t_2 \phi_{xy} + s_1 s_2 \phi_{yy} + s_1 n_2 \phi_{yz} + n_1 t_2 \phi_{xz} + n_1 s_2 \phi_{yz} + n_1 n_2 \phi_{zz}$$

$$\phi_{\bar{x}\bar{z}} = t_1 t_3 \phi_{xx} + t_1 s_3 \phi_{xy} + t_1 n_3 \phi_{xz} + s_1 t_3 \phi_{xy} + s_1 s_3 \phi_{yy} + s_1 n_3 \phi_{yz} + n_1 t_3 \phi_{xz} + n_1 s_3 \phi_{yz} + n_1 n_3 \phi_{zz}$$

$$\phi_{\bar{y}\bar{y}} = t_2^2 \phi_{xx} + s_2^2 \phi_{yy} + n_2^2 \phi_{zz} + 2(s_2 t_2 \phi_{xy} + n_2 t_2 \phi_{xz} + n_2 s_2 \phi_{yz})$$

$$\phi_{\bar{y}\bar{z}} = t_2 t_3 \phi_{xx} + t_2 s_3 \phi_{xy} + t_2 n_3 \phi_{xz} + s_2 t_3 \phi_{xy} + s_2 s_3 \phi_{yy} + s_2 n_3 \phi_{yz} + n_2 t_3 \phi_{xz} + n_2 s_3 \phi_{yz} + n_2 n_3 \phi_{zz}$$

$$\phi_{\bar{x}\bar{x}\bar{z}} = t_1^2(t_3 \phi_{xxx} + s_3 \phi_{xxy} + n_3 \phi_{xxz}) + s_1^2(t_3 \phi_{xyy} + s_3 \phi_{yyy} + n_3 \phi_{yyz}) + n_1^2(t_3 \phi_{xzz} + s_3 \phi_{yzz} + n_3 \phi_{zzz}) + 2(s_1 t_1 (t_3 \phi_{xxy} + s_3 \phi_{xyy} + n_3 \phi_{xyz}) + n_1 t_1 (t_3 \phi_{xxz} + s_3 \phi_{xyz} + n_3 \phi_{xzz}) + n_1 s_1 (t_3 \phi_{xyz} + s_3 \phi_{yyz} + n_3 \phi_{yzz}))$$

$$\phi_{\bar{x}\bar{y}\bar{z}} = t_1t_2(t_3\phi_{xxx} + s_3\phi_{xxy} + n_3\phi_{xxz}) + (t_1s_2 + s_1t_2)$$
$$\times (t_3\phi_{xxy} + s_3\phi_{xyy} + n_3\phi_{xyz}) + s_1s_2(t_3\phi_{xyy} + s_3\phi_{yyy} + n_3\phi_{yyz})$$
$$+ (t_1n_2 + n_1t_2)(t_3\phi_{xxz} + s_3\phi_{xyz} + n_3\phi_{xzz})$$
$$+ n_1n_2(t_3\phi_{xzz} + s_3\phi_{yzz} + n_3\phi_{zzz}) + (s_1n_2 + n_1s_2)$$
$$\times (t_3\phi_{xyz} + s_3\phi_{yyz} + n_3\phi_{yzz})$$

$$\phi_{\bar{x}\bar{z}\bar{z}} = t_1t_3(t_3\phi_{xxx} + s_3\phi_{xxy} + n_3\phi_{xxz}) + (t_1s_3 + s_1t_3)$$
$$\times (t_3\phi_{xxy} + s_3\phi_{xyy} + n_3\phi_{xyz}) + s_1s_3(t_3\phi_{xyy} + s_3\phi_{yyy} + n_3\phi_{yyz})$$
$$+ (t_1n_3 + n_1t_3)(t_3\phi_{xxz} + s_3\phi_{xyz} + n_3\phi_{xzz})$$
$$+ n_1n_3(t_3\phi_{xzz} + s_3\phi_{yzz} + n_3\phi_{zzz}) + (s_1n_3 + n_1s_3)$$
$$\times (t_3\phi_{xyz} + s_3\phi_{yyz} + n_3\phi_{yzz})$$

$$\phi_{\bar{y}\bar{y}\bar{z}} = t_2^2(t_3\phi_{xxx} + s_3\phi_{xxy} + n_3\phi_{xxz}) + s_2^2(t_3\phi_{xyy} + s_3\phi_{yyy} + n_3\phi_{yyz})$$
$$+ n_2^2(t_3\phi_{xzz} + s_3\phi_{yzz} + n_3\phi_{zzz}) + 2(s_2t_2(t_3\phi_{xxy} + s_3\phi_{xyy} + n_3\phi_{xyz})$$
$$+ n_2t_2(t_3\phi_{xxz} + s_3\phi_{xyz} + n_3\phi_{xzz}) + n_2s_2(t_3\phi_{xyz} + s_3\phi_{yyz} + n_3\phi_{yzz}))$$

$$\phi_{\bar{y}\bar{z}\bar{z}} = t_2t_3(t_3\phi_{xxx} + s_3\phi_{xxy} + n_3\phi_{xxz}) + (t_2s_3 + s_2t_3)$$
$$\times (t_3\phi_{xxy} + s_3\phi_{xyy} + n_3\phi_{xyz}) + s_2s_3(t_3\phi_{xyy} + s_3\phi_{yyy} + n_3\phi_{yyz})$$
$$+ (t_2n_3 + n_2t_3)(t_3\phi_{xxz} + s_3\phi_{xyz} + n_3\phi_{xzz})$$
$$+ n_2n_3(t_3\phi_{xzz} + s_3\phi_{yzz} + n_3\phi_{zzz}) + (s_2n_3 + n_2s_3)$$
$$\times (t_3\phi_{xyz} + s_3\phi_{yyz} + n_3\phi_{yzz})$$

Source Elements

The most common elements used in ship flows are source elements which are used to model the displacement effect of a body. Elements used to model the lift effect such as vortices or dipoles are also employed if lift plays a significant role, e.g. in yawed ships for maneuvering.

Point Source

1. *Two-dimensional case*
   The coordinates of the source are (x_q, z_q). The distance between source point and field point (x, y) is $r = \sqrt{(x - x_q)^2 + (z - z_q)^2}$. The potential induced at the field point is then:

$$\phi = \frac{\sigma}{2\pi}\ln r = \frac{\sigma}{4\pi}\ln((x - x_q)^2 + (z - z_q)^2)$$

This yields the velocities:

$$\vec{v} = \begin{Bmatrix} \phi_x \\ \phi_z \end{Bmatrix} = \frac{\sigma}{2\pi r^2} \begin{Bmatrix} x - x_q \\ z - z_q \end{Bmatrix}$$

The absolute value of the velocity is then:

$$v = \frac{\sigma}{2\pi r^2} \sqrt{(x - x_q)^2 + (z - z_q)^2} = \frac{\sigma}{2\pi r}$$

The absolute value of the velocity is thus the same for all points on a radius r around the point source. The direction of the velocity is pointing radially away from the source point and the velocity decreases with distance as $1/r$. Thus the flow across each concentric ring around the source point is constant. The element can be physically interpreted as a source of water which constantly pours water flowing radially in all directions. σ is the strength of this source. For negative σ, the element acts like a sink with water flowing from all directions into the center. Figure A.1 illustrates the effect of the element.

Higher derivatives are:

$$\phi_{xx} = -\phi_{zz} = \frac{\sigma}{2\pi} \frac{1}{r^2} - \phi_x \cdot \frac{2(x - x_q)}{r^2}$$

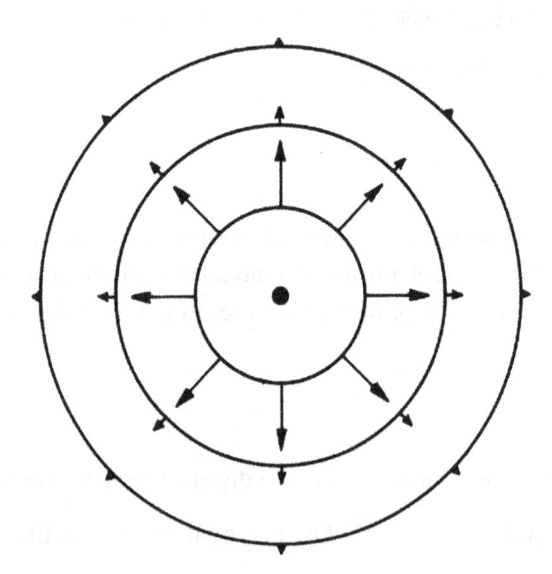

Figure A.1:
Effect of a point source

$$\phi_{xz} = -\phi_x \cdot \frac{2(z - z_q)}{r^2}$$

$$\phi_{xxz} = -\phi_{zzz} = -2 \cdot \left(\frac{(x - x_q)}{r^2} \phi_{xz} + \frac{(z - z_q)}{r^2} \phi_{xx} \right)$$

$$\phi_{xzz} = -2 \cdot \left(\frac{(x - x_q)}{r^2} \phi_{zz} + \frac{(z - z_q)}{r^2} \phi_{xz} \right)$$

2. *Three-dimensional case*
The corresponding expressions in three dimensions are:

$$r = \sqrt{(x - x_q)^2 + (y - y_q)^2 + (z - z_q)^2}$$

$$\phi = -\sigma \frac{1}{4\pi r}$$

$$\phi_x = \sigma \frac{1}{2\pi r^3}(x - x_q)$$

$$\phi_y = \sigma \frac{1}{2\pi r^3}(y - y_q)$$

$$\phi_z = \sigma \frac{1}{2\pi r^3}(z - z_q)$$

$$\phi_{xx} = (-3\phi_x(x - x_q) - \phi)/r^2$$

$$\phi_{xy} = (-3\phi_x(y - y_q))/r^2$$

$$\phi_{xz} = (-3\phi_x(z - z_q))/r^2$$

$$\phi_{yy} = (-3\phi_y(y - y_q) - \phi)/r^2$$

$$\phi_{yz} = (-3\phi_y(z - z_q))/r^2$$

$$\phi_{xxz} = -(2(\phi/r^2) + 5\phi_{xx})\mathrm{d}z/r^2$$

$$\phi_{xyz} = -5\phi_{xy}\mathrm{d}z/r^2$$

$$\phi_{xzz} = -5\phi_{xz}\mathrm{d}z/r^2 - 3\phi_x/r^2$$

$$\phi_{yyz} = -(2(\phi/r^2) + 5\phi_{yy})\mathrm{d}z/r^2$$

$$\phi_{yzz} = -5\phi_{yz}\mathrm{d}z/r^2 - 3\phi_y/r^2$$

Regular First-Order Panel

1. *Two-dimensional case*

For a panel of constant source strength we formulate the potential in a local coordinate system. The origin of the local system lies at the center of the panel. The panel lies on the local x-axis; the local z-axis is perpendicular to the panel pointing outward. The panel extends from $x = -d$ to $x = d$. The potential is then

$$\phi = \int_{-d}^{d} \frac{\sigma}{2\pi} \cdot \ln \sqrt{(x - x_q)^2 + z^2 \mathrm{d}x_q}$$

With the substitution $t = x - x_q$ this becomes:

$$\phi = \frac{1}{2} \int_{x-d}^{x+d} \frac{\sigma}{2\pi} \cdot \ln(t^2 + z^2)\mathrm{d}t$$

$$= \frac{\sigma}{4\pi} \left[t \ln (t^2 + z^2) + 2z \arctan \frac{t}{z} - 2t \right]_{x-d}^{x+d}$$

Additive constants can be neglected, giving:

$$\phi = \frac{\sigma}{4\pi} \left(x \ln \frac{r_1}{r_2} + d \ln(r_1 r_2) + z2 \arctan \frac{2\mathrm{d}z}{x^2 + z^2 - d^2} + 4d \right)$$

with $r_1 = (x + d)^2 + z^2$ and $r_2 = (x - d)^2 + z^2$. The derivatives of the potential (still in local coordinates) are:

$$\phi_x = \frac{\sigma}{2\pi} \cdot \frac{1}{2} \ln \frac{r_1}{r_2}$$

$$\phi_z = \frac{\sigma}{2\pi} \cdot \arctan \frac{2\mathrm{d}z}{x^2 + z^2 - d^2}$$

$$\phi_{xx} = \frac{\sigma}{2\pi} \cdot \left(\frac{x + d}{r_1} - \frac{x - d}{r_2} \right)$$

$$\phi_{xz} = \frac{\sigma}{2\pi} \cdot z \cdot \left(\frac{1}{r_1} - \frac{1}{r_2} \right)$$

$$\phi_{xxz} = \frac{\sigma}{2\pi} \cdot (-2z) \cdot \left(\frac{x + d}{r_1^2} - \frac{x - d}{r_2^2} \right)$$

$$\phi_{xzz} = \frac{\sigma}{2\pi} \cdot \left(\frac{(x+d)^2 - z^2}{r_1^2} - \frac{(x-d)^2 - z^2}{r_2^2} \right)$$

ϕ_x cannot be evaluated (is singular) at the corners of the panel. For the center point of the panel itself ϕ_z is:

$$\phi_z(0,0) = \lim_{z \to 0} \phi_z(0,z) = \frac{\sigma}{2}$$

If the ATAN2 function in Fortran is used for the general expression of ϕ_z, this is automatically fulfilled.

2. *Three-dimensional case*
   In three dimensions the corresponding expressions for an arbitrary panel are rather complicated. Let us therefore consider first a simplified case, namely a plane rectangular panel of constant source strength (Fig. A.2). We denote the distances of the field point to the four corner points by:

$$r_1 = \sqrt{x^2 + y^2 + z^2}$$

$$r_2 = \sqrt{(x-\ell)^2 + y^2 + z^2}$$

$$r_3 = \sqrt{x^2 + (y-h)^2 + z^2}$$

$$r_4 = \sqrt{(x-\ell)^2 + (y-h)^2 + z^2}$$

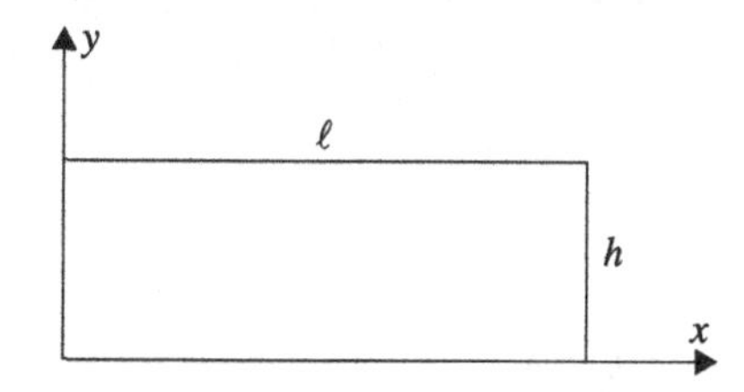

Figure A.2:
Simple rectangular flat panel of constant strength; origin at center of panel

The potential is:

$$\phi = -\frac{\sigma}{4\pi}\int_0^h\int_0^\ell \frac{1}{\sqrt{(x-\xi)^2+(y-\eta)^2+z^2}}\,d\xi\,d\eta$$

The velocity in the x direction is:

$$\frac{\partial\phi}{\partial x} = \frac{\sigma}{4\pi}\int_0^h\int_0^\ell \frac{x-\xi}{\sqrt{(x-\xi)^2+(y-\eta)^2+z^2}^{\,3}}\,d\xi\,d\eta$$

$$= \frac{\sigma}{4\pi}\int_0^h -\frac{1}{\sqrt{(x-\ell)^2+(y-\eta)^2+z^2}} + \frac{1}{\sqrt{x^2+(y-\eta)^2+z^2}}\,d\eta$$

$$= \frac{\sigma}{4\pi}\ln\frac{(r_3-(y-h))(r_1-y)}{(r_2-y)(r_4-(y-h))}$$

The velocity in the y direction is, in similar fashion:

$$\frac{\partial\phi}{\partial y} = \frac{\sigma}{4\pi}\ln\frac{(r_2-(x-\ell))(r_1-x)}{(r_3-x)(r_4-(x-\ell))}$$

The velocity in the z direction is:

$$\frac{\partial\phi}{\partial y} = \frac{\sigma}{4\pi}\int_0^h\int_0^\ell \frac{z}{\sqrt{(x-\xi)^2+(y-\eta)^2+z^2}^{\,3}}\,d\xi\,d\eta$$

$$= \frac{\sigma}{4\pi}\int_0^h -\frac{z(x-\ell)}{((y-\eta)^2+z^2)\sqrt{(x-\ell)^2+(y-\eta)^2+z^2}}$$

$$+ \frac{zx}{((y-\eta)^2+z^2)\sqrt{x^2+(y-\eta)^2+z^2}}\,d\eta$$

Substituting:

$$t = \frac{\eta-y}{\sqrt{x^2+(\eta-y)^2+z^2}}$$

yields:

$$\frac{\partial\phi}{\partial z}=\frac{\sigma}{4\pi}\left[\int_{-y/r_2}^{(h-y)/r_4}\frac{-z(x-\ell)}{z^2+(x-\ell)^2t^2}\,\mathrm{d}t+\int_{-y/r_1}^{(h-y)/r_3}\frac{zx}{z^2+x^2t^2}\,\mathrm{d}t\right]$$

$$=\frac{\sigma}{4\pi}\left[-\arctan\frac{x-\ell}{z}\frac{h-y}{r_4}+\arctan\frac{x-\ell}{z}\frac{-y}{r_2}-\arctan\frac{x}{z}\frac{-y}{r_1}+\arctan\frac{x}{z}\frac{h-y}{r_3}\right]$$

The derivation used:

$$\int\frac{1}{\sqrt{x^2+a^2}}\,\mathrm{d}x=\ln(x+\sqrt{x^2+a^2})+C$$

$$\int\frac{x}{\sqrt{x^2+a^2}^3}\,\mathrm{d}x=-\frac{1}{\sqrt{x^2+a^2}}+C$$

$$\int\frac{1}{\sqrt{x^2+a^2}^3}\,\mathrm{d}x=\frac{x}{a^2\sqrt{x^2+a^2}}+C$$

$$\int\frac{1}{\sqrt{a+bx^2}}\,\mathrm{d}x=\frac{1}{\sqrt{ab}}\arctan\frac{bx}{\sqrt{ab}}\quad\text{for}\quad b>0$$

The numerical evaluation of the induced velocities has to consider some special cases. As an example: the finite accuracy of computers can lead to problems for the above given expression of the x component of the velocity, when for small values of x and z the argument of the logarithm is rounded off to zero. Therefore, for $(\sqrt{x^2+z^2}\ll y)$ the term r_1-y must be substituted by the approximation $(x^2+z^2)/2x$. The other velocity components require similar special treatment.

Hess and Smith (1964) pioneered the development of boundary element methods in aeronautics, thus also laying the foundation for most subsequent work for BEM applications to ship flows. Their original panel used constant source strength over a plane polygon, usually a quadrilateral. This panel is still the most popular choice in practice.

The velocity is again given in a local coordinate system (Fig. A.3). For quadrilaterals of unit source strength, the induced velocities are:

$$\frac{\partial\phi}{\partial x}=\frac{y_2-y_1}{d_{12}}\ln\left(\frac{r_1+r_2-d_{12}}{r_1+r_2+d_{12}}\right)+\frac{y_3-y_2}{d_{23}}\ln\left(\frac{r_2+r_3-d_{23}}{r_2+r_3+d_{23}}\right)$$

$$+\frac{y_4-y_3}{d_{34}}\ln\left(\frac{r_3+r_4-d_{34}}{r_3+r_4+d_{34}}\right)+\frac{y_1-y_4}{d_{41}}\ln\left(\frac{r_4+r_1-d_{41}}{r_4+r_1+d_{41}}\right)$$

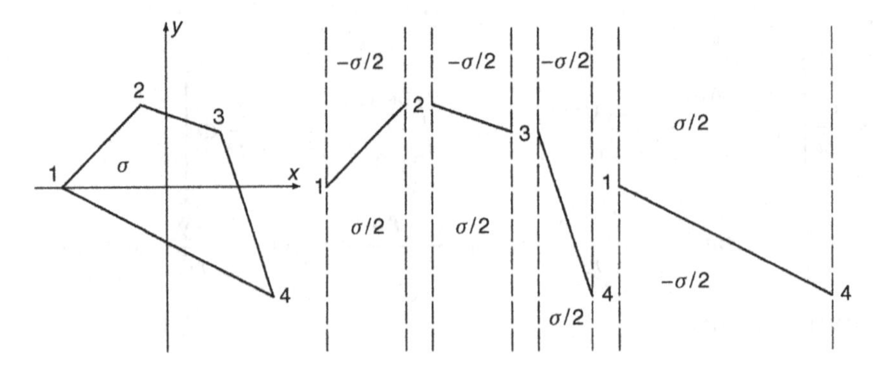

Figure A.3:
A quadrilateral flat panel of constant strength is represented by Hess and Smith as superposition of four semi-infinite strips

$$\frac{\partial\phi}{\partial y} = \frac{x_2 - x_1}{d_{12}} \ln\left(\frac{r_1 + r_2 - d_{12}}{r_1 + r_2 + d_{12}}\right) + \frac{x_3 - x_2}{d_{23}} \ln\left(\frac{r_2 + r_3 - d_{23}}{r_2 + r_3 + d_{23}}\right)$$
$$+ \frac{x_4 - x_2}{d_{34}} \ln\left(\frac{r_3 + r_4 - d_{34}}{r_3 + r_4 + d_{34}}\right) + \frac{x_1 - x_4}{d_{41}} \ln\left(\frac{r_4 + r_1 - d_{41}}{r_4 + r_1 + d_{41}}\right)$$

$$\frac{\partial\phi}{\partial z} = \arctan\left(\frac{m_{12}e_1 - h_1}{zr_1}\right) - \arctan\left(\frac{m_{12}e_2 - h_2}{zr_2}\right)$$
$$+ \arctan\left(\frac{m_{23}e_2 - h_2}{zr_2}\right) - \arctan\left(\frac{m_{23}e_2 - h_3}{zr_3}\right)$$
$$+ \arctan\left(\frac{m_{34}e_3 - h_3}{zr_3}\right) - \arctan\left(\frac{m_{34}e_4 - h_4}{zr_4}\right)$$
$$+ \arctan\left(\frac{m_{41}e_4 - h_4}{zr_4}\right) - \arctan\left(\frac{m_{41}e_1 - h_1}{zr_1}\right)$$

x_i, y_i are the local coordinates of the corner points i, r_i the distance of the field point (x, y, z) from the corner point i, d_{ij} the distance of the corner point i from the corner point $j, m_{ij} = (y_j - y_i)/(x_j - x_i), e_i = z^2 + (x - x_i)^2$ and $h_i = (y - y_i)(x - x_i)$. For larger distances between field point and panel, the velocities are approximated by a multipole expansion consisting of a point source and a point quadrupole. For large distances the point source alone approximates the effect of the panel.

For real ship geometries, four corners on the hull often do not lie in one plane. The panel corners are then constructed to lie within one plane approximating the four points on the actual hull: the normal on the panel is determined from the cross-product of the two 'diagonal' vectors. The center of the panel is determined by simple averaging of the coordinates of the four corners. This point and the normal define the plane of the panel. The four points on the hull are projected normally on this plane. The panels thus created do not form a closed body. As long as the gaps are small, the resulting errors are negligible compared to other sources of errors, e.g. the assumption of constant strength, constant pressure, constant normal over each panel, or enforcing the boundary condition only in one point of the panel. Hess and Smith (1964) comment on this issue:

> 'Nevertheless, the fact that these openings exist is sometimes disturbing to people hearing about the method for the first time. It should be kept in mind that the elements are simply devices for obtaining the surface source distribution and that the polyhedral body... has no direct physical significance, in the sense that the flow eventually calculated is not the flow about the polyhedral-type body. Even if the edges of the adjacent elements are coincident, the normal velocity is zero at only one point of each element. Over the remainder of the element there is flow through it. Also, the computed velocity is infinite on the edges of the elements, whether these are coincident or not.'

Jensen Panel

Jensen (1988) developed a panel of the same order of accuracy, but much simpler to program, which avoids the evaluation of complicated transcendental functions and in it implementation relies largely on just a repeated evaluation of point source routines. As the original publication is little known and difficult to obtain internationally, the theory is repeated here. The approach requires, however, closed bodies. Then the velocities (and higher derivatives) can be computed by simple numerical integration if the integrands are transformed analytically to remove singularities. In the formulae for this element, $\vec{n}$ is the unit normal pointing outward from the body into the fluid, $\oint$ the integral over S excluding the immediate neighborhood of $\vec{x}_q$, and ∇ the Nabla operator with respect to $\vec{x}$.

1. *Two-dimensional case*

   A Rankine source distribution on a closed body induces the following potential at a field point $\vec{x}$:

$$\phi(\vec{x}) = \int_S \sigma(\vec{x}_q)G(\vec{x},\vec{x}_q)\ \mathrm{d}S$$

   S is the surface contour of the body, σ the source strength, and $G(\vec{x},\vec{x}_q) = (1/2\pi)\ \ln|\vec{x}-\vec{x}_q|$ is the Green function (potential) of a unit point source. Then the induced normal velocity component is:

$$v_n(\vec{x}) = \vec{n}(\vec{x})\nabla\phi(\vec{x}) = \oint\sigma(\vec{x}_q)\vec{n}(\vec{x})\nabla G(\vec{x},\vec{x}_q)\ \mathrm{d}S + \frac{1}{2}\sigma(\vec{x}_q)$$

   Usually the normal velocity is given as boundary condition. Then the important part of the solution is the tangential velocity on the body:

$$v_t(\vec{x}) = \vec{t}(\vec{x})\nabla\phi(\vec{x}) = \oint\sigma(\vec{x}_q)\vec{t}(\vec{x})\nabla G(\vec{x},\vec{x}_q)\ \mathrm{d}S$$

Without further proof, the tangential velocity (circulation) induced by a distribution of point sources of the same strength at point $\vec{x}_q$ vanishes:

$$\oint_{S(\vec{x})} \nabla G(\vec{x}, \vec{x}_q)\vec{t}(\vec{x})\, dS = 0$$

Exchanging the designations $\vec{x}$ and $\vec{x}_q$ and using $\nabla G(\vec{x}, \vec{x}_q) = -\nabla G(\vec{x}_q, \vec{x})$, we obtain:

$$\oint_s \nabla G(\vec{x}, \vec{x}_q)\vec{t}(\vec{x}_q) dS = 0$$

We can multiply the integrand by $\sigma(\vec{x})$ – which is a constant as the integration variable is $\vec{x}_q$ – and subtract this zero expression from our initial integral expression for the tangential velocity:

$$v_t(\vec{x}) = \oint_S \sigma(\vec{x}_q)\vec{t}(\vec{x})\nabla G(\vec{x}, \vec{x}_q) dS - \underbrace{\oint_S \sigma(\vec{x})\nabla G(\vec{x}, \vec{x}_q)\vec{t}(\vec{x}_q) dS}_{=0}$$

$$= \oint_S \left[\sigma(\vec{x}_q)\vec{t}(\vec{x}) - \sigma(\vec{x})\vec{t}(\vec{x}_q)\right] \nabla G(\vec{x}, \vec{x}_q) dS$$

For panels of constant source strength, the integrand in this formula tends to zero as $\vec{x} \to \vec{x}_q$, i.e. at the previously singular point of the integral. Therefore this expression for v_t can be evaluated numerically. Only the length ΔS of the contour panels and the first derivatives of the source potential for each $\vec{x}, \vec{x}_q$ combination are required.

2. *Three-dimensional case*
The potential at a field point $\vec{x}$ due to a source distribution on a closed body surface S is:

$$\phi(\vec{x}) = \int_S \sigma(\vec{x}_q) G(\vec{x}, \vec{x}_q) dS$$

σ is the source strength and $G(\vec{x}, \vec{x}_q) = -(4\pi|\vec{x} - \vec{x}_q|)^{-1}$ is the Green function (potential) of a unit point source. Then the induced normal velocity component on the body is:

$$v_n(\vec{x}) = \vec{n}(\vec{x})\nabla\phi(\vec{x}) = \oint_S \sigma(\vec{x}_q)\vec{n}(\vec{x})\nabla G(\vec{x}, \vec{x}_q)\, dS + \frac{1}{2}\sigma(\vec{x}_q)$$

Usually the normal velocity is prescribed by the boundary condition. Then the important part of the solution is the velocity in the tangential directions $\vec{t}$ and $\vec{s}$. $\vec{t}$ can be chosen arbitrarily, $\vec{s}$ forms a right-handed coordinate system with $\vec{n}$ and $\vec{t}$. We will treat here only the velocity in the t direction, as the velocity in the s direction has the same form. The original, straightforward form is:

$$v_t(\vec{x}) = \vec{t}(\vec{x})\nabla\phi(\vec{x}) = \oint_S \sigma(\vec{x}_q)\vec{t}(\vec{x})\nabla G(\vec{x}, \vec{x}_q) dS$$

A source distribution of constant strength on the surface S of a sphere does not induce a tangential velocity on S:

$$\oint_S \vec{t}(\vec{x})\nabla G(\vec{x},\vec{k})\mathrm{d}S = 0$$

for $\vec{x}$ and $\vec{k}$ on S. The sphere is placed touching the body tangentially at the point $\vec{x}$. The center of the sphere must lie within the body. (The radius of the sphere has little influence on the results within wide limits. A rather large radius is recommended.) Then every point $\vec{x}_q$ on the body surface can be projected to a point $\vec{k}$ on the sphere surface by passing a straight line through $\vec{k}, \vec{x}_q$, and the sphere's center. This projection is denoted by $\vec{k} = P(\vec{x}_q)$. $\mathrm{d}S$ on the body is projected on $\mathrm{d}S$ on the sphere. R denotes the relative size of these areas: $\mathrm{d}S = R\ \mathrm{d}S$. Let R be the radius of the sphere and $\vec{c}$ be its center. Then the projection of $\vec{x}_q$ is:

$$P(\vec{x}_q) = \frac{\vec{x}_q - \vec{c}}{|\vec{x}_q - \vec{c}|} R + \vec{c}$$

The area ratio R is given by:

$$R = \frac{\vec{n}\cdot(\vec{x}_q - \vec{c})}{|\vec{x}_q - \vec{c}|}\left(\frac{R}{|\vec{x}_q - \vec{c}|}\right)^2$$

With these definitions, the contribution of the sphere ('fancy zero') can be transformed into an integral over the body surface:

$$\oint_S \vec{t}(\vec{x})\nabla G(\vec{x}, P(\vec{x}_q))R\ \mathrm{d}S = 0$$

We can multiply the integrand by $\sigma(\vec{x})$ — which is a constant as the integration variable is $\vec{x}_q$ — and subtract this zero expression from our original expression for the tangential velocity:

$$\begin{aligned} v_t(\vec{x}) &= \oint_S \sigma(\vec{x}_q)\vec{t}(\vec{x})\nabla G(\vec{x},\vec{x}_q)\mathrm{d}S - \underbrace{\oint_S \sigma(\vec{x})\,\vec{t}(\vec{x})\nabla G(\vec{x}, P(\vec{x}_q))R\ \mathrm{d}S}_{=0} \\ &= \oint_S \left[\sigma(\vec{x}_q)\vec{t}(\vec{x})\nabla G(\vec{x},\vec{x}_q) - \sigma(\vec{x})\vec{t}(\vec{x})\nabla G(\vec{x}, P(\vec{x}_q))R\right]\mathrm{d}S \end{aligned}$$

For panels of constant source strength, the integrand in this expression tends to zero as $\vec{x}\to\vec{x}_q$, i.e. at the previously singular point of the integral. Therefore this expression for v_t can be evaluated numerically.

Higher-Order Panel

The panels considered so far are 'first-order' panels, i.e. halving the grid spacing will halve the error in approximating a flow (for sufficiently fine grids). Higher-order panels (these are invariably second-order panels) will quadratically decrease the error for grid refinement. Second-order panels need to be at least quadratic in shape and linear in source distribution. They give much better results for simple geometries which can be described easily by analytical terms, e.g. spheres or Wigley parabolic hulls. For real ship geometries, first-order panels are usually sufficient and may even be more accurate for the same effort, as higher-order panels require more care in grid generation and are prone to 'overshoot' in regions of high curvature as in the aftbody. For some applications, however, second derivatives of the potential are needed on the hull and these are evaluated simply by second-order panels, but not by first-order panels.

1. *Two-dimensional case*

   We want to compute derivatives of the potential at a point (x, y) induced by a given curved portion of the boundary. It is convenient to describe the problem in a local coordinate system (Fig. A.4). The x- or ξ-axis is tangential to the curve and the perpendicular projections on the x-axis of the ends of the curve lie equal distances d to the right and the left of the origin. The y- or η-axis is normal to the curve. The arc length along the curve is denoted by s, and a general point on the curve is (ξ, η). The distance between (x, y) and (ξ, η) is:

$$r = \sqrt{(x-\xi)^2 + (y-\eta)^2}$$

The velocity induced at (x, y) by a source density distribution $\sigma(s)$ along the boundary curve is:

$$\nabla\phi = \frac{1}{2\pi}\int_{-d}^{d}\begin{Bmatrix} x-\xi \\ y-\eta \end{Bmatrix}\frac{\sigma(s)}{r^2}\frac{ds}{d\xi}d\xi$$

The boundary curve is defined by $\eta = \eta(\xi)$. In the neighborhood of the origin, the curve has a power series:

$$\eta = c\xi^2 + d\xi^3 + \cdots$$

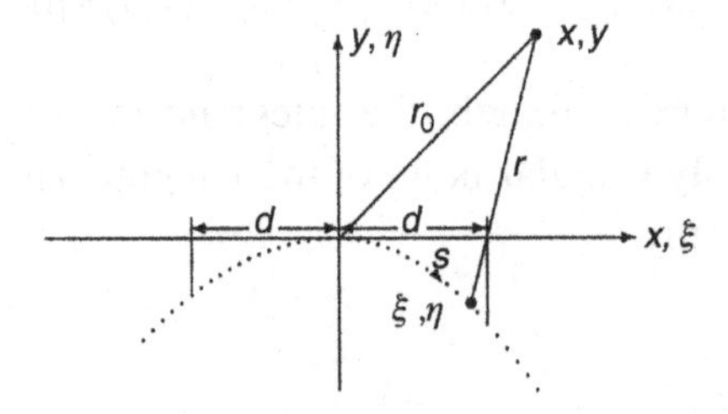

Figure A.4:
Coordinate system for higher-order panel (two-dimensional)

There is no term proportional to ξ, because the coordinate system lies tangentially to the panel. Similarly the source density has a power series:

$$\sigma(s) = \sigma^{(0)} + \sigma^{(1)}s + \sigma^{(2)}s^2 + \cdots$$

Then the integrand in the above expression for ∇_ϕ can be expressed as a function of ξ and then expanded in powers of ξ. The resulting integrals can be integrated to give an expansion for ∇_ϕ in powers of d. However, the resulting expansion will not converge if the distance of the point (x, y) from the origin is less than d. Therefore, a modified expansion is used for the distance r:

$$r^2 = \left[(x-\xi)^2 + y^2\right] - 2y\eta + \eta^2 = r_f^2 - 2y\eta + \eta^2$$

$r_f = \sqrt{(x-\xi)^2 + y^2}$ is the distance (x, y) from a point on the flat element. Only the latter terms in this expression for r^2 are expanded:

$$r^2 = r_f^2 - 2yc\xi^2 + O(\xi^3)$$

Powers $O(\xi^3)$ and higher will be neglected from now on:

$$\frac{1}{r^2} = \frac{1}{r_f^2 - 2yc\xi^2} \cdot \frac{r_f^2 + 2yc\xi^2}{r_f^2 + 2yc\xi^2} = \frac{1}{r_f^2} + \frac{2yc\xi^2}{r_f^4}$$

$$\frac{1}{r^4} = \frac{1}{r_f^4} + \frac{4yc\xi^2}{r_f^6}$$

The remaining parts of the expansion are straightforward:

$$s = \int_0^\xi \sqrt{1 + \left(\frac{d\eta}{d\xi}\right)^2} d\xi = \int_0^\xi \sqrt{1 + (2c\xi)^2} d\xi$$

$$\approx \int_0^\xi 1 + 2c^2\xi^2 d\xi = \xi + \frac{2}{3}c^2\xi^3$$

Combine this expression for s with the power series for $\sigma(s)$:

$$\sigma(s) = \sigma^{(0)} + \sigma^{(1)}\xi + \sigma^{(2)}\xi^2$$

Combine the expression of s with the above expression for $1/r^2$:

$$\frac{1}{r^2}\frac{ds}{d\xi} = \left(\frac{1}{r_f^2} + \frac{2cy\xi^2}{r_f^4}\right)(1 + 2c^2\xi^2) = \left(\frac{1}{r_f^2} + \frac{2cy\xi^2}{r_f^4} + \frac{2c^2\xi^2}{r_f^2}\right)$$

Now the integrands in the expression for $\nabla\phi$ can be evaluated:

$$\begin{aligned}
\phi_x &= \frac{1}{2\pi}\int_{-d}^{d} \sigma \frac{(x-\xi)}{r^2}\frac{ds}{d\xi}\,d\xi \\
&= \frac{1}{2\pi}\int_{-d}^{d} (\sigma^{(0)} + \sigma^{(1)}\xi + \sigma^{(2)}\xi^2)(x-\xi)\left(\frac{1}{r_f^2} + \frac{2cy\xi^2}{r_f^4} + \frac{2c^2\xi^2}{r_f^2}\right)d\xi \\
&= \frac{1}{4\pi}\left[\phi_x^{(0)}\sigma^{(0)} + \phi_x^{(1)}\sigma^{(1)} + c\phi_x^{(c)}\sigma^{(0)} + \phi_x^{(2)}(\sigma^{(2)} + 2c^2\sigma^{(0)})\right]
\end{aligned}$$

$$\phi_x^{(0)} = \int_{-d}^{d} \frac{2(x-\xi)}{r_f^2}\,d\xi = \int_{x-d}^{x+d} \frac{2t}{t^2+y^2}\,dt = \left[\ln(t^2+y^2)\right]_{x-d}^{x+d} = \ln(r_1^2/r_2^2)$$

with $r_1 = \sqrt{(x+d)^2 + y^2}$ and $r_2 = \sqrt{(x-d)^2 + y^2}$.

$$\phi_x^{(1)} = \int_{-d}^{d} \frac{2\xi(x-\xi)}{r_f^2}\,d\xi = 2\int_{x-d}^{x+d} \frac{t(x-t)}{t^2+y^2}\,dt = x\phi_x^{(0)} + y\phi_y^{(0)} - 4d$$

$$\phi_x^{(c)} = \int_{-d}^{d} \frac{4(x-\xi)y\xi^2}{r_f^4}\,d\xi = 4y\int_{x-d}^{x+d} \frac{t(t-x)^2}{(t^2+y^2)^2}\,dt - 2\phi_y^{(1)} + \frac{(2d)^3 xy}{r_1^2 r_2^2}$$

$$\phi_x^{(2)} = \int_{-d}^{d} \frac{2(x-\xi)\xi^2}{r_f^2}\,d\xi = 2\int_{x-d}^{x+d} \frac{t(t-x)^2}{t^2+y^2}\,dt = x\phi_x^{(1)} + y\phi_y^{(1)}$$

Here the integrals were transformed with the substitution $t = (x - \xi)$.

$$\phi_y = \frac{1}{2\pi}\int_{-d}^{d} \sigma \frac{(y-\eta)}{r^2}\frac{\mathrm{d}s}{\mathrm{d}\xi}\,\mathrm{d}\xi$$

$$= \frac{1}{2\pi}\int_{-d}^{d} (\sigma^{(0)} + \sigma^{(1)}\xi + \sigma^{(2)}\xi^2)(y - c\xi^2) \times \left(\frac{1}{r_f^2} + \frac{2cy\xi^2}{r_f^4} + \frac{2c^2\xi^2}{r_f^2}\right)\mathrm{d}\xi$$

$$= \frac{1}{4\pi}\left[\phi_y^{(0)}\sigma^{(0)} + \phi_y^{(1)}\sigma^{(1)} + c\phi_y^{(c)}\sigma^{(0)} + \phi_y^{(2)}(\sigma^{(2)} + 2c^2\sigma^{(0)})\right]$$

$$\phi_y^{(0)} = \int_{-d}^{d} \frac{2y}{r_f^2}\,\mathrm{d}\xi = 2\int_{x-d}^{x+d} \frac{y}{t^2+y^2}\,\mathrm{d}t$$

$$= 2\left[\arctan\frac{t}{y}\right]_{x-d}^{x+d} = 2\ \arctan\frac{2dy}{x^2+y^2-d^2}$$

$$\phi_y^{(1)} = \int_{-d}^{d} \frac{2\xi y}{r_f^2}\,\mathrm{d}\xi = 2y\int_{x-d}^{x+d} \frac{(x-t)}{t^2+y^2}\,\mathrm{d}t = x\phi_y^{(0)} - y\phi_x^{(0)}$$

$$\phi_y^{(c)} = \int_{-d}^{d} 4y^2\frac{\xi^2}{r_f^4} - \frac{2\xi^2}{r_f^2}\,\mathrm{d}\xi$$

$$= 4y^2\int_{x-d}^{x+d} \frac{(t-x)^2}{(t^2+y^2)^2}\,\mathrm{d}t - 2\int_{x-d}^{x+d} \frac{(t-x)^2}{t^2+y^2}\,\mathrm{d}t$$

$$= 2\phi_x^{(1)} - 4d^3\frac{x^2-y^2-d^2}{r_1^2 r_2^2}$$

$$\phi_y^{(2)} = \int_{-d}^{d} \frac{2y\xi^2}{r_f^2}\,\mathrm{d}\xi = 2y\int_{x-d}^{x+d} \frac{(t-x)^2}{t^2+y^2}\,\mathrm{d}t = x\phi_y^{(1)} - y\phi_x^{(1)}$$

The original formulae for the first derivatives of these higher-order panels were analytically equivalent, but less suited for programming involving more arithmetic operations than the formulae given here. Higher derivatives of the potential are given in Bertram (1999).

2. *Three-dimensional case*

   The higher-order panels are parabolic in shape with a bi-linear source distribution on each panel. The original procedure of Hess was modified by Hughes and Bertram (1995) to also include higher derivatives of the potential. The complete description of the formulae used to determine the velocity induced by the higher-order panels would be rather lengthy.

So only the general procedure is described here. The surface of the ship is divided into panels as in a first-order panel method. However, the surface of each panel is approximated by a parabolic surface, as opposed to a flat surface. The geometry of a panel in the local panel coordinate system is described as:

$$\zeta = C + A\xi + B\eta + P\xi^2 + 2Q\xi\eta + R\eta^2$$

The ξ-axis and η-axis lie in the plane tangential to the panel at the panel control point, and the ζ-axis is normal to this plane. This equation can be written in the form:

$$\zeta - \zeta_0 = P(\xi - \xi_0)^2 + 2Q(\xi - \xi_0)(\eta - \eta_0) + R(\eta - \eta_0)^2$$

The point (ξ_0, η_0, ζ_0) is used as a collocation point and origin of the local panel coordinate system. In the local panel coordinate system, terms depending on A, B, and C do not appear in the formulae for the velocity induced by a source distribution on the panel. R and P represent the local curvatures of the ship in the two coordinate directions, Q the local 'twist' in the ship form.

The required input consists of the coordinates of panel corner points lying on the body surface and information concerning how the corner points are connected to form the panels. In our implementation, each panel is allowed to have either three or four sides. The first and third sides of the panel should be (nearly) parallel. Otherwise, the accuracy of the panels deteriorates. For a given panel, the information available to determine the coefficients $A \ldots R$ consists of the three or four panel corner points of the panel and the corner points of the panels which border the panel in question. For a quadrilateral panel with neighboring panels on all sides, eight 'extra' vertex points are provided by the corner points of the adjacent panels (Fig. A.5). For triangular panels and panels lying on the edges of the body, fewer extra vertex points will be available. The panel corner points will be listed as x_i, y_i, z_i, $i = 1 \ldots 4$, and the extra vertex points as $\tilde{x}_j$, $\tilde{y}_j$, $\tilde{z}_j$, $j = 1 \ldots N_v$, where N_v is the number of extra vertex points ($3 < N_v < 8$). For triangular panels four corner points are also specified, but either the first and second or the third and fourth corner points are identical (i.e. the first or third side of the panel has zero length). The curved panel is required to pass exactly through all of its corner points and to pass as closely as possible to the extra vertex points of the neighboring panels. The order in which the corner points are specified is important, in that this determines whether the normal vector points into the fluid domain or into the body. In our method, the corner points should be ordered clockwise when viewed from the fluid domain, so that the normal vector points into the fluid domain.

The source strength on each panel is represented by a bi-linear distribution, as opposed to a constant distribution as in a first-order method:

$$\sigma(\xi, \eta) = \sigma_0 + \sigma_x(\xi - \xi_0) + \sigma_y(\eta - \eta_0)$$

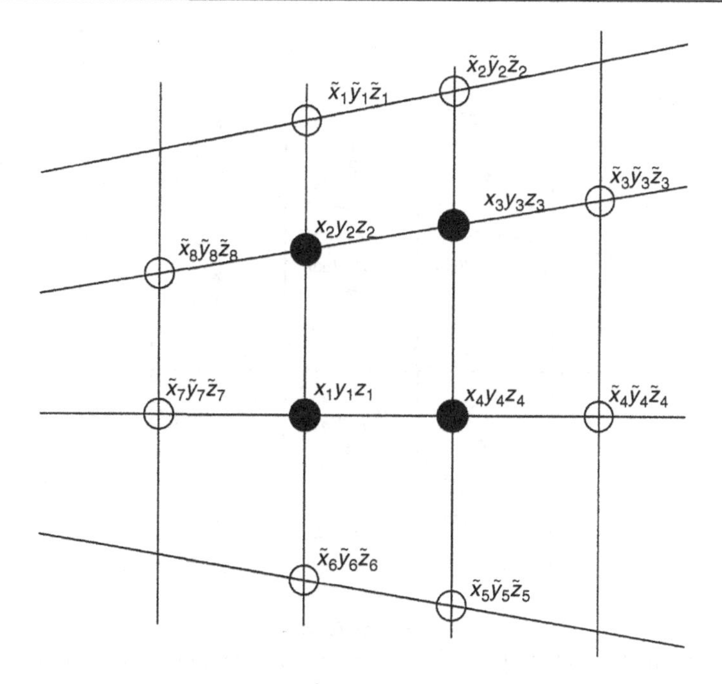

Figure A.5:
Additional points used for computing the local surface curvature of a panel (seen from the fluid domain)

σ_x and σ_y are the slopes of the source strength distribution in the ξ and η directions respectively. In the system of linear equations set up in this method, only the value of the source strength density at the collocation point on each panel, the σ_0 term, is solved for directly. The derivatives of the source density (σ_x and σ_y terms) are expressed in terms of the source strength density at the panel collocation point and at the collocation points of panels bordering the panel in question. First the collocation points of the adjacent panels are transformed into the local coordinate system of the panel in question. Then the above equation for the source strength is fitted in a least squares sense to the values of source density at the collocation points of the adjacent panels to determine σ_0, σ_x, and σ_y. For a four-sided panel which does not lie on a boundary of the body, four adjacent panel collocation points will be available for performing the least squares fit (Fig. A.6). In other cases only three or possibly two adjacent panels will be available. The procedure expresses the unknown source strength derivatives in terms of the source density at the collocation point of the adjacent panels. If the higher-order terms are set to zero, the element reduces to the regular first-order panel. A corresponding option is programmed in our version of the panel.

Vortex Elements

Vortex elements are useful to model lifting flows, e.g. in the lifting-line method for propellers and foils (see Section 2.3, Chapter 2).

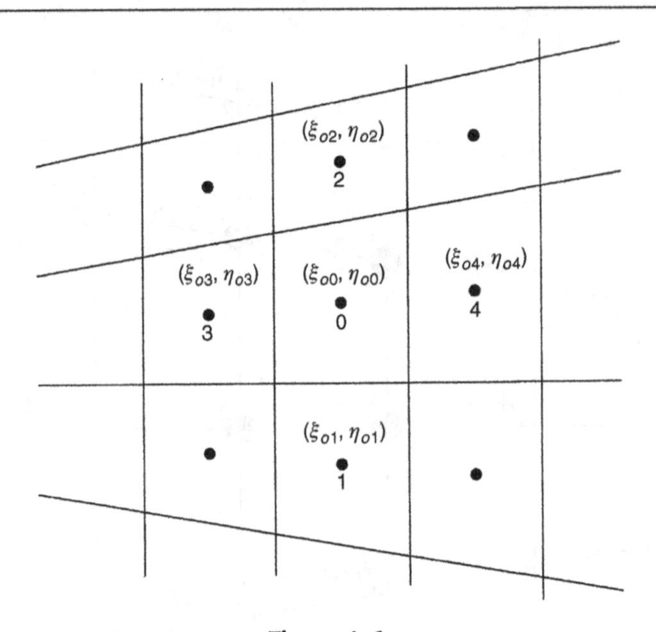

Figure A.6:
Adjacent panels used in the least-squares fit for the source density derivatives

1. *Two-dimensional case*
   Consider a vortex of strength Γ at x_w, z_w and a field point x, z. Denote $\Delta x = x - x_w$ and $\Delta z = z - z_w$. The distance between the two points is $r = \sqrt{\Delta x^2 + \Delta z^2}$. The potential and velocities induced by this vortex are:

$$\phi = -\frac{\Gamma}{2\pi} \arctan \frac{z - z_w}{x - x_w}$$

$$\phi_x = \frac{\Gamma}{2\pi} \frac{\Delta z}{r^2}$$

$$\phi_z = -\frac{\Gamma}{2\pi} \frac{\Delta x}{r^2}$$

The absolute value of the velocity is then $(\Gamma/2\pi)1/r$, i.e. the same for each point on a concentric ring around the center x_w, z_w. The velocity decays with distance to the center. So far, the vortex has the same features as the source. The difference is the direction of the velocity. The vortex induces velocities that are always tangential to the concentric ring (Fig. A.7), while the source produced radial velocities. The formulation given here produces counter-clockwise velocities for positive Γ.

The strength of the vortex is the 'circulation'. In general, the circulation is defined as the integral of the tangential velocities about any closed curve. For the definition given above, this integral about any concentric ring will indeed yield Γ as a result.

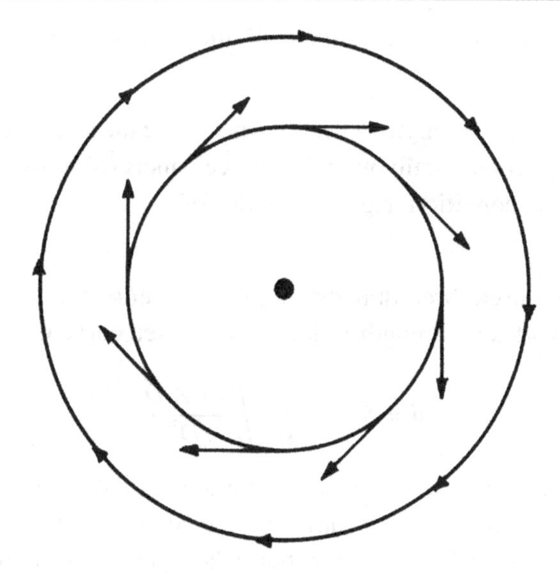

Figure A.7:
Velocities induced by vortex

This point vortex of strength Γ leads to similar expressions for velocities and higher derivatives as a point source of strength σ. One can thus express one by the other as follows:

$$\frac{2\pi}{\Gamma}\Phi_x = \frac{2\pi}{\sigma}\phi_z$$

$$\frac{2\pi}{\Gamma}\Phi_z = -\frac{2\pi}{\sigma}\phi_x$$

$$\frac{2\pi}{\Gamma}\Phi_{xx} = \frac{2\pi}{\sigma}\phi_{xz}$$

$$\frac{2\pi}{\Gamma}\Phi_{xz} = -\frac{2\pi}{\sigma}\phi_{xx}$$

$$\frac{2\pi}{\Gamma}\Phi_{xxz} = \frac{2\pi}{\sigma}\phi_{xzz}$$

$$\frac{2\pi}{\Gamma}\Phi_{xzz} = -\frac{2\pi}{\sigma}\phi_{xxz}$$

Φ is the potential of the vortex and ϕ the potential of the source. The same relations hold for converting between vortex panels and source panels of constant strength. It is thus usually not necessary to program vortex elements separately. One can rather call the source subroutines with a suitable rearrangement of the output parameters in the call of the subroutine.

A vortex panel of constant strength – i.e. all panels have the same strength – distributed on the body coinciding geometrically with the source panels (of individual strength) enforces automatically a Kutta condition, e.g. for a hydrofoil.

2. *Three-dimensional case*

The most commonly used three-dimensional vortex element is the horseshoe vortex. A three-dimensional vortex of strength Γ, lying on a closed curve C, induces a velocity field:

$$\vec{v} = \nabla\phi = \frac{\Gamma}{4\pi}\int_C \frac{d\vec{s}\times\vec{D}}{D^3}$$

We use the special case that a horseshoe vortex lies in the plane $y = y_w =$ const., from $x = -\infty$ to $x = x_w$. Arbitrary cases may be derived from this case using a coordinate transformation. The vertical part of the horseshoe vortex runs from $z = z_l$ to $z = z_2$ (Fig. A.8). Consider a field point (x, y, z). Then: $\Delta x = x - x_w$, $\Delta y = y - y_w$, $\Delta z_1 = z - z_1$, $\Delta z_2 = z - z_2$, $t_1 = \sqrt{\Delta x^2 + \Delta y^2 + \Delta z_1^2}$ and $t_2 = \sqrt{\Delta x^2 + \Delta y^2 + \Delta z_2^2}$.

The horseshoe vortex then induces the following velocity:

$$\vec{v} = \frac{\Gamma}{4\pi}\left[\left(\frac{\Delta z_1}{t_1} - \frac{\Delta z_2}{t_2}\right)\frac{1}{\Delta x^2 + \Delta y^2}\begin{Bmatrix} -\Delta y \\ \Delta x \\ 0 \end{Bmatrix} + \left(1 - \frac{\Delta x}{t_1}\right)\right.$$
$$\left.\times\frac{1}{\Delta y^2 + \Delta z_1^2}\begin{Bmatrix} 0 \\ -\Delta z_1 \\ \Delta y \end{Bmatrix} i - \left(1 - \frac{\Delta x}{t_2}\right)\frac{1}{\Delta y^2 + \Delta z_2^2}\begin{Bmatrix} 0 \\ sdz_2 \\ \Delta y \end{Bmatrix}\right]$$

The derivation used $\int (t^2 + a^2)^{-3/2} dt = t/(a^2\sqrt{t^2 + a^2})$. For $\Delta x^2 + \Delta y^2 \ll |\Delta z_1||\Delta z_2|$ or $\Delta y^2 + \Delta z_1^2 \ll \Delta x^2$ special formulae are used. Bertram (1992) gives details and expressions for higher derivatives.

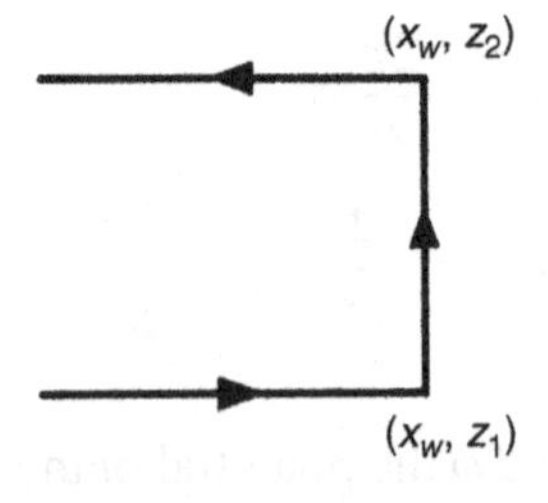

Figure A.8:
Horseshoe vortex

Dipole Elements

Point Dipole

The dipole (or doublet) is the limit of a source and sink of equal strength brought together along some direction (usually x) keeping the product of distance and source strength constant. The result is formally the same as differentiating the source potential in the required direction. The strength of a dipole is usually denoted by m. Again, r denotes the distance between field point $\vec{x}$ and the dipole at $\vec{x}_d$. We consider a dipole in the x direction here. We define $\Delta\vec{x} = \vec{x} - \vec{x}_d$.

1. *Two-dimensional case*

   The potential and derivatives for a dipole in the x direction are:

$$\phi = \frac{m}{2\pi r^2}\Delta x$$

$$\phi_x = \frac{m}{2\pi r^2} - 2\phi\frac{\Delta x}{r^2}$$

$$\phi_z = -2\phi\frac{\Delta z}{r^2}$$

$$\phi_{xx} = \left(-6 + 8\frac{\Delta x^2}{r^2}\right)\cdot\frac{\phi}{r^2}$$

$$\phi_{xz} = -2\frac{\Delta z\cdot\phi_x + \Delta x\cdot\phi_z}{r^2}$$

The streamlines created by this dipole are circles (Fig. A.9).

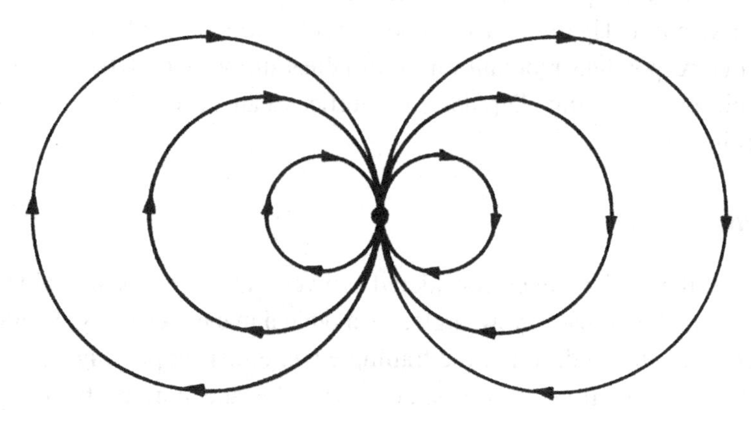

Figure A.9:
Velocities induced by point dipole

2. *Three-dimensional case*
The three-dimensional point dipole in the x direction is correspondingly given by:

$$\phi = -\frac{m}{4\pi r^3}\Delta x$$

$$\phi_x = -\frac{m}{4\pi r^3} - 3\phi\frac{\Delta x}{r^2}$$

$$\phi_y = -3\phi\frac{\Delta y}{r^2}$$

$$\phi_z = -3\phi\frac{\Delta z}{r^2}$$

$$\phi_{xx} = \frac{-5\phi_x\Delta x - 4\phi}{r^2}$$

$$\phi_{xy} = \frac{-5\phi_x\Delta y + 2\Delta y\cdot(-m/4\pi r^3)}{r^2}$$

$$\phi_{xz} = \frac{-5\phi_x\Delta z + 2\Delta z\cdot(-m/4\pi r^3)}{r^2}$$

$$\phi_{yy} = \frac{-5\phi_x\Delta y - 3\phi}{r^2}$$

$$\phi_{yz} = \frac{-5\phi_z\Delta y}{r^2}$$

$$\phi_{xx} = \frac{-5\phi_x\Delta x - 4\phi}{r^2}$$

The expressions for the dipole can be derived formally by differentiation of the corresponding source expression in x. Therefore usually source subroutines (also for distributed panels) are used with a corresponding redefinition of variables in the parameter list of the call. This avoids double programming. Dipoles like vortices can be used (rather equivalently) to generate lift in flows.

Thiart Element

The ship including the rudder can be considered as a vertical foil of considerable thickness and extremely short span. For a steady yaw angle, i.e. a typical maneuvering application, one would certainly enforce a Kutta condition at the trailing edge, either employing vortex or dipole elements. For harmonic motions in waves, i.e. a typical seakeeping problem, one should similarly employ a Kutta condition, but this is often omitted. If a Kutta condition is employed in frequency-domain computations, the wake will oscillate harmonically in strength. This can be modeled by discrete dipole elements of constant strength, but for high frequencies this approach requires many elements. The use of special elements which consider the oscillating strength analytically is more efficient and accurate, but also more complicated. Such a 'Thiart element' has been developed by Professor Gerhard Thiart of Stellen-bosch University and is described in detail in Bertram (1998), and Bertram and Thiart (1998). The oscillating ship creates a vorticity. The problem is similar to that of an oscillating airfoil. The circulation is assumed constant within the ship. Behind the ship, vorticity is shed downstream with ship speed V. Then:

$$\left(\frac{\partial}{\partial t} - V\frac{\partial}{\partial x}\right)\gamma(x, z, t) = 0$$

γ is the vortex density, i.e. the strength distribution for a continuous vortex sheet. The following distribution fulfills the above condition:

$$\gamma(x, z, t) = \mathrm{Re}(\widehat{\gamma}_a(z)\cdot \mathrm{e}^{i(\omega_e/V)(x-x_a)}\cdot \mathrm{e}^{i\omega_e t}) \quad \text{for } x \leq x_a$$

Here $\widehat{\gamma}_a$ is the vorticity density at the trailing edge x_a (stern of the ship). We continue the vortex sheet inside the ship at the symmetry plane $y = 0$, assuming a constant vorticity density:

$$\gamma(x, z, t) = \mathrm{Re}(\widehat{\gamma}_a(z)\cdot \mathrm{e}^{i\omega_e t}) \quad \text{for } x_a \leq x \leq x_f$$

x_f is the leading edge (forward stem of the ship). This vorticity density is spatially constant within the ship.

A vortex distribution is equivalent to a dipole distribution if the vortex density γ and the dipole density m are coupled by:

$$\gamma = \frac{\mathrm{d}m}{\mathrm{d}x}$$

The potential of an equivalent semi-infinite strip of dipoles is then obtained by integration. This potential is given (except for a so far arbitrary 'strength' constant) by:

$$\Phi(x, y, z) = \mathrm{Re}\left(\int_{-\infty}^{x_f}\int_{z_m}^{z_o} \widehat{m}(\xi)\frac{y}{r^3}\,\mathrm{d}\xi\,\mathrm{d}\zeta\,\mathrm{e}^{i\omega_e t}\right) = \mathrm{Re}(\varphi(x, y, z)\cdot \mathrm{e}^{i\omega_e t})$$

with $r = \sqrt{(x-\xi)^2 + y^2 + (z-\zeta)^2}$ and:

$$\widehat{m}(\xi) = \begin{cases} x_f - \xi & \text{for } x_a \leq \xi \leq x_f \\ \dfrac{V}{i\omega_e}(1 - \mathrm{e}^{i(\omega_e/V)}(\xi - x_a)) + (x_f - x_a) & \text{for } -\infty \leq \xi \leq x_a \end{cases}$$

It is convenient to write φ as:

$$\varphi(x,y,z) = y\int_{z_n}^{z_o}\int_{x_a}^{x_f}\frac{x-\xi}{r^3}\,\mathrm{d}\xi\mathrm{d}\zeta + (x_f - x)\int_{z_m}^{z_o}\int_{x_a}^{x_f}\frac{y}{r^3}\,\mathrm{d}\xi\,\mathrm{d}\zeta$$

$$+\left(\frac{V}{i\omega_e} + (x_f - x_a)\right)\int_{z_u}^{z_o}\int_{-\infty}^{x_a}\frac{y}{r^3}\,\mathrm{d}\xi\,\mathrm{d}\zeta$$

$$-\left(\frac{V}{i\omega_e}\mathrm{e}^{-i\omega_e x_a/V}\right)y\int_{z_u}^{z_o}\int_{-\infty}^{x_a}\mathrm{e}^{i\omega_e\xi/V}\frac{1}{r^3}\,\mathrm{d}\xi\,\mathrm{d}\zeta$$

The velocity components and higher derivatives are then derived by differentiation of Φ, which can be reduced to differentiation of φ. The exact formulae are given in Bertram (1998), and Bertram and Thiart (1998). The expressions involve integrals with integrands of the form 'arbitrary smooth function' · 'harmonically oscillating function'. These are accurately and efficiently evaluated using a modified Simpson's method developed by Söding:

$$\int_{x_1}^{x_1+2h} f(x)\mathrm{e}^{ikx}\mathrm{d}x = \frac{\mathrm{e}^{ikx_1}}{k}\left[\mathrm{e}^{2ikh}\left(\frac{0.5f_1 - 2f_2 + 1.5f_3}{kh} - i\left(f_3 - \frac{\Delta^2 f}{k^2h^2}\right)\right)\right.$$

$$\left. + \frac{1.5f_1 - 2f_2 + 0.5f_3}{kh} + i\left(f_1 - \frac{\Delta^2 f}{h^2k^2}\right)\right]$$

where $f_1 = f(x_1)$, $f_2 = f(x_1 + h)$, $f_3 = f(x_1 + 2h)$, $\Delta^2 f = f_1 - 2f_2 + f_3$.

Special Techniques

Desingularization

The potential and its derivatives become singular directly on a panel, i.e. infinite terms appear in the usual formulae which prevent straightforward evaluation. For the normal velocity, this singularity can be removed analytically for the collocation point on the panel itself, but the resulting special treatment makes parallelization of codes difficult. When the element is placed somewhat outside the domain of the problem (Fig. A.10), it is 'desingularized', i.e. the singularity is removed. This has several advantages:

- In principle the same expression can be evaluated everywhere. This facilitates parallel algorithms in numerical evaluation and makes the code generally shorter and easier.
- Numerical experiments show that desingularization improves the accuracy as long as the depth of submergence is not too large.

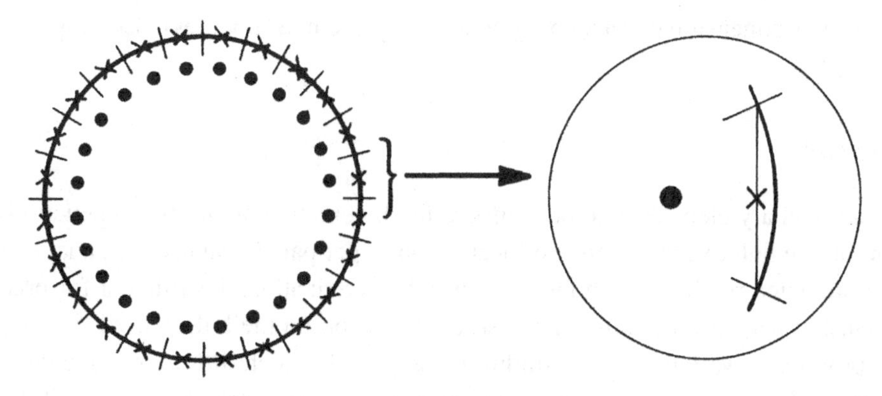

Figure A.10:
Desingularization

The last point surprised some mathematicians. Desingularization results in a Fredholm integral equation of the first kind. (Otherwise a Fredholm equation of second kind results.) This can lead in principle to problems with uniqueness and existence of solutions, which in practice manifest themselves first by an ill-conditioned matrix for the unknowns (source strengths or directly potential). For engineers, the problems are directly apparent without going into mathematical classification:

- If the individual elements (sources) are too far from the collocation points, they will all have almost the same influence. Then they will not be able to represent arbitrary local flow patterns.
- If the individual elements are somewhat removed, the individual sharp local steepness in flow pattern (singularity) will smooth out rapidly, forming a relatively smooth flow distribution which can relatively smoothly approximate arbitrary flows.
- If the individual elements are very close, an uneven cobblestone flow distribution results due to the discontinuity between the individual elements.

Thus the desingularization distance has to be chosen appropriately within a bandwidth to yield acceptable results. The distance should be related to the grid size. As the grid becomes finer, the desingularized solution approaches the conventional singular-element solution. Fortunately, several researchers have shown that the results are relatively insensitive to the desingularization distance, as long as this ranges between 0.5 and 2 typical grid spacings.

The historical development of desingularization of boundary element methods is reviewed in Cao et al. (1991), and Raven (1998).

Desingularization is used in many 'fully non-linear' wave resistance codes in practice for the free-surface elements. Sometimes it is also used for the hull elements, but here narrow pointed bows introduce difficulties often requiring special effort in grid generation. Some close-fit routines for two-dimensional seakeeping codes (strip-method modules) also employ desingularization.

Patch Method

Traditional boundary element methods enforce the kinematic condition (no-penetration condition) on the hull exactly at one collocation point per panel, usually the panel center. The resistance predicted by these methods is for usual discretizations insufficient for practical requirements, at least if conventional pressure integration on the hull is used. Söding (1993) therefore proposed a variation of the traditional approach which differs in some details from the conventional approach. Since his approach also uses flat segments on the hull, but not as distributed singularities, he called the approach 'patch' method to distinguish it from the usual 'panel' methods.

For double-body flows the resistance in an ideal fluid should be zero. This allows the comparison of the accuracy of various methods and discretizations as the non-zero numerical resistance is then purely due to discretization errors. For double-body flows, the patch method reduces the error in the resistance by one order of magnitude compared to ordinary first-order panel methods, without increasing the computational time or the effort in grid generation. However, higher derivatives of the potential or the pressure directly on the hull cannot be computed as easily as for a regular panel method.

The patch method basically introduces three changes to ordinary panel methods:

- 'Patch condition'
  Panel methods enforce the no-penetration condition on the hull exactly at one collocation point per panel. The 'patch condition' states that the integral of this condition over one patch of the surface is zero. This averaging of the condition corresponds to the techniques used in finite element methods.
- Pressure integration
  Potentials and velocities are calculated at the patch corners. Numerical differentiation of the potential yields an average velocity. A quadratic approximation for the velocity using the average velocity and the corner velocities is used in pressure integration. The unit normal is still considered constant.
- Desingularization
  Single point sources are submerged to give a smoother distribution of the potential on the hull. As desingularization distance between patch center and point source, the minimum of 10% of the patch length, 50% of the normal distance from patch center to a line of symmetry is recommended.

Söding (1993) did not investigate the individual influence of each factor, but the higher-order pressure integration and the patch condition contribute approximately the same.

The patch condition states that the flow through a surface element (patch) (and not just at its center) is zero. Desingularized Rankine point sources instead of panels are used as elementary solutions. The potential of the total flow is:

$$\phi = -Vx + \sum_i \sigma_i \varphi_i$$

σ is the source strength, φ is the potential of a Rankine point source, r is the distance between source and field point. Let M_i be the outflow through a patch (outflow = flow from interior of the body into the fluid) induced by a point source of unit strength.

1. *Two-dimensional case*

   The potential of a two-dimensional point source is:

$$\varphi = \frac{1}{4\pi} \ln r^2$$

   The integral zero-flow condition for a patch is:

$$-V \cdot n_x \cdot l + \sum_i \sigma_i M_i = 0$$

   n_x is the x component of the unit normal (from the body into the fluid), l the patch area (length). The flow through a patch is invariant of the coordinate system. Consider a local coordinate system x, z (Fig. A.11). The patch extends in this coordinate system from $-s$ to s. The flow through the patch is:

$$M = -\int_{-s}^{s} \phi_z \mathrm{d}x$$

   A Rankine point source of unit strength induces at x, z the vertical velocity:

$$\phi_z = \frac{1}{2\pi} \frac{z - z_q}{(x - x_q)^2 + (z - z_q)^2}$$

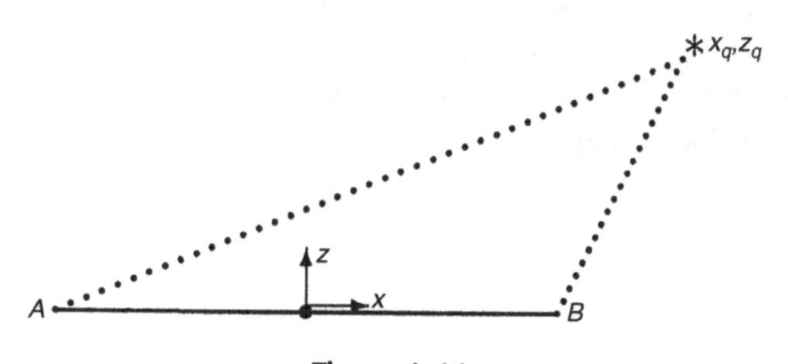

Figure A.11:
Patch in 2d

Since $z = 0$ on the patch, this yields:

$$M = \int_{-s}^{s} \frac{1}{2\pi} \frac{z_q}{(x - x_q)^2 + z_q^2} \mathrm{d}x = \frac{1}{2\pi} \arctan \frac{l z_q}{x_q^2 + z_q^2 - s^2}$$

The local z_q transforms from the global coordinates:

$$z_q = -n_x \cdot (\bar{x}_q - \bar{x}_c) - n_z \cdot (\bar{z}_q - \bar{z}_c)$$

$\bar{x}_c$, $\bar{z}_c$ are the global coordinates of the patch center, $\bar{x}_q, \bar{z}_q$ of the source.

From the value of the potential ϕ at the corners A and B, the average velocity within the patch is found as:

$$\bar{\vec{v}} = \frac{\phi_B - \phi_A}{|\vec{x}_B - \vec{x}_A|} \cdot \frac{\vec{x}_B - \vec{x}_A}{|\vec{x}_B - \vec{x}_A|}$$

i.e. the absolute value of the velocity is:

$$\frac{\Delta\phi}{\Delta s} = \frac{\phi_B - \phi_A}{|\vec{x}_B - \vec{x}_A|}$$

The direction is tangential to the body; the unit tangential is $(\vec{x}_B - \vec{x}_A)/|\vec{x}_B - \vec{x}_A|$. The pressure force on the patch is:

$$\Delta\vec{f} = \vec{n} \int p\mathrm{d}l = \vec{n} \frac{\rho}{2} \left(V^2 \cdot l - \int \vec{v}^2 \mathrm{d}l \right)$$

$\vec{v}$ is not constant! To evaluate this expression, the velocity within the patch is approximated by:

$$\vec{v} = a + bt + ct^2$$

t is the tangential coordinate directed from A to B. $\vec{v}_A$ and $\vec{v}_B$ are the velocities at the patch corners. The coefficients a, b, and c are determined from the conditions:

- The velocity at $t = 0$ is $\vec{v}_A$: $a = \vec{v}_A$.
- The velocity at $t = 1$ is $\vec{v}_B$: $a + b + c = \vec{v}_B$.
- The average velocity (integral over one patch) is $\bar{\vec{v}}$: $a + 1/2b + 1/3c = \bar{\vec{v}}$.

This yields:

$$a = \vec{v}_A$$

$$b = 6\bar{\vec{v}} - 4\vec{v}_A - 2\vec{v}_B$$

$$c = -6\bar{\vec{v}} + 3\vec{v}_A + 3\vec{v}_B$$

Using the above quadratic approximation for $\vec{v}$, the integral of $\bar{\vec{v}}^2$ over the patch area is found after some lengthy algebraic manipulations as:

$$\int \vec{v}^2 \mathrm{d}l = l \int_0^1 \vec{v}^2 \mathrm{d}t = l \cdot \left(a^2 + ab + \frac{1}{3}(2ac + b^2) + \frac{1}{2}bc + \frac{1}{5}c^2 \right)$$

$$= l \cdot \left(\bar{\vec{v}}^2 + \frac{2}{15}((\vec{v}_A - \bar{\vec{v}}) + (\vec{v}_B - \bar{\vec{v}}))^2 - \frac{1}{3}(\vec{v}_A - \bar{\vec{v}})(\vec{v}_B - \bar{\vec{v}}) \right)$$

Thus the force on one patch is:

$$\Delta \vec{f} = -\vec{n} \cdot l \cdot \left((\bar{\vec{v}}^2 - V^2) + \frac{2}{15}((\vec{v}_A - \bar{\vec{v}}) + (\vec{v}_B - \bar{\vec{v}}))^2 - \frac{1}{3}((\vec{v}_A - \bar{\vec{v}})(\vec{v}_B - \bar{\vec{v}})) \right)$$

2. *Three-dimensional case*

The potential of a three-dimensional source is:

$$\varphi = -\frac{1}{4\pi |\vec{x} - \vec{x}_q|}$$

Figure A.12 shows a triangular patch ABC and a source S. Quadrilateral patches may be created by combining two triangles. The zero-flow condition for this patch is:

$$-V \frac{(\vec{a} \times \vec{b})_1}{2} + \sum_i \sigma_i M_i = 0$$

The first term is the volume flow through ABC due to the uniform flow; the index 1 indicates the x component (of the vector product of two sides of the triangle). The flow M through a patch ABC induced by a point source of unit strength is $-\alpha/(4\pi)$. α is the solid angle in which ABC is seen from S. The rules of spherical geometry give α as the sum of the angles between each pair of planes SAB, SBC, and SCA minus π:

$$\alpha = \beta_{SAB,SBC} + \beta_{SBC,SCA} + \beta_{SCA,SAB} - \pi$$

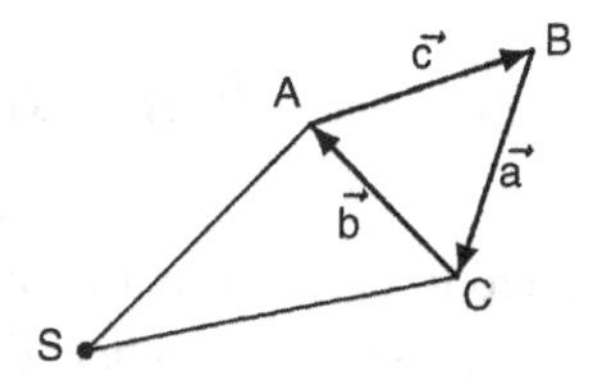

Figure A.12:
Source point S and patch ABC

where, e.g.,

$$\beta_{SAB,SBC} = \arctan\frac{-[(\vec{A}\times\vec{B})\times(\vec{B}\times\vec{C})]\cdot\vec{B}}{(\vec{A}\times\vec{B})\cdot(\vec{B}\times\vec{C})|\vec{B}|}$$

Here $\vec{A}$, $\vec{B}$, $\vec{C}$ are the vectors pointing from the source point S to the panel corners A, B, C. The solid angle may be approximated by A^*/d^2 if the distance d between patch center and source point exceeds a given limit. A^* is the patch area projected on a plane normal to the direction from the source to the patch center:

$$\vec{d} = \frac{1}{3}(\vec{A}+\vec{B}+\vec{C})$$

$$A^* = \frac{1}{2}(\vec{a}\times\vec{b})\,\frac{\vec{d}}{d}$$

With known source strengths σ_i, one can determine the potential ϕ and its derivatives $\nabla\phi$ at all patch corners. From the ϕ values at the corners A, B, C, the average velocity within the triangle is found as:

$$\bar{\vec{v}} = \overline{\nabla\phi} = \frac{\phi_A-\phi_C}{\vec{n}_{AB}^2}\vec{n}_{AB} + \frac{\phi_B-\phi_A}{\vec{n}_{AC}^2}\vec{n}_{AC}$$

$$\vec{n}_{AB} = \vec{b} - \frac{\vec{c}\cdot\vec{b}}{\vec{c}^2}\vec{c} \quad \text{and} \quad \vec{n}_{AC} = \vec{c} - \frac{\vec{b}\cdot\vec{c}}{\vec{b}^2}\vec{b}$$

With known $\bar{\vec{v}}$ and corner velocities $\vec{v}_A$, $\vec{v}_B$, $\vec{v}_C$, the pressure force on the triangle can be determined:

$$\Delta\vec{f} = \vec{n}\int p\,\mathrm{d}A = \vec{n}\frac{\rho}{2}\left(V^2\cdot A - \int \vec{v}^2\mathrm{d}A\right)$$

where $\vec{v}$ is *not* constant! $A = {}^1/_2|\vec{a}\times\vec{b}|$ is the patch area. To evaluate this equation, the velocity within the patch is approximated by:

$$\vec{v} = \bar{\vec{v}} + (\vec{v}_A - \bar{\vec{v}})(2r^2 - r) + (\vec{v}_B - \bar{\vec{v}})(2s^2 - s) + (\vec{v}_C - \bar{\vec{v}})(2t^2 - t)$$

r is the 'triangle coordinate' directed to patch corner A: $r = 1$ at A and $r = 0$ at the line BC. s and t are the corresponding 'triangle coordinates' directed to B resp. C. Using this quadratic v formula, the integral of $\vec{v}^2$ over the triangle area is found after some algebraic manipulations as:

$$\int \vec{v}^2 \mathrm{d}A = A \cdot \left[\bar{\vec{v}}^2 + \frac{1}{30}(\vec{v}_A - \bar{\vec{v}})^2 + \frac{1}{30}(\vec{v}_B - \bar{\vec{v}})^2 + \frac{1}{30}(\vec{v}_C - \bar{\vec{v}})^2 - \frac{1}{90}(\vec{v}_A - \bar{\vec{v}})\,(\vec{v}_B - \bar{\vec{v}}) - \frac{1}{90}(\vec{v}_B - \bar{\vec{v}})\,(\vec{v}_C - \bar{\vec{v}}) - \frac{1}{90}(\vec{v}_C - \bar{\vec{v}})(\vec{v}_A - \bar{\vec{v}}) \right]$$

APPENDIX B

Numerical Examples for BEM

Two-Dimensional Flow Around a Body in Infinite Fluid

One of the most simple applications of boundary element methods is the computation of the potential flow around a body in an infinite fluid. The inclusion of a rigid surface is straightforward in this case and leads to the double-body flow problem which will be discussed at the end of this chapter.

Theory

We consider a submerged body of arbitrary (but smooth) shape moving with constant speed V in an infinite fluid domain. For inviscid and irrotational flow, this problem is equivalent to a body being fixed in an inflow of constant speed. For testing purposes, we may select a simple geometry like a circle (cylinder of infinite length) as a body.

For the assumed ideal fluid, there exists a velocity potential ϕ such that $\vec{v} = \nabla\phi$. For the considered ideal fluid, continuity gives Laplace's equation, which holds in the whole fluid domain:

$$\Delta\phi = \phi_{xx} + \phi_{zz} = 0$$

In addition, we require the boundary condition that water does not penetrate the body's surface (hull condition). For an inviscid fluid, this condition can be reformulated requiring just vanishing normal velocity on the body:

$$\vec{n}\cdot\nabla\phi = 0$$

$\vec{n}$ is the inward unit normal vector on the body hull. This condition is mathematically a Neumann condition as it involves only derivatives of the unknown potential.

Once a potential and its derivatives have been determined, the forces on the body can be determined by direct pressure integration:

$$f_1 = \int_S p n_1 \, \mathrm{d}S$$

$$f_2 = \int_S p n_2 \, \mathrm{d}S$$

S is the wetted surface. p is the pressure determined from Bernoulli's equation:

$$p = \frac{\rho}{2}(V^2 - (\nabla\phi)^2)$$

The force coefficients are then:

$$C_x = \frac{f_1}{\frac{\rho}{2}V^2 S}$$

$$C_z = \frac{f_2}{\frac{\rho}{2}V^2 S}$$

Numerical Implementation

The velocity potential ϕ is approximated by uniform flow superimposed by a finite number N of elements. These elements are in the sample program DOUBL2D desingularized point sources inside the body (Fig. A.10). The choice of elements is rather arbitrary, but the most simple elements are selected here for teaching purposes.

We formulate the potential ϕ as the sum of parallel uniform flow (of speed V) and a residual potential which is represented by the elements:

$$\phi = -Vx + \sum \sigma_i \varphi$$

σ_i is the strength of the ith element, φ the potential of an element of unit strength. The index i for φ is omitted for convenience but it should be understood in the equations below that φ refers to the potential of only the ith element.

Then the Neumann condition on the hull becomes:

$$\sum_{i=1}^{N} \sigma_i(\vec{n} \cdot \nabla\varphi) = V n_1$$

This equation is fulfilled on N collocation points on the body, thus forming a linear system of equations in the unknown element strengths σ_i. Once the system is solved, the velocities and pressures are determined on the body.

The pressure integral for the x force is evaluated approximately by:

$$\int_S pn_1 \, \mathrm{d}S \approx \sum_{i=1}^{N} p_i n_{1,i} s_i$$

The pressure p_i and the inward normal on the hull n_i are taken constant over each panel. s_i is the area of one segment.

For double-body flow, an 'element' consists of a source at $z = z_q$ and its mirror image at $z = -z_q$. Otherwise, there is no change in the program.

Two-Dimensional Wave Resistance Problem

The extension of the theory for a two-dimensional double-body flow problem to a two-dimensional free-surface problem with optional shallow-water effect introduces these main new features:

- 'fully non-linear' free-surface treatment
- shallow-water treatment
- treatment of various element types in one program.

While the problem is purely academical as free-surface steady flows for ships in reality are always strongly three-dimensional, the two-dimensional problem is an important step in understanding the three-dimensional problem. Various techniques have in the history of development always been tested and refined first in the much faster and easier two-dimensional problem, before being implemented in three-dimensional codes. The two-dimensional problem is thus an important stepping stone for researchers and a useful teaching example for students.

Theory

We consider a submerged body of arbitrary (but smooth) shape moving with constant speed V under the free surface in water of constant depth. The depth may be infinite or finite. For inviscid and irrotational flow, this problem is equivalent to a body being fixed in an inflow of constant speed.

We extend the theory simply repeating the previously discussed conditions and focusing on the new conditions. Laplace's equation holds in the whole fluid domain. The boundary conditions are:

- Hull condition: water does not penetrate the body's surface.
- Kinematic condition: water does not penetrate the water surface.

- Dynamic condition: there is atmospheric pressure at the water surface.
- Radiation condition: waves created by the body do not propagate ahead.
- Decay condition: far ahead of and below the body, the flow is undisturbed.
- Open-boundary condition: waves generated by the body pass unreflected any artificial boundary of the computational domain.
- Bottom condition (shallow-water case): no water flows through the sea bottom.

The decay condition replaces the bottom condition if the bottom is at infinity, i.e. in the usual infinite fluid domain case.

The wave resistance problem features two special problems requiring an iterative solution:

1. A non-linear boundary condition appears on the free surface.
2. The boundaries of water (waves) are not a priori known.

The iteration starts by approximating:

- the unknown wave elevation by a flat surface
- the unknown potential by the potential of uniform parallel flow.

In each iterative step, wave elevation and potential are updated yielding successively better approximations for the solution of the non-linear problem.

The equations are formulated here in a right-handed Cartesian coordinate system with x pointing forward towards the 'bow' and z pointing upward. For the assumed ideal fluid, there exists a velocity potential ϕ such that $\vec{v} = \nabla\phi$. The velocity potential ϕ fulfills Laplace's equation in the whole fluid domain:

$$\Delta\phi = \phi_{xx} + \phi_{zz} = 0$$

The hull condition requires vanishing normal velocity on the body:

$$\vec{n}\cdot\nabla\phi = 0$$

$\vec{n}$ is the inward unit normal vector on the body hull.

The kinematic condition (no penetration of water surface) gives at $z = \zeta$:

$$\nabla\phi\cdot\nabla\zeta = \phi_z$$

For simplification, we write $\zeta(x, z)$ with $\zeta_z = \partial\zeta/\partial z = 0$.

The dynamic condition (atmospheric pressure at water surface) gives at $z = \zeta$:

$$\frac{1}{2}(\nabla\phi)^2 + gz = \frac{1}{2}V^2$$

with $g = 9.81$ m/s^2. Combining the dynamic and kinematic boundary conditions eliminates the unknown wave elevation $z = \zeta$:

$$\frac{1}{2}\nabla\phi\cdot\nabla(\nabla\phi)^2 + g\phi_z = 0$$

This equation must still be fulfilled at $z = \zeta$. If we approximate the potential ϕ and the wave elevation ζ by arbitrary approximations Φ and $\bar{\zeta}$, linearization about the approximated potential gives at $z = \zeta$:

$$\nabla\Phi\cdot\nabla\left(\frac{1}{2}(\nabla\Phi)^2 + \nabla\Phi\cdot\nabla(\phi - \Phi)\right) + \nabla(\phi - \Phi)\cdot\nabla\left(\frac{1}{2}(\nabla\Phi)^2\right) + g\phi_z = 0$$

Φ and $\phi - \Phi$ are developed in a Taylor expansion about $\bar{\zeta}$. The Taylor expansion is truncated after the linear term. Products of $\zeta - \bar{\zeta}$ with derivatives of $\phi - \Phi$ are neglected. This yields at $z = \bar{\zeta}$:

$$\nabla\Phi\cdot\nabla\left(\frac{1}{2}(\nabla\Phi)^2 + \nabla\Phi\cdot\nabla(\phi - \Phi)\right) + \nabla(\phi - \Phi)\cdot\nabla\left(\frac{1}{2}(\nabla\Phi)^2\right) + g\phi_z$$
$$+ \left[\frac{1}{2}\nabla\Phi\cdot\nabla(\nabla\Phi)^2 + g\Phi_z\right]_z (\zeta - \bar{\zeta}) = 0$$

A consistent linearization about Φ and $\bar{\zeta}$ substitutes ζ by an expression depending solely on $\bar{\zeta}$, $\Phi(\bar{\zeta})$ and $\phi(\bar{\zeta})$. For this purpose, the original expression for ζ is also developed in a truncated Taylor expansion and written at $z = \bar{\zeta}$:

$$\zeta = -\frac{1}{2g}(-(\nabla\Phi)^2 + 2\nabla\Phi\cdot\nabla\phi + 2\nabla\Phi\cdot\nabla\Phi_z(\zeta - \bar{\zeta}) - V^2)$$

$$\zeta - \bar{\zeta} = \frac{-\frac{1}{2}(2\nabla\Phi\cdot\nabla\phi - (\nabla\Phi)^2 - V^2) - g\bar{\zeta}}{g + \nabla\Phi\cdot\nabla\Phi_z}$$

Substituting this expression in our equation for the free-surface condition gives the consistently linearized boundary condition at $z = \bar{\zeta}$:

$$\nabla\Phi\nabla[-(\nabla\Phi)^2 + \nabla\Phi\cdot\nabla\phi] + \frac{1}{2}\nabla\phi\nabla(\nabla\Phi)^2 + g\phi_z + \frac{\left[\frac{1}{2}\nabla\Phi\nabla(\nabla\Phi)^2 + g\Phi_z\right]_z}{g + \nabla\Phi\cdot\nabla\Phi_z}$$
$$\times\left(-\frac{1}{2}[-(\nabla\Phi)^2 + 2\nabla\Phi\cdot\nabla\phi - V^2] - g\bar{\zeta}\right) = 0$$

The denominator in the last term becomes zero when the vertical particle acceleration is equal to gravity. In fact, the flow becomes unstable already at 0.6 to $0.7g$ both in reality and in numerical computations.

It is convenient to introduce the following abbreviations:

$$\vec{a} = \frac{1}{2}\nabla((\nabla\Phi)^2) = \begin{Bmatrix} \Phi_x\Phi_{xx} + \Phi_z\Phi_{xz} \\ \Phi_x\Phi_{xz} + \Phi_z\Phi_{zz} \end{Bmatrix}$$

$$B = \frac{\left[\frac{1}{2}\nabla\Phi\nabla(\nabla\Phi)^2 + g\Phi_z\right]_z}{g + \nabla\Phi\cdot\nabla\Phi_z} = \frac{[\nabla\Phi\vec{a} + g\Phi_z]_z}{g + a_2}$$
$$= \frac{1}{g + a_2}(\Phi_x^2\Phi_{xxz} + \Phi_z^2\Phi_{zzz} + g\Phi_{zz} + 2[\Phi_x\Phi_z\Phi_{xzz} + \Phi_{xz}\cdot a_1 + \Phi_{zz}\cdot a_2])$$

Then the boundary Φ condition at $z = \overline{\zeta}$ becomes:

$$2(\vec{a}\nabla\phi + \Phi_x\Phi_z\phi_{xz}) + \Phi_x^2\phi_{xx} + \Phi_z^2\phi_{zz} + g\phi_z - B\nabla\Phi\nabla\phi$$
$$= 2\vec{a}\nabla\Phi - B\left(\frac{1}{2}((\nabla\Phi)^2 + V^2) - g\overline{\zeta}\right)$$

The non-dimensional error in the boundary condition at each iteration step is defined by:

$$\varepsilon = \max(|\vec{a}\nabla\Phi + g\Phi_z|)/(gV)$$

where 'max' means the maximum value of all points at the free surface.

For given velocity, Bernoulli's equation determines the wave elevation:

$$z = \frac{1}{2g}\left(V^2 - (\nabla\phi)^2\right)$$

The first step of the iterative solution is the classical linearization around uniform flow. To obtain the classical solutions for this case, the above equation should also be linearized as:

$$z = \frac{1}{2g}(V^2 + (\nabla\Phi)^2 - 2\nabla\Phi\nabla\phi)$$

However, it is computationally simpler to use the non-linear equation.

The bottom, radiation, and open-boundary conditions are fulfilled by the proper arrangement of elements as described below. The decay condition – like the Laplace equation – is automatically fulfilled by all elements.

Once a potential has been determined, the force on the body in the x direction can be determined by direct pressure integration:

$$f_1 = \int_S pn_1\, dS$$

S is the wetted surface. p is the pressure determined from Bernoulli's equation:

$$p = \frac{\rho}{2}\left(V^2 - (\nabla\phi)^2\right)$$

The force in the x direction, f_1, is the (negative) wave resistance. The non-dimensional wave resistance coefficient is:

$$C_w = -f_1 / \left(\frac{\rho}{2} V^2 S\right)$$

Numerical Implementation

The velocity potential ϕ is approximated by uniform flow superimposed by a finite number of elements. These elements are, in the sample program SHAL2D:

- desingularized point source clusters above the free surface
- desingularized point sources inside the body.

The choice of elements is rather arbitrary, but very simple elements have been selected for teaching purposes.

The height of the elements above the free surface is not corrected in SHAL2D. For usual discretizations (10 elements per wave length) and moderate speeds, this procedure should work without problems. For finer discretizations (as often found for high speeds), problems occur which require a readjustment of the panel layer. However, in most cases it is sufficient to adjust the source layer just once after the first iteration and then 'freeze' it.

We formulate the potential ϕ as the sum of parallel uniform flow (of speed V) and a residual potential which is represented by the elements:

$$\phi = -Vx + \sum \sigma_i \varphi$$

σ_i is the strength of the ith element, φ the potential of an element of unit strength. The expression 'element' refers to one source (cluster) and all its mirror images. If the collocation point and source center are sufficiently far from each other, e.g. three times the grid spacing, the source cluster may be substituted by a single point source. This accelerates the computation without undue loss of accuracy.

Then the no-penetration boundary condition on the hull becomes:

$$\sum \sigma_i (\vec{n} \cdot \nabla\varphi) = V n_1$$

The linearized free-surface condition becomes:

$$\sum \sigma_i(2(\vec{a}\nabla\varphi + \Phi_x\Phi_z\varphi_{xz}) + \Phi_x^2\varphi_{xx} + \Phi_z^2\varphi_{zz} + g\varphi_z - B\nabla\Phi\nabla\varphi)$$
$$= 2(\vec{a}\nabla\Phi + a_1 V) - B\left(\frac{1}{2}((\nabla\Phi)^2 + V^2) - g\overline{\zeta} + V\Phi_x\right)$$

These two equations form a linear system of equations in the unknown element strengths σ_i. Once the system is solved, the velocities (and higher derivatives of the potential) are determined on the water surface. Then the error ε is determined.

For shallow water, mirror images of elements at the ocean bottom are used. This technique is similar to the mirror imaging at the still waterplane used for double-body flow.

The radiation and open-boundary conditions are fulfilled using 'staggered grids'. This technique adds an extra row of panels at the downstream end of the computational domain and an extra row of collocation points at the upstream end (Fig. 3.11). For equidistant grids, this can also be interpreted as shifting or staggering the grid of collocation points vs. the grid of elements, hence the name 'staggered grid'. However, this name is misleading as for non-equidistant grids or three-dimensional grids with quasi-streamlined grid lines, adding an extra row at the ends is not the same as shifting the whole grid.

The pressure integral for the x force is evaluated approximately by:

$$\int_S pn_1 \, \mathrm{d}S \approx \sum_{i=1}^{N_B} p_i n_{1,i} s_i$$

N_B is the number of elements on the hull. The pressure p_i, and the inward normal on the hull, n_i, are taken constant over each panel. s_i is the area of one segment.

Three-Dimensional Wave Resistance Problem

The extension of the theory for a two-dimensional submerged body to a three-dimensional surface-piercing ship free to trim and sink introduces these main new features:

- surface-piercing hulls
- dynamic trim and sinkage
- transom stern
- Kutta condition for multihulls.

The theory outlined here is the theory behind the STEADY code (Hughes and Bertram 1995). The code is a typical representative of a state-of-the-art 'fully non-linear' wave resistance code of the 1990s.

Theory

We consider a ship moving with constant speed V in water of constant depth. The depth and width may be infinite and are in fact assumed to be so in most cases. For inviscid and irrotational flow, this problem is equivalent to a ship being fixed in an inflow of constant speed.

For the considered ideal fluid, continuity gives Laplace's equation, which holds in the whole fluid domain. A unique description of the problem requires further conditions on all boundaries of the fluid resp. the modeled fluid domain:

- Hull condition: water does not penetrate the ship's surface.
- Transom stern condition: for ships with a transom stern, we assume that the flow separates and the transom stern is dry. Atmospheric pressure is then enforced at the edge of the transom stern.
- Kinematic condition: water does not penetrate the water surface.
- Dynamic condition: there is atmospheric pressure at the water surface.
- Radiation condition: waves created by the ship do not propagate ahead. (This condition is not valid for transcritical depth Froude numbers when the flow becomes unsteady and soliton waves are pulsed ahead. But ships are never designed for such speeds.)
- Decay condition: far away from the ship, the flow is undisturbed.
- Open-boundary condition: waves generated by the ship pass unreflected any artificial boundary of the computational domain.
- Equilibrium: the ship is in equilibrium, i.e. trim and sinkage are changed such that the dynamic vertical force and the trim moment are counteracted.
- Bottom condition (shallow-water case): no water flows through the sea bottom.
- Kutta condition (for multihulls): at the end of each side floater the flow separates smoothly. This is approximated by setting the y velocity to zero.

The decay condition replaces the bottom condition if the bottom is at infinity, i.e. in the usual infinite fluid domain case.

The problem is solved using boundary elements (in the case of STEADY higher-order panels on the ship hull, point source clusters above the free surface). The wave resistance problem features two special problems requiring an iterative solution approach:

1. A non-linear boundary condition appears on the free surface.
2. The boundaries of water (waves) and ship (trim and sinkage) are not a priori known.

The iteration starts by approximating:

- the unknown wave elevation by a flat surface
- the unknown potential by the potential of uniform parallel flow
- the unknown position of the ship by the position of the ship at rest.

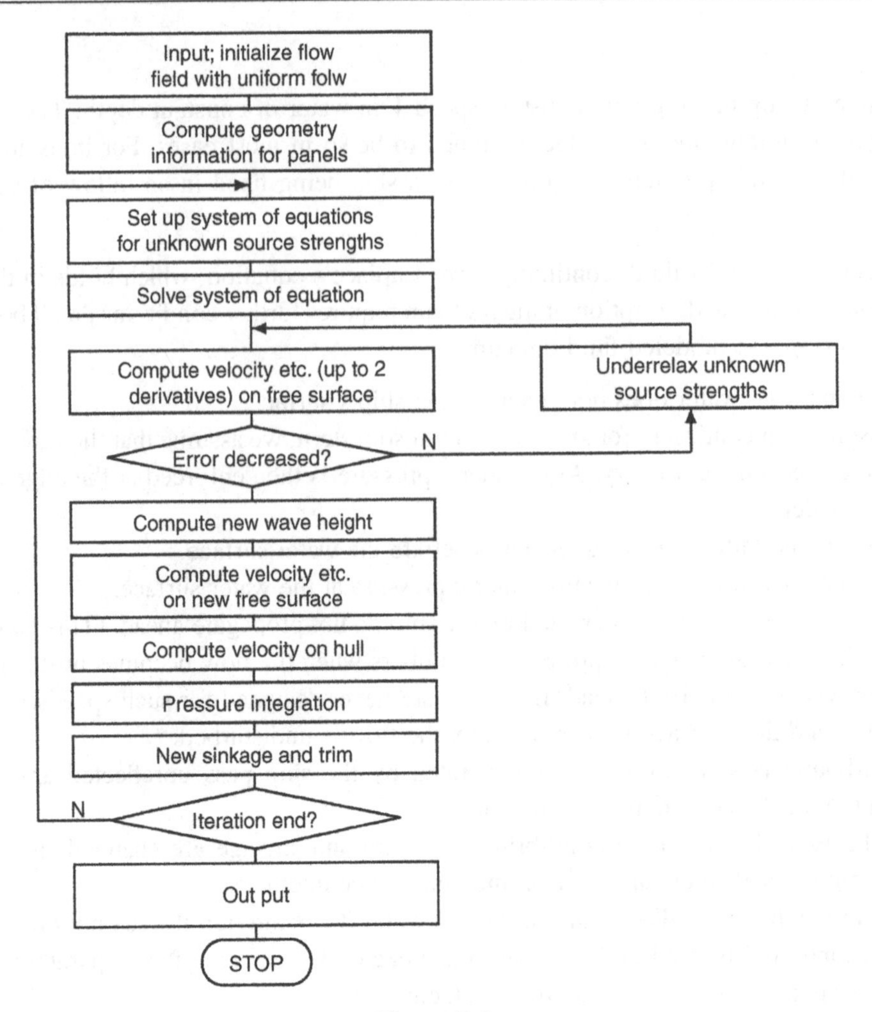

Figure B.1:
Flow chart of iterative solution

In each iterative step, wave elevation, potential, and position are updated, yielding successively better approximations for the solution of the non-linear problem (Fig. B.1).

The equations are formulated here in a right-handed Cartesian coordinate system with x pointing forward towards the bow and z pointing upward. The moment about the y-axis (and the trim angle) are positive clockwise (bow immerses for positive trim angle).

For the assumed ideal fluid, there exists a velocity potential ϕ such that $\vec{v} = \nabla\phi$. The velocity potential ϕ fulfills Laplace's equation in the whole fluid domain:

$$\Delta\phi = \phi_{xx} + \phi_{yy} + \phi_{zz} = 0$$

A unique solution requires the formulation of boundary conditions on all boundaries of the modeled fluid domain.

The hull condition (no penetration of ship hull) requires that the normal velocity on the hull vanishes:

$$\vec{n} \cdot \nabla\phi = 0$$

$\vec{n}$ is the inward unit normal vector on the ship hull.

The transom stern condition (atmospheric pressure at the edge of the transom stern $z = z_T$) is derived from Bernoulli's equation:

$$\frac{1}{2}(\nabla\phi)^2 + gz_T = \frac{1}{2}V^2$$

with $g = 9.81$ m/s^2. This condition is non-linear in the unknown potential. We assume that the water flows at the stern predominantly in the x direction, such that the y and z components are negligible. This leads to the linear condition:

$$\phi_x + \sqrt{V^2 - 2gz_T} = 0$$

For points above the height of stagnation $V^2/2g$, this condition leads to a negative term in the square root. For these points, stagnation of horizontal flow is enforced instead. Both cases can be combined as:

$$\phi_x + \sqrt{\max(0, V^2 - 2gz_T)} = 0$$

The Kutta condition is originally a pressure condition, thus also non-linear. However, the obliqueness of the flow induced at the end of each side floater is so small that a simplification can be well justified. We then enforce just zero y velocity (Joukowski condition):

$$\phi_y = 0$$

The kinematic condition (no penetration of water surface) gives at $z = \zeta$:

$$\nabla\phi \cdot \nabla\zeta = \phi_z$$

For simplification, we write $\zeta(x, y, z)$ with $\zeta_z = \partial\zeta/\partial z = 0$.

The dynamic condition (atmospheric pressure at water surface) gives at $z = \zeta$:

$$\frac{1}{2}(\nabla\phi)^2 + gz = \frac{1}{2}V^2$$

Combining the dynamic and kinematic boundary conditions and linearizing consistently yields again at $z = \overline{\zeta}$:

$$\begin{aligned} &2(\vec{a}\nabla\phi + \Phi_x\Phi_y\phi_{xy} + \Phi_x\Phi_z\phi_{xz} + \Phi_y\Phi_z\phi_{yz}) + \Phi_x^2\phi_{xx} \\ &\quad + \Phi_y^2\phi_{yy} + \Phi_z^2\phi_{zz} + g\phi_z - B\nabla\Phi\nabla\phi = 2\vec{a}\nabla\Phi - B\left(\frac{1}{2}((\nabla\Phi)^2 + V^2) - g\overline{\zeta}\right) \end{aligned}$$

with

$$\vec{a} = \frac{1}{2}\nabla((\nabla\Phi)^2) = \begin{Bmatrix} \Phi_x\Phi_{xx} + \Phi_y\Phi_{xy} + \Phi_z\Phi_{xz} \\ \Phi_x\Phi_{xy} + \Phi_y\Phi_{yy} + \Phi_z\Phi_{yz} \\ \Phi_x\Phi_{xz} + \Phi_y\Phi_{yz} + \Phi_z\Phi_{zz} \end{Bmatrix}$$

$$B = \frac{\left[\frac{1}{2}\nabla\Phi\nabla(\nabla\Phi)^2 + g\Phi_z\right]_z}{g + \nabla\Phi\cdot\nabla\Phi_z} = \frac{[\nabla\Phi\vec{a} + g\Phi_z]_z}{g + a_3}$$

$$= \frac{1}{g + a_3}\Big(\Phi_x^2\Phi_{xxz} + \Phi_y^2\Phi_{yyz} + \Phi_z^2\Phi_{zzz} + g\Phi_{zz} + 2[\Phi_x\Phi_y\Phi_{xyz}$$

$$+ \Phi_x\Phi_z\Phi_{xzz} + \Phi_y\Phi_z\Phi_{yzz} + \Phi_{xz}\cdot a_1 + \Phi_{yz}\cdot a_2 + \Phi_{zz}\cdot a_3]\Big)$$

The bottom, radiation, and open-boundary conditions are fulfilled by the proper arrangement of elements as described below. The decay condition – like the Laplace equation – is automatically fulfilled by all elements.

Once a potential has been determined, the forces can be determined by direct pressure integration on the wetted hull. The forces are corrected by the hydrostatic forces at rest. (The hydrostatic x force and y moment should be zero, but are non-zero due to discretization errors. The discretization error is hoped to be reduced by subtracting the value for the hydrostatic force):

$$f_1 = \int_S pn_1 \, dS - \int_{S_0} p_s n_1 \, dS$$

$$f_3 = \int_S pn_3 \, dS - \int_{S_0} p_s n_3 \, dS$$

$$f_5 = \int_S p(zn_1 - xn_3) \, dS - \int_{S_0} p_s(zn_1 - xn_3) \, dS$$

S is the actually wetted surface. S_0 is the wetted surface of the ship at rest. $p_s = -\rho gz$ is the hydrostatic pressure, where ρ is the density of water. p is the pressure determined from Bernoulli's equation:

$$p = \frac{\rho}{2}\left(V^2 - (\nabla\phi)^2\right) - \rho gz$$

The force in the x direction, f_1, is the (negative) wave resistance. The non-dimensional wave resistance coefficient is:

$$C_w = -f_1 \Big/ \left(\frac{\rho}{2}V^2 S\right)$$

The z force and y moments are used to adjust the position of the ship. We assume small changes of the position of the ship. Δz is the deflection of the ship (positive, if the ship surfaces) and $\Delta\theta$ is the trim angle (positive if bow immerses) (Fig. B.2).

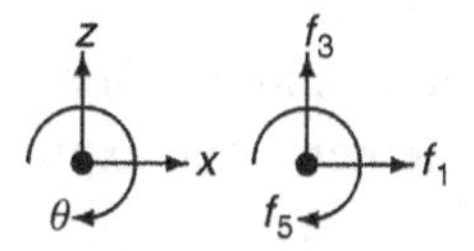

Figure B.2:
Coordinate system; x points towards bow, origin is usually amidships in still waterline; relevant forces and moment

For given Δz and $\Delta\theta$, the corresponding z force and y moment (necessary to enforce this change of position) are:

$$\begin{Bmatrix} f_3 \\ f_5 \end{Bmatrix} = \begin{bmatrix} A_{WL}\cdot\rho\cdot g & -A_{WL}\cdot\rho\cdot g\cdot x_{WL} \\ -A_{WL}\cdot\rho\cdot g\cdot x_{WL} & I_{WL}\cdot\rho\cdot g \end{bmatrix} \begin{Bmatrix} \Delta z \\ \Delta\theta \end{Bmatrix}$$

A_{WL} is the area, I_{WL} the moment of inertia, and x_{WL} the center of the still water-plane. I_{WL} and x_{WL} are taken relative to the origin, which we put amidships. Inversion of this matrix gives an equation of the form:

$$\begin{Bmatrix} \Delta z \\ \Delta\theta \end{Bmatrix} = \begin{bmatrix} a_{11} & a_{12} \\ a_{21} & a_{22} \end{bmatrix} \begin{Bmatrix} f_3 \\ f_5 \end{Bmatrix}$$

The coefficients a_{ij} are determined once in the beginning by inverting the matrix for the still waterline. Then during each iteration the position of the ship is changed by Δz and $\Delta\theta$ giving the final sinkage and trim when converged. The coefficients should actually change as the ship trims and sinks and thus its actual waterline changes from the still waterline. However, this error just slows down the convergence, but (for convergence) does not change the final result for trim and sinkage.

Numerical Implementation

The velocity potential ϕ is approximated by parallel flow superimposed by a finite number of elements. These elements are, for STEADY higher-order panels lying on the ship surface, linear panels (constant strength) in a layer above part of the free surface, and vortex elements lying on the local center plane of any side floater. However, the choice of elements is rather arbitrary. If just wave resistance computations are performed, first-order elements are sufficient and actually preferable due to their greater robustness.

The free-surface elements are again usually 'desingularized'. We place them approximately one panel length above the still-water plane ($z = 0$).

We formulate the potential ϕ as the sum of parallel uniform flow (of speed V) and a residual potential which is represented by the elements:

$$\phi = -Vx + \sum \sigma_i \varphi$$

σ_i is the strength of the ith element, φ the potential of an element of unit strength. The index i for φ is omitted for convenience but it should be understood in the equations below that φ refers to the potential of only the ith element. The expression 'element' refers to one panel or vortex and all its mirror images.

Then the no-penetration boundary condition on the hull becomes:

$$\sum \sigma_i(\vec{n}\cdot\nabla\varphi) = Vn_1$$

The Kutta condition becomes:

$$\sum \sigma_i\varphi_y = 0$$

The transom stern condition becomes:

$$\sum \sigma_i\varphi_x = V - \sqrt{\max(0, V^2 - 2gz_T)}$$

The linearized free surface condition then becomes:

$$\sum \sigma_i(2(\vec{a}\,\nabla\varphi + \Phi_x\Phi_y\varphi_{xy} + \Phi_x\Phi_z\varphi_{xz} + \Phi_y\Phi_z\varphi_{yz}) + \Phi_x^2\varphi_{xx} + \Phi_y^2\varphi_{yy}$$
$$+ \Phi_z^2\varphi_{zz} + g\varphi_z - B\nabla\Phi\nabla\varphi) = 2(\vec{a}\,\nabla\Phi + a_1V) - B\left(\frac{1}{2}((\nabla\Phi)^2 + V^2) - g\overline{\zeta} + V\Phi_x\right)$$

These four equations form a linear system of equations in the unknown element strengths σ_i. Once the system is solved, the velocities (and higher derivatives of the potential) are determined on the water surface and the error ε is determined. A special refinement accelerates and stabilizes to some extent the iteration process: if the error ε_{i+1} in iteration step i+1 is larger than the error ε_i in the previous ith step the source strengths are underrelaxed:

$$\sigma_{i+1,\text{new}} = \frac{\sigma_{i+1,\text{old}}\cdot\varepsilon_i + \sigma_i\cdot\varepsilon_{i+1}}{\varepsilon_i + \varepsilon_{i+1}}$$

Velocities and errors are evaluated again with the new source strengths. If the error is decreased the computation proceeds, otherwise the underrelaxation is repeated. If four repetitions still do not improve the error compared to the previous step, the computation is stopped. In this case, no converged non-linear solution can be found. This is usually the case if breaking waves appear in the real flow at a location of a collocation point.

Mirror images of panels are used (Fig. B.3):

1. In the y direction with respect to the center plane $y = 0$.
2. For shallow water in the z direction with respect to the water bottom $z = z_{\text{bottom}}: z' = -2|z_{\text{bottom}}| - z$.

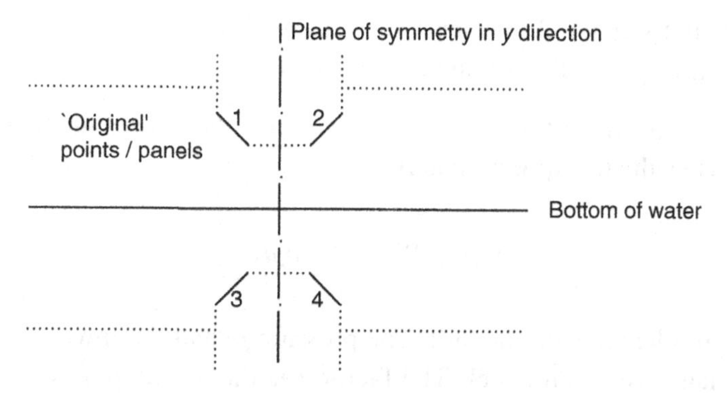

Figure B.3:
Mirror images of panels are used

The computation of the influence of one element on one collocation point uses the fact that the influence of a panel at A on a point at B has the same absolute value and opposite sign as a panel at B on a point at A. Actually, mirror images of the collocation point are produced and the influence of the original panel is computed for all mirror points. Then the sign of each influence is changed according to Table B.1.

The radiation and open-boundary conditions are fulfilled using 'staggered grids' as for the two-dimensional case. No staggering in the y direction is necessary.

For equidistant grids and collocation points along lines of $y =$ const., this can also be interpreted as shifting or staggering the grid of collocation points vs. the grid of elements, hence the name 'staggered grid'. However, for three-dimensional grids around surface-piercing

Table B.1: Sign for derivatives of potential due to interchanging source and collocation point; mirror image number as in Fig. B.4

	1	2	3	4
ϕ_x	+	+	+	+
ϕ_y	+	−	+	−
ϕ_z	+	+	+	+
ϕ_{xx}	+	+	+	+
ϕ_{xy}	+	−	+	−
ϕ_{xz}	+	+	+	+
ϕ_{yy}	+	+	+	+
ϕ_{yz}	+	−	+	−
ϕ_{xxz}	+	+	+	+
ϕ_{xyz}	+	−	+	−
ϕ_{xzz}	+	+	+	+
ϕ_{yyz}	+	+	+	+
ϕ_{yzz}	+	−	+	−

ships the grids are not staggered in a strict sense as, with the exception of the very ends, collocation points always lie directly under panel centers.

The pressure integral for the x force – the procedure for the z force and the y moment are corresponding – is evaluated approximately by:

$$\int_S pn_1 \, dS \approx 2\sum_{i=1}^{N_B} p_i n_{1,i} s_i$$

N_B is the number of elements on the hull. The pressure p_i and the inward normal on the hull n_i are taken constant over each panel. The factor 2 is due to the port/starboard symmetry.

The non-linear solution makes it necessary to discretize the ship above the still waterline. The grid can then be transformed (regenerated) such that it always follows the actually wetted surface of the ship. However, this requires fully automatic grid generation, which is difficult on complex ship geometries prefering to discretize a ship initially to a line $z =$ const. above the free waterline. Then the whole grid can trim and sink relative to the free surface, as the grids on free surface and ship do not have to match. Then in each step, the actually wetted part of the ship grid has to be determined. The wetted area of each panel can be determined as follows.

A panel is subdivided into triangles. Each triangle is formed by one side of the panel and the panel center. Bernoulli's equation correlates the velocity in a panel to a height z_w where the pressure would equal atmospheric pressure:

$$z_w = \frac{1}{2g}(V^2 - (\nabla\phi)^2)$$

If z_w lies above the highest point of the triangle, s_i is taken as the triangle area. If z_w lies below the lowest point of the triangle, $s_i = 0$. If z_w lies between the highest and the lowest point of the triangle, the triangle is partially submerged and pierces the water surface (Fig. B.4).

In this case, the line z_w divides the triangle into a subtriangle ABC and the remaining trapezoid. If the triangle ABC is submerged (left case) s_i is taken to the area of ABC, otherwise to the triangle area minus ABC. The value of z in the pressure integral (e.g. for the hydrostatic contribution) is taken from the center of the submerged partial area.

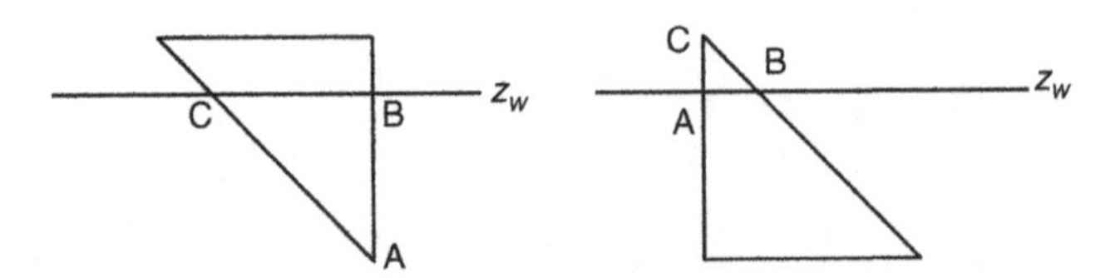

Figure B.4:
Partially submerged triangle with subtriangle ABC submerged (left) or surfaced (right)

If a panel at the upper limit of discretization is completely submerged, the discretization was chosen too low. The limit of upper discretization is given for the trimmed ship by:

$$z = m_{\mathrm{sym}}x + n_{\mathrm{sym}}$$

Strip Method Module (Two-Dimensional)

Strip methods as discussed in Section 4.4.2, Chapter 4, are the standard tool in evaluating ship seakeeping. An essential part of each strip method is the computation of hydrodynamic masses, damping, and exciting forces for each strip. This computation was traditionally based on conformal mapping techniques, where an analytical solution for a semicircle was transformed to a shape resembling a ship section. This technique is not capable of reproducing complex shapes as found in the forebody of modern ships, where possibly cross-section may consist of unconnected parts for bulbous bow and upper stem. Numerical 'close-fit' methods became available with the advent of computers in naval architecture and are now widely used in practice. In the following, one example of such a close-fit method to solve the two-dimensional strip problem is presented. The Fortran source code for the method is available on the internet (www.bh.com/companions/0750648511).

We compute the radiation and diffraction problems for a two-dimensional cross-section of arbitrary shape in harmonic, elementary waves. As usual, we assume an ideal fluid. Then there exists a velocity potential ϕ such that the partial derivatives of this potential in an arbitrary direction yield the velocity component of the flow in that direction. We neglect all non-linear effects in our computations. The problem is formulated in a coordinate system as shown in Fig. B.5. Indices y, z, and t denote partial derivatives with respect to these variables.

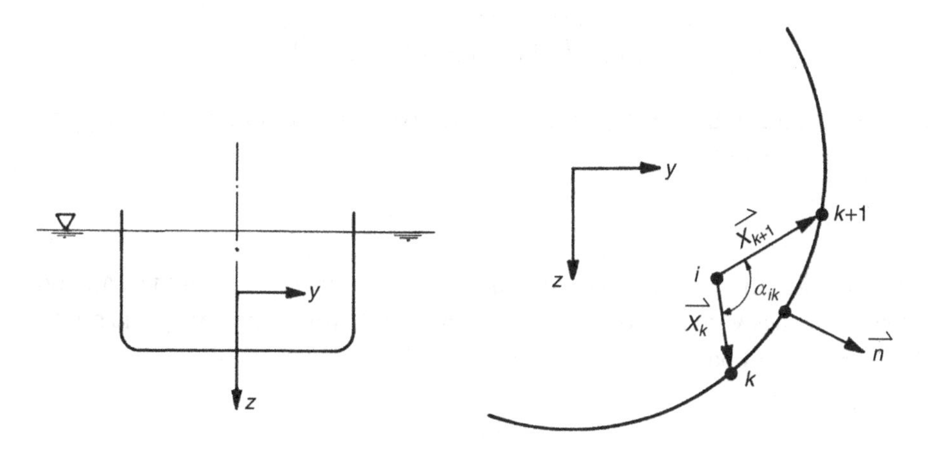

Figure B.5:
Flow chart of iterative solution

We solve the problem in the frequency domain. The two-dimensional seakeeping potentials will then be harmonic functions oscillating with encounter frequency ω_e:

$$\phi(y,z,t) = \mathrm{Re}(\hat{\phi}(y,z)e^{i\omega_e t})$$

The potential must fulfill the Laplace equation:

$$\phi_{yy} + \phi_{zz} = 0$$

in the whole fluid domain ($z < 0$) subject to the following boundary conditions:

1. Decaying velocity with water depth:

$$\lim_{z\to\infty} \nabla\phi = 0$$

2. There is atmospheric pressure everywhere on the free surface $z = \zeta$ (dynamic condition). Then Bernoulli's equation yields:

$$\phi_t + \frac{1}{2}(\nabla\phi)^2 - g\zeta = 0$$

3. There is no flow through the free surface (kinematic condition), i.e. the local vertical velocity of a particle coincides with the rate of change of the surface elevation in time:

$$\phi_z = \zeta_t$$

4. Differentiation of the dynamic condition with respect to time and combination with the kinematic condition yields:

$$\phi_{tt} + \phi_y\phi_{yt} + \phi_z\phi_{zt} - g\phi_z = 0$$

This expression can be developed in a Taylor expansion around $z = 0$. Omitting all non-linear terms then yields:

$$\phi_{tt} - g\phi_z = 0$$

5. There is no flow through the body contour, i.e. the normal velocity of the water on the body contour coincides with the normal velocity of the hull (or, respectively, the relative normal velocity between body and water is zero):

$$\vec{n}\cdot\nabla\phi = \vec{n}\cdot\vec{v}$$

Here $\vec{v}$ is the velocity of the body, $\vec{n}$ is the outward unit normal vector.

6. Waves created by the body must radiate away from the body:

$$\lim_{|y|\to\infty} \phi = \mathrm{Re}(\hat{\varphi} e^{-kz} e^{i(\omega_e t - k|y|)})$$

$\hat{\varphi}$ here is a yet undetermined, but constant, amplitude.

Using the harmonic time dependency of the potential, we can reformulate the Laplace equation and all relevant boundary conditions such that only the time-independent complex amplitude of the potential $\hat{\phi}$ appears:

Laplace equation:

$$\hat{\phi}_{yy} + \hat{\phi}_{zz} = 0 \quad \text{for } z < 0$$

Decay condition:

$$\lim_{z\to\infty} \nabla\hat{\phi} = 0$$

Combined free-surface condition:

$$\frac{\omega_e^2}{g}\hat{\phi} + \hat{\phi}_z = 0 \quad \text{at } z = 0$$

The body boundary condition here is explicitly given for the radiation problem of the body in heave motion. This will serve as an example. The other motions (sway, roll) and the diffraction problem are treated in a very similar fashion. The body boundary condition for heave is then:

$$\vec{n}\nabla\hat{\phi} = i\omega_e n_2$$

n_2 is the z component of the (two-dimensional) normal vector $\vec{n}$.

The radiation condition for $\hat{\phi}$ is derived by differentiation of the initial radiation condition for ϕ with respect to y and z, respectively. The resulting two equations allow the elimination of the unknown constant amplitude $\hat{\varphi}$, yielding:

$$i\hat{\phi}_z = \mathrm{sign}\,(y)\cdot\hat{\phi}_y$$

$\hat{\phi}$ can be expressed as the superposition of a finite number n of point source potentials (see Appendix A). The method described here uses desingularized sources located (a small distance) inside the body and above the free surface. The grid on the free surface extends to a sufficient distance to both sides depending on the wavelength of the created wave. Due to symmetry, sources at y_i, z_i should have the same strength as sources at $-y_i$, z_i. (For sway and roll motion, we have antisymmetrical source strength.) We then exploit symmetry and use source pairs as elements to represent the total potential:

$$\hat{\phi}(y,z) = \sum_{i=1}^{n} \sigma_i \varphi_i$$

$$\varphi_i = \frac{1}{4\pi} \ln[(y-y_i)^2 + (z-z_i)^2] + \frac{1}{4\pi} \ln[(y+y_i)^2 + (z-z_i)^2]$$

This formulation automatically fulfills the Laplace equation and the decay condition. The body, free surface, and radiation conditions are fulfilled numerically by adjusting the element strengths σ_i appropriately. We enforce these conditions only on points $y_i > 0$. Due to symmetry, they will then also be fulfilled automatically for $y_i < 0$.

The method described here uses a patch method to numerically enforce the boundary conditions (see Appendix A). The body boundary condition is then integrated over one patch, e.g. between the points k and $k + 1$ on the contour (Fig. B.5):

$$\sum_{i=1}^{n} \sigma_i \int_{P_k}^{P_{k+1}} \nabla \varphi_i \vec{n}_k \, \mathrm{d}S = i\omega_e \int_{P_k}^{P_{k+1}} n_2 \, \mathrm{d}S$$

As n_2 can be expressed as $n_2 = \mathrm{d}y/\mathrm{d}s$, this yields:

$$\sum_{i=1}^{n} \sigma_i \int_{P_k}^{P_{k+1}} \nabla \varphi_i \vec{n}_k \, \mathrm{d}S = i\omega_e (y_{k+1} - y_k)$$

The integral on the l.h.s. describes the flow per time (flux) through the patch (contour section) under consideration due to a unit source at y_i, z_i and its mirror image. The flux for just the source without its image corresponds to the portion of the angle α_{ik} (Fig. B.5):

$$\int_{P_k}^{P_{k+1}} \nabla \varphi_i \vec{n}_k \, \mathrm{d}S = \frac{\alpha_{ik}}{2\pi}$$

Correspondingly we write for the elements formed by a pair of sources:

$$\int_{P_k}^{P_{k+1}} \nabla \varphi_i \vec{n}_k \, \mathrm{d}S = \frac{\alpha_{ik}^{+}}{2\pi} + \frac{\alpha_{ik}^{-}}{2\pi}$$

The angle α_{ik}^{+} is determined by:

$$\alpha_{ik} = \arctan \left(\frac{\vec{x}_{k+1} \times \vec{x}_k}{\vec{x}_{k+1} \cdot \vec{x}_k} \right)_1$$

The index 1 here denotes the x component of the vector.

The other numerical conditions can be formulated in an analogous way. The number of patches corresponds to the number of elements. The patch conditions then form a system of linear equations for the unknown element strengths σ_i, which can be solved straightforwardly. Once the element strengths are known, the velocity can be computed everywhere. The pressure integration for the patch method, then yields the forces on the section. The forces can then again be decomposed into exciting forces (for diffraction) and radiation forces expressed as added mass and damping coefficients analogous to the decomposition described in Section 4.4, Chapter 4. The method has been encoded in the Fortran routines HMASSE and WERREG (see www.bh.com/companions/0750648511).

Rankine Panel Method in the Frequency Domain

Theory

The seakeeping method is limited theoretically to $\tau > 0.25$. In practice, accuracy problems may occur for $\tau < 0.4$. The method does not treat transom sterns. The theory given is that behind the FREDDY code (Bertram 1998).

We consider a ship moving with mean speed V in a harmonic wave of small amplitude h with $\tau = V\omega_e/g > 0.25$. ω_e is the encounter frequency, $g = 9.81$ m/s^2. The resulting (linearized) seakeeping problems are similar to the steady wave resistance problem described previously and can be solved using similar techniques.

The fundamental field equation for the assumed potential flow is again Laplace's equation. In addition, boundary conditions are postulated:

1. No water flows through the ship's surface.
2. At the trailing edge of the ship, the pressures are equal on both sides. (Kutta condition.)
3. A transom stern is assumed to remain dry. (Transom condition.)
4. No water flows through the free surface. (Kinematic free surface condition.)
5. There is atmospheric pressure at the free surface. (Dynamic free surface condition.)
6. Far away from the ship, the disturbance caused by the ship vanishes.
7. Waves created by the ship move away from the ship. For $\tau > 0.25$, waves created by the ship propagate only downstream. (Radiation condition.)
8. Waves created by the ship should leave artificial boundaries of the computational domain without reflection. They may not reach the ship again. (Open-boundary condition.)
9. Forces on the ship result in motions. (Average longitudinal forces are assumed to be counteracted by corresponding propulsive forces, i.e. the average speed V remains constant.)

Note that this verbal formulation of the boundary conditions coincides virtually with the formulation for the steady wave resistance problem.

All coordinate systems here are right-handed Cartesian systems. The inertial *Oxyz* system moves uniformly with velocity *V*. *x* points in the direction of the body's mean velocity *V*, *z* points vertically upwards. The $\underline{Oxyz}$ system is fixed at the body and follows its motions. When the body is at rest position, $\underline{x}$, $\underline{y}$, $\underline{z}$ coincide with *x*, *y*, *z*. The angle of encounter μ between body and incident wave is defined such that $\mu = 180°$ denotes head sea and $\mu = 90°$ beam sea.

The body has six degrees of freedom for rigid body motion. We denote corresponding to the degrees of freedom:

u_1 surge motion of $\underline{O}$ in the *x* direction, relative to *O*

u_2 heave motion of $\underline{O}$ in the *y* direction, relative to *O*

u_3 heave motion of $\underline{O}$ in the *z* direction, relative to *O*

u_4 angle of roll = angle of rotation around the *x*-axis

u_5 angle of pitch = angle of rotation around the *y*-axis

u_6 angle of yaw = angle of rotation around the *z*-axis

The motion vector $\vec{u}$ and the rotational motion vector $\vec{\alpha}$ are given by:

$$\vec{u} = \{u_1, u_2, u_3\}^T \quad \text{and} \quad \vec{\alpha} = \{u_4, u_5, u_6\}^T = \{\alpha_1, \alpha_2, \alpha_3\}^T$$

All motions are assumed to be small, of order $O(h)$. Then for the three angles α_i, the following approximations are valid: $\sin(\alpha_i) = \tan(\alpha_i) = \alpha_i$, $\cos(\alpha_i) = 1$.

The relation between the inertial and the hull-bound coordinate system is given by the linearized transformation equations:

$$\vec{x} = \underline{\vec{x}} + \vec{\alpha} \times \underline{\vec{x}} + \vec{u}$$

$$\underline{\vec{x}} = \vec{x} - \vec{\alpha} \times \vec{x} - \vec{u}$$

Let $\vec{v} = \vec{v}(\vec{x})$ be any velocity relative to the *Oxyz* system and $\underline{\vec{v}} = \underline{\vec{v}}(\underline{x})$ the velocity relative to the $\underline{Oxyz}$ system where $\vec{x}$ and $\underline{\vec{x}}$ describe the same point.

Then the velocities transform:

$$\vec{v} = \underline{\vec{v}} + \vec{\alpha} \times \underline{\vec{v}} + (\vec{\alpha}_t \times \underline{\vec{x}} + \vec{u}_t)$$

$$\underline{\vec{v}} = \vec{v} - \vec{\alpha} \times \vec{v} + (\vec{\alpha}_t \times \vec{x} + \vec{u}_t)$$

The differential operators ∇_x and $\nabla_{\underline{x}}$ transform:

$$\nabla_x = \{\partial/\partial x, \partial/\partial y, \partial/\partial z\}^T = \nabla_{\underline{x}} + \vec{\alpha} \times \nabla_{\underline{x}}$$

$$\nabla_{\underline{x}} = \{\partial/\partial \underline{x}, \partial/\partial \underline{y}, \partial/\partial \underline{z}\}^T = \nabla_x - \vec{\alpha} \times \nabla_x$$

Using a three-dimensional truncated Taylor expansion, a scalar function transforms from one coordinate system into the other:

$$f(\vec{x}) = f(\vec{\underline{x}}) + (\vec{\alpha} \times \vec{\underline{x}} + \vec{u})\nabla_{\underline{x}} f(\vec{\underline{x}})$$

$$f(\vec{\underline{x}}) = f(\vec{x}) - (\vec{\alpha} \times \vec{x} + \vec{u})\nabla_x f(\vec{x})$$

Correspondingly we write:

$$\nabla_x f(\vec{x}) = \nabla_{\underline{x}} f(\vec{\underline{x}}) + ((\vec{\alpha} \times \vec{\underline{x}} + \vec{u})\nabla_{\underline{x}})\nabla_{\underline{x}} f(\vec{\underline{x}})$$

$$\nabla_{\underline{x}} f(\vec{\underline{x}}) = \nabla_x f(\vec{x}) + ((\vec{\alpha} \times \vec{x} + \vec{u})\nabla_x)\nabla_x f(\vec{x})$$

A perturbation formulation for the potential is used:

$$\phi^{\text{total}} = \phi^{(0)} + \phi^{(1)} + \phi^{(2)} + \cdots$$

$\phi^{(0)}$ is the part of the potential which is independent of the wave amplitude h. It is the solution of the steady wave resistance problem described in the previous section (where it was denoted by just ϕ). $\phi^{(1)}$ is proportional to h, $\phi^{(2)}$ proportional to h^2, etc. Within a theory of first order (linearized theory), terms proportional to h^2 or higher powers of h are neglected. For reasons of simplicity, the equality sign is used here to denote equality of low-order terms only, i.e. $A = B$ means $A = B + O(h^2)$.

We describe both the z-component of the free surface ζ and the potential in a first-order formulation. $\phi^{(1)}$ and $\zeta^{(1)}$ are time harmonic with ω_e, the frequency of encounter:

$$\begin{aligned}\phi^{\text{total}}(x, y, z; t) &= \phi^{(0)}(x, y, z) + \phi^{(1)}(x, y, z; t) \\ &= \phi^{(0)}(x, y, z) + \text{Re}(\hat{\phi}^{(1)}(x, y, z)e^{i\omega_e t})\end{aligned}$$

$$\begin{aligned}\zeta^{\text{total}}(x, y; t) &= \zeta^{(0)}(x, y) + \zeta^{(1)}(x, y; t) \\ &= \zeta^{(0)}(x, y) + \text{Re}(\hat{\zeta}^{(1)}(x, y)e^{i\omega_e t})\end{aligned}$$

Correspondingly the symbol ^ is used for the complex amplitudes of all other first-order quantities, such as motions, forces, pressures, etc.

The superposition principle can be used within a linearized theory. Therefore the radiation problems for all six degrees of freedom of the rigid-body motions and the diffraction problem are solved separately. The total solution is a linear combination of the solutions for each independent problem.

The harmonic potential $\phi^{(1)}$ is divided into the potential of the incident wave $\phi^{(w)}$, the diffraction potential ϕ^d, and six radiation potentials:

$$\phi^{(1)} = \phi^d + \phi^w + \sum_{i=1}^{6} \phi^i u_i$$

It is convenient to decompose ϕ^w and ϕ^d into symmetrical and antisymmetrical parts to take advantage of the (usual) geometrical symmetry:

$$\phi^w(x,y,z) = \underbrace{\frac{\phi^w(x,y,z) + \phi^w(x,-y,z)}{2}}_{\phi^{w,s}} + \underbrace{\frac{\phi^w(x,y,z) - \phi^w(x,-y,z)}{2}}_{\phi^{w,a}}$$

$$\phi^d = \phi^{d,s} + \phi^{d,a} = \phi^7 + \phi^8$$

Thus:

$$\phi^{(1)} = \phi^{w,s} + \phi^{w,a} + \sum_{i=1}^{6} \phi^i u_i + \phi^7 + \phi^8$$

The conditions satisfied by the steady flow potential $\phi^{(0)}$ are repeated here without further comment.

The particle acceleration in the steady flow is:	$\vec{a}^{(0)} = (\nabla\phi^{(0)}\nabla)\nabla\phi^{(0)}$
We define an acceleration vector $\vec{a}^g$	$\vec{a}^g = \vec{a}^{(0)} + \{0,0,g\}^T$
For convenience I introduce an abbreviation:	$B = \frac{1}{a_3^g}\frac{\partial}{\partial z}(\nabla\phi^{(0)}\vec{a}^g)$
In the whole fluid domain:	$\Delta\phi^{(0)} = 0$
At the steady free surface:	$\nabla\phi^{(0)}\vec{a}^g = 0$
	$\frac{1}{2}(\nabla\phi^{(0)})^2 + g\zeta^{(0)} = \frac{1}{2}V^2$
On the body surface:	$\underline{n}(\underline{x})\nabla\phi^{(0)}(\underline{x}) = 0$

Also, suitable radiation and decay conditions are observed.

The linearized potential of the incident wave on water of infinite depth is expressed in the inertial system:

$$\phi^w = \mathrm{Re}\left(-\frac{igh}{\omega}\mathrm{e}^{-ik(x\cos\mu - y\sin\mu)-kz}\mathrm{e}^{i\omega_e t}\right) = \mathrm{Re}(\hat{\phi}^w \mathrm{e}^{i\omega_e t})$$

$\omega = \sqrt{gk}$ is the frequency of the incident wave, $\omega_e = |\omega - kV\cos\mu|$ the frequency of encounter, k is the wave number. The derivation of the expression for ϕ^w assumes a linearization around $z = 0$. The same formula will be used now in the seakeeping computations, although the average boundary is at the steady wave elevation, i.e. different near the ship. This may be an inconsistency, but the diffraction potential should compensate this 'error'.

We write the complex amplitude of the incident wave as:

$$\hat{\phi}^w = -\frac{igh}{\omega}e^{\vec{x}\vec{d}} \quad \text{with} \quad \vec{d} = \{-ik\cos\mu,\ ik\sin\mu,\ -kz\}^T$$

At the free surface ($z = \zeta^{\text{total}}$) the pressure is constant, namely atmospheric pressure ($p = p_0$):

$$\frac{D(p-p_0)}{Dt} = \frac{\partial(p-p_0)}{\partial t} + (\nabla\phi^{\text{total}}\nabla)(p-p_0) = 0$$

Bernoulli's equation gives at the free surface ($z = \zeta^{\text{total}}$) the dynamic boundary condition:

$$\phi_t^{\text{total}} + \frac{1}{2}(\nabla\phi^{\text{total}})^2 + g\zeta^{\text{total}} + \frac{p}{\rho} = \frac{1}{2}V^2 + \frac{p_0}{\rho}$$

The kinematic boundary condition gives at $z = \zeta^{\text{total}}$:

$$\frac{D\zeta^{\text{total}}}{Dt} = \frac{\partial}{\partial t}\zeta^{\text{total}} + (\nabla\phi^{\text{total}}\nabla)\zeta^{\text{total}} = \phi_z^{\text{total}}$$

Combining the above three equations yields at $z = \zeta^{\text{total}}$:

$$\phi_{tt}^{\text{total}} + 2\nabla\phi^{\text{total}}\nabla\phi_t^{\text{total}} + \nabla\phi^{\text{total}}\nabla\left(\frac{1}{2}\nabla\phi^{\text{total}}\right)^2 + g\phi_z^{\text{total}} = 0$$

Formulating this condition in $\phi^{(0)}$ and $\phi^{(1)}$ and linearizing with regard to instationary terms gives at $z = \zeta^{\text{total}}$:

$$\phi_{tt}^{(1)} + 2\nabla\phi^{(0)}\nabla\phi_t^{(1)} + \nabla\phi^{(0)}\nabla\left(\frac{1}{2}(\nabla\phi^{(0)})^2 + \nabla\phi^{(1)}\nabla\phi^{(0)}\right)$$
$$+\nabla\phi^{(1)}\nabla\left(\frac{1}{2}(\nabla\phi^{(0)})^2\right) + g\phi_z^{(0)} + g\phi_z^{(1)} = 0$$

We develop this equation in a linearized Taylor expansion around $\zeta^{(0)}$ using the abbreviations $\vec{a}, \vec{a}^g$, and B for steady flow contributions. This yields at $z = \zeta^0$:

$$\phi_{tt}^{(1)} + 2\nabla\phi^{(0)}\nabla\phi_t^{(1)} + \nabla\phi^{(0)}\vec{a}^g + \nabla\phi^{(0)}(\nabla\phi^{(0)}\nabla)\nabla\phi^{(1)} + \nabla\phi^{(1)}(\vec{a} + \vec{a}^g) + Ba_3^g\zeta^{(1)} = 0$$

The steady boundary condition can be subtracted, yielding:

$$\phi_{tt}^{(1)} + 2\nabla\phi^{(0)}\nabla\phi_t^{(1)} + \nabla\phi^{(0)}(\nabla\phi^{(0)}\nabla)\nabla\phi^{(1)} + \nabla\phi^{(1)}(\vec{a} + \vec{a}^g) + Ba_3^g\zeta^{(1)} = 0$$

$\zeta^{(1)}$ will now be substituted by an expression depending solely on $\zeta^{(0)}, \phi^{(0)}(\zeta^{(0)})$ and $\phi^{(1)}(\zeta^{(0)})$. To this end, Bernoulli's equation is also developed in a Taylor expansion. Bernoulli's equation yields at $z = \zeta^{(0)} + \zeta^{(1)}$:

$$\phi_t^{\text{total}} + \frac{1}{2}(\nabla\phi^{\text{total}})^2 + g\zeta^{\text{total}} = \frac{1}{2}V^2$$

A truncated Taylor expansion gives at $z = \zeta^{(0)}$:

$$\phi_t^{(1)} + \frac{1}{2}(\nabla\phi^{\text{total}})^2 + g\zeta^{(0)} - \frac{1}{2}V^2 + (\nabla\phi^{\text{total}}\phi_z^{\text{total}} + g)\zeta^{(1)} = 0$$

Formulating this condition in $\phi^{(0)}$ and $\phi^{(1)}$, linearizing with regard to instationary terms and subtracting the steady boundary condition yields:

$$\phi_t^{(1)} + \nabla\phi^{(0)}\nabla\phi^{(1)} + a_3^g\zeta^{(1)} = 0$$

This can be reformulated as:

$$\zeta^{(1)} = \frac{\phi_t^{(1)} + \nabla\phi^{(0)}\nabla\phi^{(1)}}{a_3^g}$$

By inserting this expression in the free-surface condition and performing the time derivatives leaving only complex amplitudes, the free-surface condition at $z = \zeta^{(0)}$ becomes:

$$(-\omega_e^2 + Bi\omega_e)\hat{\phi}^{(1)} + ((2i\omega_e + B)\nabla\phi^{(0)} + \vec{a}^{(0)} + \vec{a}^g)\nabla\hat{\phi}^{(1)} + \nabla\phi^{(0)}(\nabla\phi^{(0)}\nabla)\nabla\hat{\phi}^{(1)} = 0$$

The last term in this condition is explicitly written:

$$\begin{aligned}\nabla\phi^{(0)}(\nabla\phi^{(0)}\nabla)\nabla\hat{\phi}^{(1)} = {} & (\phi_x^{(0)})^2\phi_{xx}^{(1)} + (\phi_y^{(0)})^2\phi_{yy}^{(1)} + (\phi_z^{(0)})^2\phi_{zz}^{(1)} + 2\cdot(\phi_x^{(0)}\phi_y^{(0)}\phi_{xy}^{(1)} \\ & + \phi_x^{(0)}\phi_z^{(0)}\phi_{xz}^{(1)} + \phi_y^{(0)}\phi_z^{(0)}\phi_{yz}^{(1)})\end{aligned}$$

Complications in formulating the kinematic boundary condition on the body's surface arise from the fact that the unit normal vector is conveniently expressed in the body-fixed coordinate system, while the potential is usually given in the inertial system. The body surface is defined in the body-fixed system by the relation $\underline{S}(\underline{\vec{x}}) = 0$.

Water does not penetrate the body's surface, i.e. relative to the body-fixed coordinate system the normal velocity is zero, at $\underline{S}(\underline{\vec{x}}) = 0$:

$$\underline{\vec{n}}(\underline{\vec{x}}) \cdot \underline{\vec{v}}(\underline{\vec{x}}) = 0$$

$\underline{\vec{n}}$ is the inward unit normal vector. The velocity transforms into the inertial system as:

$$\underline{\vec{v}}(\underline{\vec{x}}) = \vec{v}(\vec{x}) - \vec{\alpha} \times \vec{v}(\vec{x}) - (\vec{\alpha}_t \times \vec{x} + \vec{u}_t)$$

where $\vec{x}$ is the inertial system description of the same point as $\underline{\vec{x}}$. $\vec{v}$ is expressed as the sum of the derivatives of the steady and the first-order potential:

$$\vec{v}(\vec{x}) = \nabla\phi^{(0)}(\vec{x}) + \nabla\phi^{(1)}(\vec{x})$$

For simplicity, the subscript x for the ∇ operator is dropped. It should be understood that from now on the argument of the ∇ operator determines its type, i.e. $\nabla\phi(\vec{x}) = \nabla_x\phi(\vec{x})$ and $\nabla\phi(\underline{\vec{x}}) = \nabla_{\underline{x}}\phi(\underline{\vec{x}})$. As $\phi^{(1)}$ is of first order small, $\phi^{(1)}(\underline{\vec{x}}) = \phi^{(1)}(\vec{x}) = \phi^{(1)}$.

The r.h.s. of the above equation for $\vec{v}(\vec{x})$ transforms back into the hull-bound system:

$$\vec{v}(\vec{x}) = \nabla\phi^{(0)}(\underline{\vec{x}}) + ((\vec{\alpha} \times \underline{\vec{x}} + \vec{u})\nabla)\nabla\phi^{(0)}(\underline{\vec{x}}) + \nabla\phi^{(1)}$$

Combining the above equations and omitting higher-order terms yields:

$$\underline{\vec{n}}(\underline{\vec{x}})(\nabla\phi^{(0)}(\underline{\vec{x}}) - \vec{\alpha} \times \nabla\phi^{(0)} + ((\vec{\alpha} \times \underline{\vec{x}} + \vec{u})\nabla)\nabla\phi^{(0)} + \nabla\phi^{(1)} - (\vec{\alpha}_t \times \vec{x} + \vec{u}_t)) = 0$$

This boundary condition must be fulfilled at any time. The steady terms give the steady body-surface condition as mentioned above. Because only terms of first order are left, we can exchange $\vec{x}$ and $\underline{\vec{x}}$ at our convenience. Using some vector identities we derive:

$$\underline{\vec{n}}\nabla\hat{\phi}^{(1)} + \hat{\vec{u}}[(\underline{\vec{n}}\nabla)\nabla\phi^{(0)} - i\omega_e\underline{\vec{n}}] + \hat{\vec{\alpha}}[\underline{\vec{n}} \times \nabla\phi^{(0)} + \underline{\vec{x}} \times ((\underline{\vec{n}}\nabla)\nabla\phi^{(0)} - i\omega_e\underline{\vec{n}})] = 0$$

where all derivatives of potentials can be taken with respect to the inertial system.

With the abbreviation $\vec{m} = (\underline{\vec{n}}\nabla)\nabla\phi^{(0)}$ the boundary condition at $\underline{S}(\underline{\vec{x}}) = 0$ becomes:

$$\underline{\vec{n}}\nabla\hat{\phi}^{(1)} + \hat{\vec{u}}(\vec{m} - i\omega_e\underline{\vec{n}}) + \hat{\vec{\alpha}}(\underline{\vec{x}} \times (\vec{m} - i\omega_e\underline{\vec{n}}) + \underline{\vec{n}} \times \nabla\phi^{(0)}) = 0$$

The Kutta condition requires that at the trailing edge the pressures are equal on both sides. This is automatically fulfilled for the symmetric contributions (for monohulls). Then only the antisymmetric pressures have to vanish:

$$-\rho(\phi_t^i + \nabla\phi^{(0)}\nabla\hat{\phi}^i) = 0$$

This yields on points at the trailing edge:

$$i\omega_e\hat{\phi}^i + \nabla\phi^{(0)}\nabla\hat{\phi}^i = 0$$

Diffraction and radiation problems for unit amplitude motions are solved independently as described in the next section. After the potentials $\hat{\phi}^i$ ($i = 1 \ldots 8$) have been determined, only the motions u_i remain as unknowns.

The forces $\vec{F}$ and moments $\vec{M}$ acting on the body result from the body's weight and from integrating the pressure over the instantaneous wetted surface S. The body's weight $\vec{G}$ is:

$$\vec{G} = \{0, 0, -mg\}^T$$

m is the body's mass. (In addition, a propulsive force counteracts the resistance. This force could be included in a similar fashion as the weight. However, resistance and propulsive force are assumed to be negligibly small compared to the other forces.)

$\vec{F}$ and $\vec{M}$ are expressed in the inertial system ($\vec{n}$ is the inward unit normal vector):

$$\vec{F} = \int_S (p(\vec{x}) - p_0)\vec{n}(\vec{x})\, dS + \vec{G}$$

$$\vec{M} = \int_S (p(\vec{x}) - p_0)(\vec{x} \times \vec{n}(\vec{x}))\, dS + \vec{x}_g \times \vec{G}$$

$\vec{x}_g$ is the center of gravity. The pressure is given by Bernoulli's equation:

$$p(\vec{x}) - p_0 = -\rho\left(\frac{1}{2}(\nabla\phi^{\text{total}}(\vec{x}))^2 - \frac{1}{2}V^2 + gz + \phi_t^{\text{total}}(\vec{x})\right)$$

$$= \underbrace{-\rho\left(\frac{1}{2}(\nabla\phi^{\text{total}}(\vec{x}))^2 - \frac{1}{2}V^2 + gz\right)}_{p^{(0)}} - \underbrace{\rho(\nabla\phi^{(0)}\nabla\phi^{(1)} + \phi_t^{(1)})}_{p^{(1)}}$$

The r.h.s. of the expressions for $\vec{F}$ and $\vec{M}$ are now transformed from the inertial system to the body-fixed system. This includes a Taylor expansion around the steady position of the body. The normal vector $\vec{n}$ and the position $\vec{x}$ are readily transformed as usual:

$$\vec{x} = \underline{\vec{x}} + \vec{\alpha} \times \underline{\vec{x}} + \vec{u}$$

$$\vec{n}(\vec{x}) = \underline{\vec{n}}(\underline{\vec{x}}) + \vec{\alpha} \times \underline{\vec{n}}(\underline{\vec{x}})$$

The steady parts of the equations give:

$$\vec{F}^{(0)} = \int_{\underline{S}^{(0)}} p^{(0)}\, \underline{\vec{n}}\, d\underline{S} + \vec{G} = 0$$

$$\vec{M}^{(0)} = \int_{\underline{S}^{(0)}} p^{(0)}(\underline{\vec{x}} \times \underline{\vec{n}})\, d\underline{S} + \underline{\vec{x}}_g \times \vec{G} = 0$$

The ship is in equilibrium for steady flow. Therefore the steady forces and moments are all zero.

The first-order parts give (r.h.s. quantities are now all functions of $\underline{\vec{x}}$):

$$\vec{F}^{(1)} = \int_{\underline{S}^{(0)}} [(p^{(1)} + \nabla p^{(0)}(\vec{\alpha} \times \underline{\vec{x}} + \vec{u})] \, \underline{\vec{n}} \, d\underline{S} - \vec{\alpha} \times \vec{G}$$

$$\vec{M}^{(1)} = \int_{\underline{S}^{(0)}} [(p^{(1)} + \nabla p^{(0)}(\vec{\alpha} \times \underline{\vec{x}} + \vec{u})](\underline{\vec{x}} \times \underline{\vec{n}}) d\underline{S} - \underline{\vec{x}}_g \times (\vec{\alpha} \times \vec{G})$$

where $(\vec{\alpha} \times \underline{\vec{x}}) \times \underline{\vec{n}} + \underline{\vec{x}} \times (\vec{\alpha} \times \underline{\vec{n}}) = \vec{\alpha} \times (\underline{\vec{x}} \times \underline{\vec{n}})$ and the expressions for $\vec{F}^{(0)}$ and $\vec{M}^{(0)}$ have been used. Note: $\nabla p^{(0)} = -\rho \vec{a}^g$. The difference between instantaneous wetted surface and average wetted surface still does not have to be considered as the steady pressure $p^{(0)}$ is small in the region of difference.

The instationary pressure is divided into parts due to the incident wave, radiation and diffraction:

$$p^{(1)} = p^w + p^d + \sum_{i=1}^{6} p^i u_i$$

Again the incident wave and diffraction contributions can be decomposed into symmetrical and antisymmetrical parts:

$$p^w = p^{w,s} + p^{w,a}$$

$$p^d = p^{d,s} + p^{d,a} = p^7 + p^8$$

Using the unit motion potentials, the pressure parts p^i are derived:

$$p^i = -\rho(\phi_t^i + \nabla\phi^{(0)} \nabla\phi^i)$$

$$p^w = -\rho(\phi_t^w + \nabla\phi^{(0)} \nabla\phi^w)$$

$$p^d = -\rho(\phi_t^d + \nabla\phi^{(0)} \nabla\phi^d)$$

The individual terms in the integrals for $\vec{F}^{(1)}$ and $\vec{M}^{(1)}$ are expressed in terms of the motions u_i, using the vector identity $(\vec{\alpha} \times \underline{\vec{x}})\vec{a}^g = \vec{\alpha}(\underline{\vec{x}} \times \vec{a}^g)$:

$$\vec{F}^{(1)} = \int_{\underline{S}^{(0)}} (p^w + p^d) \, \underline{\vec{n}} \, d\underline{S} + \sum_{i=1}^{6} \left(\int_{\underline{S}^{(0)}} p^i \, \underline{\vec{n}} \, d\underline{S} \right) u_i$$
$$+ \int_{\underline{S}^{(0)}} -\rho(\vec{u}\vec{a}^g + \vec{\alpha}(\underline{\vec{x}} \times \vec{a}^g))\vec{n} \, d\underline{S} - \vec{\alpha} \times \vec{G}$$

$$\vec{M}^{(1)} = \int_{\underline{S}^{(0)}} (p^w + p^d)(\underline{\vec{x}} \times \vec{n})\mathrm{d}\underline{S} + \sum_{i=1}^{6}\left(\int_{\underline{S}^{(0)}} p^i(\vec{x} \times \underline{\vec{n}})\mathrm{d}\underline{S}\right)u_i$$

$$- \underline{\vec{x}}_g \times (\vec{\alpha} \times \vec{G}) + \int_{\underline{S}^{(0)}} -\rho(\vec{u}\,\vec{a}^g + \vec{\alpha}(\underline{\vec{x}} \times \vec{a}^g))(\vec{x} \times \vec{n})\mathrm{d}\underline{S}$$

The relation between forces, moments and motion acceleration is:

$$\vec{F}^{(1)} = m(\vec{u}_{tt} + \vec{\alpha}_{tt} \times \underline{\vec{x}}_g)$$

$$\vec{M}^{(1)} = m(\underline{\vec{x}}_g \times \vec{u}_{tt}) + I\vec{\alpha}_{tt}$$

I is the matrix of moments of inertia:

$$I = \begin{bmatrix} \Theta_{\underline{x}} & 0 & -\Theta_{\underline{xz}} \\ 0 & \Theta_{\underline{y}} & 0 \\ -\Theta_{\underline{xz}} & 0 & \Theta_{\underline{z}} \end{bmatrix}$$

where mass distribution symmetrical in y is assumed. $\Theta_{\underline{x}}$ etc. are the moments of inertia and the centrifugal moments with respect to the origin of the body-fixed $\underline{Oxyz}$ system:

$$\Theta_{\underline{x}} = \int (\underline{y}^2 + \underline{z}^2)\,dm; \quad \Theta_{\underline{x}\,\underline{y}} = \int \underline{x}\,\underline{y}\,dm; \quad \text{etc.}$$

Combining the above equations for $\vec{F}^{(1)}$ and $\vec{M}^{(1)}$ yields a linear system of equations in the unknown u_i that is quickly solved using Gauss elimination.

Numerical Implementation

Systems of equations for unknown potentials

The two unknown diffraction potentials and the six unknown radiation potentials are determined by approximating the unknown potentials by a superposition of a finite number of Rankine higher-order panels on the ship and above the free surface. For the antisymmetric cases, in addition Thiart elements (Appendix A) are arranged and a Kutta condition is imposed on collocation points at the last column of collocation points on the stern. Radiation and open-boundary conditions are fulfilled by the 'staggering' technique (adding one row of collocation points at the upstream end of the free-surface grid and one row of source elements at the downstream end of the free-surface grid). This technique only works well for $\tau > 0.4$.

Elements use mirror images at $y = 0$ and for shallow water at $z = z_{\text{bottom}}$. For the symmetrical cases, all mirror images have the same strength. For the antisymmetrical case, the mirror images on the negative y sector(s) have negative element strength of the same absolute magnitude.

Each unknown potential is then written as:

$$\hat{\phi}^i = \sum \hat{\sigma}_i \varphi$$

σ_i is the strength of the ith element, φ the potential of an element of unit strength. φ is real for the Rankine elements and complex for the Thiart elements.

The same grid on the hull may be used as for the steady problem, but the grid on the free surface should be created new depending on the wave length of the incident wave. The quantities on the new grid can be interpolated within the new grid from the values on the old grid. Outside the old grid in the far field, all quantities are set to uniform flow on the new grid.

For the boundary condition on the free surface, we introduce the following abbreviations:

$$f_q = -\omega_e^2 + i\omega_e B$$

$$f_{qx} = (2i\omega_e + B)\phi_x^{(0)} + 2a_1$$

$$f_{qy} = (2i\omega_e + B)\phi_y^{(0)} + 2a_2$$

$$f_{qz} = (2i\omega_e + B)\phi_z^{(0)} + 2a_3$$

$$f_{qxx} = \phi_x^{(0)} \cdot \phi_x^{(0)} - \phi_z^{(0)} \cdot \phi_z^{(0)}$$

$$f_{qxy} = 2 \cdot \phi_x^{(0)} \cdot \phi_y^{(0)}$$

$$f_{qxz} = 2 \cdot \phi_x^{(0)} \cdot \phi_z^{(0)}$$

$$f_{qyy} = \phi_y^{(0)} \cdot \phi_y^{(0)} - \phi_z^{(0)} \cdot \phi_z^{(0)}$$

$$f_{qyz} = 2 \cdot \phi_y^{(0)} \cdot \phi_z^{(0)}$$

Then we can write the free-surface condition for the radiation cases ($i = 1 \ldots 6$):

$$\sum \hat{\sigma}_i (f_q\varphi + f_{qx}\varphi_x + f_{qy}\varphi_y + f_{qz}\varphi_z + f_{qxx}\varphi_{xx} + f_{qxy}\varphi_{xy} + f_{qxz}\varphi_{xz} + f_{qyy}\varphi_{yy} + f_{qyz}\varphi_{yz}) = 0$$

where it has been exploited that all potentials fulfill Laplace's equation. Similarly, we get for the symmetrical diffraction problem:

$$\begin{aligned}
&\sum \hat{\sigma}_i (f_q\varphi + f_{qx}\varphi_x + f_{qy}\varphi_y + f_{qz}\varphi_z + f_{qxx}\varphi_{xx} + f_{qxy}\varphi_{xy} + f_{qxz}\varphi_{xz} + f_{qyy}\varphi_{yy} + f_{qyz}\varphi_{yz}) \\
&\quad + f_q\hat{\phi}^{w,s} + f_{qx}\hat{\phi}_x^{w,s} + f_{qy}\hat{\phi}_y^{w,s} + f_{qz}\hat{\phi}_z^{w,s} + f_{qxx}\hat{\phi}_{xx}^{w,s} + f_{qxy}\hat{\phi}_{xy}^{w,s} + f_{qxz}\hat{\phi}_{xz}^{w,s} + f_{qyy}\hat{\phi}_{yy}^{w,s} \\
&\quad + f_{qyz}\hat{\phi}_{yz}^{w,s} = 0
\end{aligned}$$

The expression for the antisymmetrical diffraction problem is written correspondingly using $\hat{\phi}^{w,a}$ on the r.h.s.

Hull condition

For the hull conditions for the eight radiation and diffraction problems, we introduce the following abbreviations, where the auxiliary variable h is used as a local variable with different meaning than further below for the system of equations for the motions:

$$\{h_1, h_2, h_3\}^T = \vec{m} \; - i\omega_e \underline{\vec{n}}$$

$$\{h_4, h_5, h_6\}^T = \underline{\vec{x}} \times (\vec{m} - i\omega_e \underline{\vec{n}}) + \underline{\vec{n}} \times \nabla\phi^{(0)}$$

$$h_7 = \nabla\hat{\phi}^{w,s} \cdot \underline{\vec{n}}$$

$$h_8 = \nabla\hat{\phi}^{w,a} \cdot \underline{\vec{n}}$$

Then the hull condition can be written for the jth case ($j = 1 \ldots 8$):

$$\sum \hat{\sigma}_i(\underline{\vec{n}} \cdot \varphi) + h_j = 0$$

The Kutta condition is simply written:

$$\sum \hat{\sigma}_i(i\omega_e\varphi + \nabla\phi^{(0)}\nabla\varphi) = 0 \qquad \text{for case } j = 2, 4, 6$$

$$\sum \hat{\sigma}_i(i\omega_e\varphi + \nabla\phi^{(0)}\nabla\varphi) + i\omega_e\hat{\phi}^{w,s} + \nabla\phi^{(0)}\nabla\hat{\phi}^{w,s} = 0 \qquad \text{for case } j = 8$$

The l.h.s. of the four systems of equations for the symmetrical cases and the l.h.s. for the four systems of equations for the antisymmetrical cases each share the same coefficients. Thus four systems of equations can be solved simultaneously using a Gauss elimination procedure.

System of equations for motions

We introduce the abbreviations:

$$\{h_1, h_2, h_3\}^T = -\rho\vec{a}^g$$

$$\{h_4, h_5, h_6\}^T = -\rho\underline{\vec{x}} \times \vec{a}^g$$

$$h_7 = p^{w,s} = -\rho(i\omega_e\hat{\phi}^{w,s} + \nabla\phi^{(0)}\nabla\hat{\phi}^{w,s})$$

$$h_8 = p^{w,a} = -\rho(i\omega_e\hat{\phi}^{w,a} + \nabla\phi^{(0)}\nabla\hat{\phi}^{w,a})$$

Recall that the instationary pressure contribution is:

$$p^i = -\rho(i\omega_e\hat{\phi}^i + \nabla\phi^{(0)}\nabla\hat{\phi}^i)$$

Then we can rewrite the conditions for $\vec{F}^{(1)}$ and $\vec{M}^{(1)}$:

$$-m(\vec{u}_{tt} + \vec{\alpha}_{tt} \times \vec{\underline{x}}_g) + \sum_{i=1}^{8}\left(\int_{\underline{S}^{(0)}} (p^i + h^i)\,\vec{\underline{n}}\;\mathrm{d}\underline{S}\right)u_i - \vec{\alpha} \times \vec{G} = 0$$

$$-m(\vec{\underline{x}}_g \times \vec{u}_{tt}) - I\vec{\alpha}_{tt}\sum_{i=1}^{8}\left(\int_{\underline{S}^{(0)}} (p^i + h^i)(\vec{\underline{x}} \times \vec{\underline{n}})\;\mathrm{d}\underline{S}\right)u_i - \vec{\underline{x}}_g \times (\vec{\alpha} \times \vec{G}) = 0$$

The weight terms $-\vec{\alpha} \times \vec{G}$ and $-\vec{\underline{x}}_g \times (\vec{\alpha} \times \vec{G})$ contribute with $W = mg$:

$$\begin{bmatrix} 0 & 0 & 0 & 0 & W & 0 \\ 0 & 0 & 0 & -W & 0 & 0 \\ 0 & 0 & 0 & 0 & 0 & 0 \\ 0 & 0 & 0 & \underline{z}_g W & 0 & 0 \\ 0 & 0 & 0 & 0 & \underline{z}_g W & 0 \\ 0 & 0 & 0 & \underline{x}_g W & & 0 \end{bmatrix} \begin{Bmatrix} u_1 \\ u_2 \\ u_3 \\ u_4 \\ u_5 \\ u_6 \end{Bmatrix}$$

The mass terms $-m(\hat{u}_{tt} + \hat{\alpha}_{tt} \times \vec{\underline{x}}_g)$ and $-m(\vec{\underline{x}}_g \times \hat{u}_{tt}) - I\hat{\alpha}_{tt}$ contribute:

$$-m\frac{\partial^2}{\partial t^2}\begin{bmatrix} 1 & 0 & 0 & 0 & \underline{z}_g & 0 \\ 0 & 1 & 0 & -\underline{z}_g & 0 & \underline{x}_g \\ 0 & 0 & 1 & 0 & -\underline{x}_g & 0 \\ 0 & -\underline{z}_g & 0 & k_{\underline{x}}^2 & 0 & -k_{\underline{xz}}^2 \\ \underline{z}_g & 0 & -\underline{x}_g & 0 & k_{\underline{y}}^2 & 0 \\ 0 & \underline{x}_g & 0 & -k_{\underline{xz}}^2 & 0 & k_{\underline{z}}^2 \end{bmatrix} \begin{Bmatrix} u_1 \\ u_2 \\ u_3 \\ u_4 \\ u_5 \\ u_6 \end{Bmatrix}$$

where the radii of inertia k have been introduced, e.g. $\Theta_{\underline{x}} = mk_{\underline{x}}^2$, etc.

References

Abbott, I. & Doenhoff, A. (1959). *A theory of wing sections.* Dover Publ.

Abt, C. & Harries, S. (2007). A new approach to integration of CAD and CFD for naval architects. *6th Conf. Computer and IT Applications in the Maritime Industries (COMPIT).* Cortona.

Allan, R. G. (2004). Tugs and towboats. *Ship Design & Construction, SNAME.*

Allison, J. (1993). Marine waterjet propulsion. *Trans. SNAME, 101*, 275–335.

Alte, B. & Baur, M. (1986). Propulsion. *Handbuch der Werften, XVIII*, Hansa, p. 132.

Andersen, P., Friesch, J. & Kappel, J. (2002). Development and full-scale evaluation of a new marine propeller type. In: *Jahrbuch Schiffbautechnische Gesellschaft.* Springer, pp. 465–476.

ANEp II-I. (1983). *Standardized wave and wind environments for NATO operational areas.* Allied Naval Engineering Publication.

Asmussen, I., Menzel, W. & Mumm, H. (1998). Schiffsschwingungen. *Handbuch der Werften, XXIV*, 75–147, Hansa-Verlag.

Bailey, D. (1976). The NPL high-speed round bilge displacement hull series. *Maritime Technology Monograph, 4.* RINA.

Bertram, V. (1992). Wellenwiderstandsberechnung für SWATH-Schiffe und Katamarane. In: *Jahrbuch Schiffbautechnische Gesellschaft*, Springer, pp. 343–351.

Bertram, V. (1998). Marching towards the numerical ship model basin. *Euromech Conf., 374*, 3–17, Poitiers.

Bertram, V. (1998). Numerical investigation of steady flow effects in three-dimensional seakeeping computations. *22nd Symp. Naval Hydrodynamics*, Wasshington.

Bertram, V. (1999). Panel methods for wave resistance computations. *31 Wegemt school 'CFD for ship and offshore design'*, WEGEMT, London.

Bertram, V. & Bentin, M. (2001). Tugs. *36th WEGEMT Summer School "Manoeuvring and Manoeuvring Devices"*, WEGEMT, London.

Bertram, V. & Couser, P. (2010). Aspects of selecting the appropriate CAD and CFD software. *9th Conf. Computer and IT Applications in the Maritime Industries (COMPIT).* Gubbio.

Bertram, V. & Jensen, G. (1994). Recent applications of computational fluid dynamics. *Ship Technology Research, 41/3*, 131–134.

Bertram, V. & Mesbahi, E. (2004). Estimating resistance and power of fast monohulls employing artificial neural nets, *4th Int. Conf. High-Performance Marine Vehicles (HIPER)*, Rome.

Bertram, V. & Thiart G. (1998). A Kutta condition for ship seakeeping computations with a Rankine panel method. *Ship Technology Research, 45*, 54–63.

Bertram, V. & Yasukawa, H. (1996). Rankine source methods for seakeeping problems. In: *Jahrbuch Schiffbautechnische Gesellschaft.* Springer, pp. 411–425.

Biran, A. (2003). *Ship Hydrostatics and Stability.* Oxford: Butterworth-Heinemann.

Blaurock, J. (1990). An appraisal of unconventional aftbody configurations and propulsion devices. *Marine Technology, 27/6*, 325.

Blendermann, W. (1993). Parameter identification of wind loads on ships. *Journal of Wind Engineering and Industrial Aerodynamics, 51*, 339–351.

Blendermann, W. (1996). *Wind loading of ships – Collected data from wind tunnel tests in uniform flow.* In: *IfS Report.* Univ. Hamburg. *574.*

Blendermann, W. (1998). Parameters of the long-term wind conditions above the oceans. *Ship Technology Research, 45*, 99–104.

Blount, D. L. & Clement, E. P. (1963). Resistance tests of a systematic series of planing hull forms. *Trans. SNAME, 71*, 491.

Blume, P. (1979). Experimentelle Bestimmung von Koeffizienten der wirksamen Rolldämpfung und ihre Anwendung zur Abschätzung extremer Rollwinkel. *Schiffstechnik, 26*, 3–19.

Bohlmann, H. J. (1989). Vorausberechnung des Bewegungsverhaltens von Ubooten. In: *Jahrbuch Schiffbautechnische Gesellschaft*. Springer.

Bohlmann, H. J. (1990). Berechnung hydrodynamischer Koeffizienten von Ubooten zur Vorhersage des Bewegungsverhaltens. In: *IfS Report. 513*. Univ. Hamburg.

Breslin, J. P. & Andersen, P. (1994). *Hydrodynamics of Ship Propellers*. Cambridge University Press.

Brix, J. (Ed.) (1993). *Manoeuvring Technical Manual*. Seehafen-Verlag.

Brix, J., Nolte, A. & Heinzel, S. (1971). Konstruktive Maßnahmen zur Verhinderung der Kavitation an der Rudersohle. *Hansa, 16*, 1579–1587.

Buhaug, Ø., Corbett, J. J., Endresen, Ø., Eyring, V., Faber, J., Hanayama, S., Lee, D. S., Lee, D., Lindstad, H., Markowska, A. Z., Mjelde, A., Nelissen, D., Nilsen, J., Pålsson, C., Winebrake, J. J., Wu, W. Q., & Yoshida, K. (2009). *Second IMO GHG study 2009*. Int. Maritime Organization (IMO), London.

Cao, Y., Schultz, W. & Beck, R. F. (1991). Three-dimensional desingularized boundary integral methods for potential problems. *Int. J. Num. Methods in Fluids, 12*, 785–803.

Carlton, J. (2007). *Marine Propellers and Propulsion*. Oxford: Butterworth-Heinemann.

Celik, F. (2007). A numerical study for effectiveness of a wake equalizing duct. *Ocean Engineering, 34*, 2138–2145.

Chang, B. C. & Blume, P. (1998). Survivability of damaged roro passenger vessels. *Ship Technology Research, 45*, 105–117.

Chapman, R. B. (1975). Free-surface effects for yawed surface piercing plates. *J. Ship Research, 20*, 125–132.

Chuang, S. L. (1967). Experiments on slamming of wedge-shaped bodies. *J. Ship Research, 11*, 190–198.

Chung, T. J. (1978). *Finite element analysis in fluid dynamics*. McGraw-Hill.

Clarke, D., Gedling, P. & Hine, G. (1983). The application of manoeuvring criteria in hull design using linear theory. *The Naval Architect*, 45–68.

Danckwardt, E. C. M. (1969). Ermittlung des Widerstandes von Frachtschiffen und Hecktrawlern beim Entwurf. *Schiffbauforschung, 8*, 124–160.

Donath, G. & Bryndum, L. (1988). Schwingungen auf Schiffen im Zusammenhang mit der Antriebsanlage. *Handbuch der Werften, XIX*, 126–200, Hansa-Verlag.

Doyle, J. & Loh, S. (1997). *Wave propagation in structures: Spectral analysis using fast discrete Fourier transforms*. Springer.

Faltinsen, O. (1993). *Sea loads on ships and offshore structures*. Cambridge University Press.

Faltinsen, O. (2005). *Hydrodynamics of high-speed marine vehicles*. Cambridge University Press.

Ferziger, J. H. & Peric, M. (1996). *Computational methods in fluid dynamics* (2nd ed.). Springer.

Foeth, E. J. (2008). Decreasing frictional resistance by air lubrication. *20th Int. Hiswa Symp. Yacht Design and Yacht Construction*, Amsterdam.

Fricke, W. (2002). Festigkeitsanalysen der Schiffskonstruktion mit der Methode der finiten Elemente. *Handbuch der Werften, XXVI*, 13–44, Hansa-Verlag.

Fritsch, M. & Bertram, V. (2002). Power prediction and hull improvement for conventional fast vessels, *3rd Int. Conf. High-Performance Marine Vehicles (HIPER)*, Bergen, pp. 167–177.

Gertler, M. (1954). A reanalysis of the original test data for the Taylor standard series. *DTMB report 806*. Bethesda: David Taylor Model Basin.

Granville, P. S. (1956). The viscous resistance of surface vessels and the skin friction of flat plates. *Trans. Society of Naval Architects and Marine Engineers, 64*, 209–240.

Grim, O. (1980). Propeller and vane wheel. *J. Ship Research, 24/4*, 203–226.

Grothues-Spork, H. (1988). Bilge vortex control devices and their benefits for propulsion. *Int. Shipb. Progr., 35*, 183–214.

Guldhammer, H. E. & Harvald, S. A. (1974). *Ship resistance, effect of form and principal dimensions*. Akademisk Forlag Copenhagen.

Hadler, J. B. (1966). The prediction of power performance of planing craft. *Trans. SNAME, 74*, 563–610.

Hansen, H. & Freund, M. (2010). Assistance tools for operational fuel efficiency. *9th Conf. Computer and IT Applications in the Maritime Industries (COMPIT)*, Gubbio.

Harries, S. & Tillig, F. (2011). Numerical hull series. *10th Conf. Computer and IT Applications in the Maritime Industries (COMPIT)*, Berlin.

Harries, S. & Vesting, F. (2010). Aerodynamic optimization of superstructures and components. *9th Conf. Computer and IT Applications in the Maritime Industries (COMPIT)*, Gubbio.

Helm, G. (1980). *Systematische Propulsions-Untersuchungen von Kleinschiffen*. Rep. 100. Hamburg: Forschungszentrum des Deutschen Schiffbaus.

Henschke, W. (1965). *Schiffbautechnisches Handbuch* (2nd ed.). Verlag Technik.

Hess, J. L. (1986). Review of the source panel technique for flow computation. In: *Innovative Numerical Methods in Engineering*. Springer, pp. 197–210.

Hess, J. L. (1990). Panel methods in computational fluid dynamics. *Annual Rev. Fluid Mech., 22*, 255–274.

Hess, J. L. & Smith A. M. O. (1964). Calculation of nonlifting potential flow about arbitrary three-dimensional bodies. *J. Ship Research, 8*, 22–44.

Holden, K., Fagerjord, O. & Frostad, R. (1980). Early design stage approach to reducing hull surface forces due to propeller cavitation. *Trans. SNAME, 88*, 403–442.

Hollenbach, U. (1998). Estimating resistance and propulsion for single-screw and twin-screw ships. *Ship Technology Research, 45*, 72–76.

Hollenbach, U. (1999). Estimating resistance and propulsion for single-screw and twin-screw ships in the preliminary design. *10th ICCAS Conf.*, Cambridge.

Hollenbach, U. & Friesch, J. (2007). Efficient hull forms – What can be gained, *1st Int. Conf. on Ship Efficiency*, Hamburg.

Holtrop, J. (1977). A statistical analysis of performance test results. *Int. Shipb. Progr., 24*, 23–28.

Holtrop, J. (1978). Statistical data for the extrapolation of model performance. *Int. Shipb. Progr, 25*, 122–126.

Holtrop, J. (1984). A statistical re-analysis of resistance and propulsion data. *Int. Shipbuilding Progress, 31*, 272–276.

Holtrop, J. & Mennen, G. G. (1978). A statistical power prediction method. *Int. Shipbuilding Progress, 25*, 253–256.

Holtrop, J. & Mennen, G. G. (1982). An approximate power prediction method. *Int. Shipbuilding Progress, 29*, 166–170.

Hoshino, T., Oshima, A., Fujita, K., Kuroiwa, T., Hayashi, F. & Yamazaki, E. (2004). Development of high-performance stator fin by using advanced panel method. *MHI Technical Review, 41/6*, 1–4.

Hughes, M. & Bertram, V. (1995). A higher-order panel method for 3-d free surface flows. *IfS Report 558*, Univ. Hamburg.

Inoue, S. & Kijima, K. (1978). The hydrodynamic derivatives on ship manoeuvrability in the trimmed condition. *ITTC, Vol. 2*, 87–92.

Isay, W. H. (1964). *Propellertheorie – Hydrodynamische Probleme*. Springer.

Isay, W. H. (1989). *Kavitation*. Hansa-Verlag.

ISO (1997). *Mechanical vibration and shock – Evaluation of human exposure to whole-body vibration*. ISO 2631.

ISO (2000). *Mechanical vibration. Guidelines for the measurement, reporting and evaluation of vibration with regard to habitability on passenger and merchant ships*. BS ISO 6954.

ISO (1996a). *Mechanical vibration of non-reciprocating machines – Measurements on rotating shafts and evaluation criteria*. ISO 7919.

ISO (1996b). *Mechanical vibration - Evaluation of machine vibration by measurements on non-rotating parts.* ISO 10816.
ISSC (1994). Int. Ship and Offshore Structures Congress, In: N. E. Jeffery, A. M. Kendrick (Eds.), St. John's.
ITTC (1990). Report of the powering performance committee. *19th Int. Towing Tank Conf.*, Madrid.
ITTC (1999). The specialist committee on unconventional propulsors. *22nd Int. Towing Tank Conf.*, Seoul.
Janson, C. E. (1996). *Potential flow panel methods for the calculation of free surface flows with lift.* PhD Thesis. Chalmers University of Technology.
Jensen, G. (1994). Moderne Schiffslinien. *Handbuch der Werften, XXII*, Hansa.
Jensen, G. (1988). Berechnung der stationären Potentialströmung um ein Schiff unter Berücksichtigung der nichtlinearen Randbedingung an der freien Wasseroberfläche. *IfS Report 484*, Univ. Hamburg.
Kashiwagi, M. (1997). Numerical seakeeping calculations based on slender ship theory. *Ship Technology Research, 44*, 167–192.
Keller, W. H. (1973). Extended diagrams for determining the resistance and required power for single-screw ships. *Int. Shipbuilding Progress, 20*, 133–142.
Kerwin, J. E. (1986). Marine propellers. *Ann. Rev. Fluid Mech., 18*, 387–403.
Kerwin, J. E., Kinnas, S. A., Lee, J. T. & Shih, W. Z. (1987). A surface panel method for the hydrodynamic analysis of ducted propellers. *Trans. SNAME, 95*, 93–122.
Kinnas, S. A. (1996). Theory and numerical methods for the hydrodynamic analysis of marine propulsors. *Advances in Marine Hydrodynamics,* 279–322, Comp. Mech. Publ.
Korobkin, A. (1996). Water impact problems in ship hydrodynamics. *Advances in Marine Hydrodynamics,* pp. 323–371, Comp. Mech. Publ.
Kose, K. (1982). On the new mathematical model of manoeuvring motions of a ship and applications. *Int. Shipb. Progr., 29*, 205–220.
Kracht, A. (1987). Kavitation an Rudern. In: *Jahrbuch Schiffbautechnische Gesellschaft,* pp. 167–179, Springer.
Kruppa, C. (1994). Wasserstrahlantriebe. *Handbuch der Werften, XXII*, 111–136, Hansa-Verlag.
Lackenby, H. (1963). The effect of shallow water on ship speed. *Shipbuilder and Marine Engineer, 70*, 446–450.
Lamb, T. (1969). A ship design procedure. *Marine Technology, 6*(4).
Lap, A. J. W. (1954). Diagrams for determining the resistance of single screw ships. *Int. Shipb. Progr.*, 179.
Larsson, L. (1997). CFD in ship design – Prospects and limitations. *Ship Technology Research, 44*, 133–154.
Larsson, L. & Baba, E. (1996). Ship resistance and flow computations. In M. Ohkusu (Ed.), *Advances in marine hydrodynamics* (pp. 1–75). Comp. Mech. Publ.
Larsson, L., Regnström, B., Broberg, L., Li, D. Q. & Janson, C. E. (1998). Failures, fantasies, and feats in the theoretical/numerical prediction of ship performance. *22. Symp. Naval Shiphydrodyn.*, Washington.
Launder, B. E. & Spalding, D. B. (1974). The numerical computation of turbulent flows. *Computer Methods in Applied Mechanics and Engineering, 3.*
Lehmann, E. (2000). *Grundzüge des Schiffbaus Bd 1.* Natural-Verlag, TU Hamburg-Harburg.
Lerbs, H. (1952). Moderately loaded propellers with a finite number of blades and an arbitrary distribution of circulation. *Trans. SNAME, 60*, 73–123.
Lerbs, H. (1955). Ergebnisse der angewandten Theorie des Schiffspropellers. In: *Jahrbuch Schiffbautechnische Gesellschaft.* Springer, pp. 163–206.
Lewis, E.V. (Ed.) (1990). *Principles of Naval Architecture.* Soc. Naval Arch. Marine Eng.
Lewis, F. M. (1929). The inertia of the water surrounding a vibrating ship. *Trans. SNAME, 37*, 1–20.
MacPherson, D. M. (1993). Reliable performance prediction: techniques using a personal computer. *Marine Technology, 30/4*, 243–257.
Menzel, W., El Moctar, O. & Mumm, H. (2008). Advanced thinking on tricky excitations. *The Naval Architect,* March 64–69.
Mewis, F. (2001). Pod drives – pros and cons. *Hansa, 138/8*, 25–30.
Mewis, F. (2009). A novel power-saving device for full-form vessels. *1st Int. Symp. Marine Propulsors (SMP'09)*, Trondheim.

Mizoguchi, S. & Tanizawa, K. (1996). Impact wave loads due to slamming – A review. *Ship Technology Research, 43*, 139–154.

Morgan, W. B. & Lin, W. C. (1998). Predicting ship hydrodynamic performance in today's world. *Naval Engineers J.*, 91–98, September.

Morino, L. (1975). Steady and oscillatory subsonic and supersonic aerodynamics around complex configurations. *AIAA J., 13/3*, 368–374.

Morino, L. & Kuo, C. C. (1974). Subsonic potential aerodynamics for complex configurations: A general theory. *AIAA J., 12/2*, 191–197.

Munk, T. (2006). *Fuel conservation through managing hull resistance*. Copenhagen: Motorship Propulsion Conf.

Munk, T. & Kane, D. (2009). *The effects of corrosion and fouling on the performance of ocean-going vessels: A naval architect perspective. Advances in Marine Antifouling Coatings and Technologies.* CRC Press.

Nakos, D. (1990). *Ship wave patterns and motions by a three-dimensional Rankine panel method.* Ph.D. thesis. MIT.

Nakos, D. & Sclavounos, P. (1990). Steady and unsteady wave patterns. *J. Fluid Mechanics, 215*, 256–288.

Newman, J. N. (1965). The exciting forces on a moving body in waves. *J. Ship Research, 10*, 190–199.

Newman, J. N. (1977). *Marine hydrodynamics*. MIT Press.

Newman, J. N. (1978). The theory of ship motions. *Adv. Appl. Mech., 18*, 222–283.

Ok, J. P. (2005). Numerical investigation of scale effects of a wake-equalizing duct. *Ship Technology Research, 52*, 34–53.

Oortmerssen, G.van (1971). A power prediction method and its application to small ships. *Int. Shipbuilding Progress, 18*, 397–415.

Östergaard, C. (1996). Schiffspropulsion. In: *Technikgeschichte des industriellen Schiffbaus in Deutschland*, Vol. 2, Ed. L.U. Scholl. Ernst Kabel Verlag, p. 65.

Parsons, M. G. (2004). Parametric design. *Ship Design & Construction, SNAME.*

Pereira, R. (1988). Simulation nichtlinearer Seegangslasten. *Schiffstechnik, 35*, 173–193.

Perez Gomez, G. & Gonzalez-Adalid, J. (1997). *Detailed Design of Ship Propellers*. Madrid: FEIN.

Peric, M. & Bertram, V. (2011). Trends in industry applications of CFD to maritime flows. *10th Conf. Computer and IT Applications in the Maritime Industries (COMPIT)*, Berlin.

Peyret, R. & Taylor, T. D. (1985). *Computational methods for fluid flow.* Springer.

Raven, H. (1996). *A solution method for the nonlinear ship wave resistance problem.* PhD Thesis. TU Delft.

Reich, Y., Bertram, V. & Friesch, J. (1997). The development of a decision support system for propeller design. *ICCAS'97*, Yokohama, 363–378.

Riegels, F. W. (1958). *Aerodynamische Profile*. R. Oldenbourg.

Salvesen, N., Tuck, E. O. & Faltinsen, O. (1970). Ship motions and sea loads. *Trans. SNAME, 78*, 250–287.

Schlichting, H. (1979). *Boundary layer theory* (7th ed.). McGraw-Hill.

Schneekluth, H. (1986). Wake equalizing duct. *The Naval Architect, 103*, 147–150, April.

Schneekluth, H. & Bertram, V. (1998). *Ship design for efficiency and economy.* Butterworth-Heinemann.

Schwanecke, H. (1963). Gedanken zur Frage der hydrodynamisch erregten Schwingungen des Propellers und der Wellenleitung. In: *Jahrbuch Schiffbautechnische Gesellschaft*. Springer, pp. 252–280.

Sharma, S. D. (1986). Kräfte am Unter- und Überwasserschiff. *Handbuch der Werften, XVIII*, 31–43.

Söding, H. (1982). Prediction of ship steering capabilities. *Schiffstechnik*, 3–29.

Söding, H. (1984). Bewertung der Manövriereigenschaften im Entwurfsstadium. In: *Jahrbuch Schiffbautechnische Gesellschaft*, pp. 179–201. Springer.

Söding, H. (1986). Kräfte am Ruder. *Handbuch der Werften, XVIII*, 47–56, Hansa-Verlag.

Söding, H. (1987). Ermittlung der Kentergefahr aus Bewegungssimulationen. *Schiffstechnik, 34*, 28–39.

Söding, H. (1992). Speed optimization in ship operation. *Ship Technology Research, 39*, 167–174.

Söding, H. (1993). CFD for manoeuvring of ships. *19th WEGEMT School*. Nantes.

Söding, H. (1993). A method for accurate force calculations in potential flow. *Ship Technology Research, 40*, 176–186.

Söding, H. (1994). Was kann und sollte unsere Forschung im nächsten Jahrzehnt leisten? *Schiff & Hafen, 46/12*, 47–51.

Söding, H. (1995). Wave resistance by Michell integral. *Ship Technology Research, 42*, 163–164.

Söding, H. (1997). Wind wave spectrum. *Ship Technology Research, 44*, 71–72.

Söding, H. (1998a). Limits of potential theory in rudder flow predictions. *Ship Technology Research, 45*, 141–155.

Söding, H. (1998b). Limits of potential theory in rudder flow predictions. *22 Symp. Naval Hydrodynamics*, Washington.

Söding, H. & Bertram, V. (1998). Schiffe im Seegang. *Handbuch der Werften, XXIV*, 151–189, Hansa-Verlag.

Sparenberg, J. A. & de Vries, J. (1987). An optimum screw propeller with endplates. *Int. Shipbuilding Progress, 34*(395), 124–133.

SSC (1995). *Hydrodynamic impact on displacement ship hulls*. Ship Structure Committee, SSC-385, SSC-385A.

Streckwall, H. (1993). Anwendung auf Propeller. In *27 Kontakt-Studium 'Numerische Methoden in der Strömungsmechanik'. IfS*, Univ. Hamburg.

Streckwall, H. (1999). Numerical techniques for propeller design. *31 WEGEMT school 'CFD for ship and offshore design'*, WEGEMT, London.

Swain, G. W. (2010). The importance of ship hull coatings and maintenance as drivers for environmental sustainability. *RINA Conf. Ship Design and Operation for Environmental Sustainability*, London.

Tanizawa, K. & Bertram, V. (1998). Slamming. *Handbuch der Werften, XXIV*, 191–210, Hansa-Verlag.

Terswiga, T. J. C. van (1996). *Waterjet-hull interaction*. Ph.D. Thesis, TU Delft, ISBN 90-75757-01-8.

Thieme, H. (1962). Zur Formgebung von Schiffsrudern. In: *Jahrbuch Schiffbautechnische Gesellschaft*. Springer, pp. 381–426.

Thompson, J. F., Warsi, Z. U. A. & Mastin, C. W. (1985). *Numerical Grid Generation*. North-Holland.

Townsin, R. L., Byrne, D., Milne, A. & Svensen, T. (1980). Speed, power and roughness: the economics of outer bottom maintenance. *Trans. RINA, 122*, 459–483.

Townsin, R. L. & Kwon, Y. J. (1983). Approximate formulae for the speed loss due to added resistance in wind and waves. *Trans. RINA, 125*, 199–207.

Van Manen, J. D. & Sentic, A. (1956). Contra-rotating propellers. *Int. Shipbuilding Progress, 3*.

Versteeg, H. K. & Malalasekera, W. (1995). *An introduction to computational fluid dynamics – The finite volume method*. Addison Wesley Longman.

Völker, H. (1974). Entwerfen von Schiffen. *Handbuch der Werften, XII*, 17, Hansa.

Von Karman, T. (1929). *The impact of seaplane floats during landing*. Washington: NACA TN 321.

Watanabe. (1986). Analytical expression of hydrodynamic impact pressure by matched asymptotic expansion technique. *Trans. West-Japan Soc. Naval Arch., 71*.

Whicker, L. F. & Fehlner, L. F. (1958). *Free stream characteristics of a family of low aspect ratio control surfaces for application to ship design*. DTMB Report 933. Washington: David Taylor Model Basin.

Wolff, K. (1981). Ermittlung der Manövriereigenschaften fünf repräsentativer Schiffstypen mit Hilfe von CPMC-Modellversuchen. *IfS Report. 412*. Univ. Hamburg.

Zou, Z. J. (1990). Hydrodynamische Kräfte am manövrierenden Schiff auf flachem Wasser bei endlicher Froudezahl. *IfS Report. 503*. Univ. Hamburg.

Index

C

D

E

F

G

P

Q

R

U

V

W

Printed in the United States
By Bookmasters